NEUTRON SCATTERING METHODS AND STUDIES

CHEMICAL ENGINEERING METHODS AND TECHNOLOGY

Additional books in this series can be found on Nova's website
under the Series tab.

Additional E-books in this series can be found on Nova's website
under the E-books tab.

BIOCHEMISTRY RESEARCH TRENDS

Additional books in this series can be found on Nova's website
under the Series tab.

Additional E-books in this series can be found on Nova's website
under the E-books tab.

CHEMICAL ENGINEERING METHODS AND TECHNOLOGY

NEUTRON SCATTERING METHODS AND STUDIES

MICHAEL J. LYONS
EDITOR

Nova Science Publishers, Inc.
New York

Copyright © 2011 by Nova Science Publishers, Inc.

For permission to use material from this book please contact us:
Telephone 631-231-7269; Fax 631-231-8175
Web Site: http://www.novapublishers.com

NOTICE TO THE READER

The Publisher has taken reasonable care in the preparation of this book, but makes no expressed or implied warranty of any kind and assumes no responsibility for any errors or omissions. No liability is assumed for incidental or consequential damages in connection with or arising out of information contained in this book. The Publisher shall not be liable for any special, consequential, or exemplary damages resulting, in whole or in part, from the readers' use of, or reliance upon, this material. Any parts of this book based on government reports are so indicated and copyright is claimed for those parts to the extent applicable to compilations of such works.

Independent verification should be sought for any data, advice or recommendations contained in this book. In addition, no responsibility is assumed by the publisher for any injury and/or damage to persons or property arising from any methods, products, instructions, ideas or otherwise contained in this publication.

This publication is designed to provide accurate and authoritative information with regard to the subject matter covered herein. It is sold with the clear understanding that the Publisher is not engaged in rendering legal or any other professional services. If legal or any other expert assistance is required, the services of a competent person should be sought. FROM A DECLARATION OF PARTICIPANTS JOINTLY ADOPTED BY A COMMITTEE OF THE AMERICAN BAR ASSOCIATION AND A COMMITTEE OF PUBLISHERS.

Additional color graphics may be available in the e-book version of this book.

LIBRARY OF CONGRESS CATALOGING-IN-PUBLICATION DATA

Neutron scattering methods and studies / editors, Michael J. Lyons.
p. cm.
  Includes index.
  ISBN 978-1-61122-521-1 (hardcover)
  1. Neutrons--Scattering. I. Lyons, Michael J.
  QC793.5.N4628N53 2010
  539.7'58--dc22
2010041303

Published by Nova Science Publishers, Inc. + *New York*

CONTENTS

PREFACE

Neutron scattering encompasses all scientific techniques whereby the deflection of neutron radiation is used as a scientific probe. Neutrons readily interact with atomic nuclei and magnetic fields from unpaired electrons, making a useful probe of both structure and magnetic order. Neutron Scattering falls into two basic categories - elastic and inelastic. This book presents research in the study of neutron scattering, including neutron diffuse scattering measurements from ionic crystals and metal crystals; small angle scattering analysis of virus-like particles for biomedical diagnostic assays; the structure and properties of different proteins by small angles neutrons scattering; neutron scattering methods in the investigation of the SC lipid membrane structure; and neutron scattering from solutions of biological macromolecules.

Chapter 1 - Diffuse scatterings data contain information about static and dynamic disorders in crystals. Neutron diffuse scattering measurements from ionic crystals, covalent crystals and metal crystals are investigated at low and room temperatures by angular-dispersive and energy-dispersive scattering methods. The intensity of the diffuse scattering changes with temperature. The oscillatory forms of the diffuse scattering are observed even from ordered crystals at room temperature. The observed diffuse scattering intensities are analyzed by including the correlation effects among thermal displacements of atoms. The values of correlation effects decrease with decrease of temperature. For most of the materials, the value of correlation effects among first nearest neighboring atoms is observed to be ~ 0.7 at room temperature. The values of correlation effects decrease rapidly with the increase of inter-atomic distances. The correlation effects do not depend much on the type of the crystal structure and crystal binding. From the parameters of Debye-Waller temperature factor and the values of correlation effects, the inter-atomic force constants among first, second and third nearest neighboring atoms are determined. A rough estimation of phonon dispersion relation has been tried from the inter-atomic force constants and crystal structure using computer simulation. The diffuse background scattering measurement is found to be very useful in optimizing annealing process during preparation of lithium-based metal oxides.

Chapter 2 - Technical advances in biotechnology have made the production of genetically engineered biomaterials commonplace. Many artificially produced biomolecules retain native binding to protein or nucleic acid targets, assemble into higher order complexes, and perform native catalytic functions.

Virus-based tools for biomedical purposes are of particular interest. Virus coat proteins can spontaneously assemble into intact virus-like particles (VLPs), with many structural features of the native virus. Specifically, the VLP displays surface antigens for host recognition, and its protective shell prevents damage to and acts as a delivery agent for its contents. Unlike native viruses, VLPs are noninfectious and usually lack nucleic acid content. These features have been used to advantage to produce vaccines such as for human papillomavirus (HPV). However, VLP vaccine development would be greatly served by a facile, high resolution method for physical characterization of VLPs in the form to be administered (i.e., in solution).

VLPs are also ideally suited as standards for corresponding infectious viral particles in clinical diagnostic assays. In particular, VLP carriers of short nucleic acid sequences show great promise as quantitative reference standards in DNA and RNA diagnostic assays. However, as for VLP vaccine development, the use of nucleic acid-filled VLPs in diagnostic testing has been hampered by a lack of physical techniques to easily measure VLP physical properties in solution.

Small-angle neutron scattering (SANS) and small-angle x-ray scattering (SAXS) are novel approaches to physical characterization of VLPs for vaccine and diagnostic standard development. SANS offers the unique capability of studying both the coat protein and the RNA or DNA of interest as separate entities, within a VLP containing both components. Both SAXS and SANS measurements are performed using VLPs in solution. This chapter examines the unique power and utility of scattering techniques for structural characterization of native viral particles, nucleic acid-free VLPs (reconstituted capsids), and recombinant VLPs carrying a foreign RNA of interest. In addition, the analysis of SANS and SAXS results will be demonstrated for modeling the structure of the coat protein and encapsulated RNA (required for vaccine development) in solution, as well as for quantifying the amount of RNA in a VLP. Quantification of nucleic acid content is critical for assessment of whether recombinant VLPs meet the requirements for a standard biomedical assay.

Chapter 3 - In this part of book, the authors presented the different steps to study the structure and properties of different proteins by small angles neutrons scattering (SANS). They have shown in this case, how to treat the neutron scattering spectra i.e the corrections of spectra. The form and structure factors used to determine respectively the structure and interactions effects. The figures presented are good witnesses and examples for these.

Chapter 4 - The mammalian skin plays a major role in the protection of the body against pathogen threats from the environment. The unique skin barrier properties are ensured by the outermost skin layer, the *stratum corneum* (SC). The SC consists of corneocytes which are embedded in a lipid matrix consisting of ceramides, free fatty acids, cholesterol, and its derivatives. The SC lipids are organized in lipid membranes arranged into a lamellar structure. Due to the special physicochemical properties of ceramides, these membranes are extremely rigid and, therefore, very poorly permeable. A number of skin diseases, namely atopic dermatitis, psoriasis or recessive X-linked ichthyosis, are related to an impaired structure of the SC lipid membrane. On the other hand, a reversible decrease in the skin barrier function connected with the changes in the SC lipid membrane organisation is often required in the transdermal drug delivery. These facts encourage looking for new approaches towards investigating the internal membrane arrangement of the SC lipid matrix.

Recently, several neutron scattering techniques have been applied in this context. Above all, the neutron scattering on unilamellar vesicles and the neutron diffraction on oriented multilamellar lipid films allow describing the organisation of SC lipid membrane models on the molecular level. These promising results open new opportunities in the skin research.

The present chapter aims to review the recent experience with the neutron scattering methods concerning the investigation of the SC lipid membrane structure.

Chapter 5 - Precipitates (2nd phase particles) are a major strengthening mechanism in many types of steels. In particular, it is the size (and distribution) and volume fraction of very fine precipitates ($\approx$<10 nm) that contribute significantly to the strengthening of these materials. This important precipitate size range coincides well with the precipitate analysis capabilities of Small Angle Neutron Scattering (SANS). The major challenges in characterizing precipitates in microalloyed steels are the relatively low volume fraction of nano size precipitates (<0.20wt%) and both the wide range of composition and sizes of precipitates that can form during the TMCP process. Given these challenges, characterizing precipitates in these types of steels requires a diversity of techniques including TEM, SEM, EDX, XRD and SANS. This paper will detail the analysis of nano sized precipitates in microalloyed steels using SANS. The precipitate data (e.g. size and distribution of the nano sized precipitates) measured with SANS was compared with other measurements techniques (TEM and XRD) for validation. In addition, the importance of the SANS precipitate data generated and how it can be used to correlate nano precipitates size and volume fraction with steel processing parameters will also be presented. This relationship can be used to identify steel processing conditions that would enhance fine precipitate evolution and ultimately the strength of the steel.

Chapter 6 - The study of biomolecules in solution is quite important in order to reproduce environmental conditions quite resembling those existing invivo. Small Angle Neutron Scattering (SANS) from proteins in solution provides information concerning size and shape, over a large length scale. Performance of SANS experiments on samples at different deuteration grades enables a deeper knowledge both of protein structure and of protein solvation shell. The chance to perform numerical simulation before carrying on a SANS experiment can lead the user to a wiser choice of experiment preparation and greatly improve the structural resolution. brief discussion about the scattering theory, isotopic substitution and thermodynamic equilibria between water and cosolvent molecules in the protein solvation shell, will be presented. This chapter will discuss theory practical aspects with protein conditions in order to guide potential users to successfully apply SANS experiments to answer peculiar biological questions involving proteins in solution.

Chapter 7 - The subject of this chapter is diffraction studies of nanocomposites, materials which obtained by the chemical synthesis (or other methods) of oxides, metals and other compounds inside porous media. The study of the physical properties of confined nanoparticles has become a very active field of research during the last decade. The reason to investigate such materials is fundamental since the confined geometry and the influence of the surface yields unusual properties as compared with the bulk. Confined nanoparticles also result in new applications, for example, in the field of catalysis, high-density magnetic memory, etc.

Chapter 8 - Small-angle scattering (SAS) of X-rays and neutrons is a powerful method for the analysis of biological macromolecules in solution. The method allows one to study low resolution structure of native particles in nearly physiological environments and to analyze dynamic processes like complex formation, assembly and folding. Thanks to recent progress in SAS instrumentation and novel analysis methods, which substantially improved the resolution and reliability of the structural models, the method has attracted renewed interest of biologists and biochemists. In this chapter devoted to biological neutron scattering

(SANS) the authors cover basics of the neutron scattering theory, SANS instrumentation and modern methods used in data analysis and modeling. Differences between X-ray and neutron scattering and advantages/shortcomings of the two techniques are demonstrated. The major advantage of SANS, *i.e.* possibility of contrast variation and specific labeling due to isotope hydrogen/deuterium exchange is comprehensively discussed. Recent applications of SANS to proteins and nucleoprotein complexes are reviewed and a test case, contrast variation analysis of a GroEL/GroES chaperonin system using modern data interpretation methods, is presented in detail.

Chapter 9 - Neutron scattering has been demonstrated to be a powerful tool for investigating physical and chemical mechanisms of biophysical processes. The present paper shows Elastic (ENS), Quasi Elastic (QENS) and Inelastic Neutron Scattering (INS) findings on a class of hydrogen-bonded systems of biophysical interest, such as homologues disaccharides (i. e. trehalose, maltose and sucrose)/water mixtures as a function of temperature and concentration.

The aim of the present work is to give an overview of dynamical properties of the investigated systems in order to characterise the dynamical transition, the diffusion and the vibrational modes of disaccharides in solution and to link the observed features to their bioprotective effectiveness. The role of hydrogen bond and fragility in these bioprotectant systems is discussed in detail. The obtained findings highlight a strong destructuring effect on the hydrogen-bonded network and a marked slowing down effect on the dynamics of water by disaccharides together with a considerable cryptobiotic action and can justify the superior capability of trehalose in respect to maltose and sucrose to encapsulate biostructures in a more rigid and protective environment.

In: Neutron Scattering Methods and Studies
Editor: Michael J. Lyons

ISBN: 978-1-61122-521-1
© 2011 Nova Science Publishers, Inc.

Chapter 1

DIFFUSE NEUTRON SCATTERING STUDY OF CORRELATION EFFECTS AMONG THERMAL DISPLACEMENTS OF ATOMS

Takashi Sakuma[1]*, Xianglian[2], Hiroyuki Uehara[1]
and Saumya R. Mohapatra[1]

[1]Institute of Applied Beam Science, Ibaraki University, Mito 310-8512, Japan
[2]College of Physics and Electronics Information, Inner Mongolia University
for the Nationalities, Tongliao 028043, China

ABSTRACT

Diffuse scatterings data contain information about static and dynamic disorders in crystals. Neutron diffuse scattering measurements from ionic crystals, covalent crystals and metal crystals are investigated at low and room temperatures by angular-dispersive and energy-dispersive scattering methods. The intensity of the diffuse scattering changes with temperature. The oscillatory forms of the diffuse scattering are observed even from ordered crystals at room temperature. The observed diffuse scattering intensities are analyzed by including the correlation effects among thermal displacements of atoms. The values of correlation effects decrease with decrease of temperature. For most of the materials, the value of correlation effects among first nearest neighboring atoms is observed to be ~ 0.7 at room temperature. The values of correlation effects decrease rapidly with the increase of inter-atomic distances. The correlation effects do not depend much on the type of the crystal structure and crystal binding. From the parameters of Debye-Waller temperature factor and the values of correlation effects, the inter-atomic force constants among first, second and third nearest neighboring atoms are determined. A rough estimation of phonon dispersion relation has been tried from the inter-atomic force constants and crystal structure using computer simulation. The diffuse background scattering measurement is found to be very useful in optimizing annealing process during preparation of lithium-based metal oxides.

* Corresponding author: e-mail address: sakuma@mx.ibaraki.ac.jp. Tel.: +81-29-228-8357; Fax: +81-29-228-8357.

Keywords: neutron diffraction, diffuse scattering, thermal vibration, correlation effects, force constant, phonon, disordered crystal, ordered crystal, solid electrolyte, nanostructure, J-PARC

1. INTRODUCTION

Information related to crystal structures and thermal vibrations of atoms can be obtained from the analysis of Bragg intensities by X-ray and neutron diffraction measurements [1-2]. Apart from sharp Bragg lines, the background pattern coming out of the diffuse scatterings is also very useful to evaluate lattice dynamical parameters. Recently the study of diffuse scattering intensity of diffraction pattern has been systematically developed and its achievements have become a center of attraction. Diffuse scattering measurement by X-ray and neutron diffraction methods comes handy to analyze disordered states and lattice dynamics in crystals and non-crystalline materials [3-8]. The static disordered arrangements in solids give rise to short-range order in crystals and the dynamical disorders to the correlation effects among thermal displacements of atoms.

Neutron scattering measurements by angular-dispersive and energy-dispersive methods are briefly described in section 2. Neutron scattering measurements have been performed on powder sample of ionic crystals, semiconductors and metals by HRPD (High Resolution Powder Diffractometer) installed at JRR-3 in Japan Atomic Energy Agency. Time of flight (TOF) neutron scattering measurement is performed on the Sirius spectrometer at Neutron Science Laboratory of High Energy Accelerator Research Organizations (KEK), Japan.

In section 3, theoretical treatment of neutron diffuse scattering intensity of ordered and disordered crystals is presented. The correlation effects among thermal displacements of atoms are taken into account in the treatment. The correlation effects are used to predict the force constants among nearest neighboring atoms. The average of the thermal displacements of atoms is obtained from the values of Debye-Waller temperature parameters and correlation effects. Force constants of first, second and third nearest neighboring atoms are calculated from Debye-Waller temperature parameters and the values of correlation effects. Phonon dispersion relation is estimated from force constants and crystal structure by computer simulation. The correlation effects by the diffuse scattering measurements and those obtained from EXAFS (Extended X-ray Absorption Fine Structure) analysis have been compared.

The theoretical expressions of the diffuse scattering intensity are applied to ionic crystals, semiconductors and metals. Temperature dependence of the oscillatory diffuse scattering is explained by the correlation effect among the thermal displacements of atoms in section 4. Solid state ionics exhibit high ionic conductivities at fairly low temperatures below their melting points. The high ionic conductivity of solid state ionics has been understood with disordered arrangements of atoms. Very strong and oscillatory diffuse scatterings have been observed from solid state ionic materials. The observed diffuse scattering intensities are analyzed by including the correlation effects among thermal displacements of atoms and disordered arrangements of atoms. The intensities of the oscillatory diffuse scattering of ionic materials are mainly explained using the correlation effects among thermal displacements of nearest neighboring atoms. From the results of diffuse scattering analysis of solid state ionics it is deduced that an oscillatory diffuse scattering intensity would appear in diffraction pattern from ordered crystals near room temperature. To investigate the temperature dependence of diffuse scattering intensities, neutron diffraction measurements are performed on ordered crystal at low and room temperatures. The theoretical treatment of the diffuse scattering intensity including the correlation effects among thermal displacements of atoms is applied to the background function in Rietveld analysis of crystals.

In section 5, neutron scattering experiments have been used to optimize the annealing process during solid state synthesis of $LiNiO_2$ and $LiMn_2O_4$. The analysis of crystallite size of $LiNiO_2$ has been calculated from the FWHM of the Bragg lines using Scherrer equation. The intensity of the Bragg lines and that of the diffuse background scattering of $LiNiO_2$ composites have been used to estimate the percentage of $LiNiO_2$ formed in the annealing process. The results are compared with those of $LiMn_2O_4$ composites.

2. NEUTRON DIFFUSE SCATTERING MEASUREMENTS

Neutron diffraction measurements at low temperature (as low as ~10 K) and at room temperature have been performed on powder sample of ionic crystals, semiconductors and metals by HRPD installed at JRR-3 of Japan Atomic Energy Agency. HRPD is an angular-dispersive type diffractometer with 64 counters and shows high resolution performance that could separate diffuse scattering intensities from Bragg lines in diffraction patterns. Powder sample is set into a vanadium container of 10 mm in diameter. Indium wire as O-ring material is used for sealing the container. Incident neutron wavelength of 1.823 Å monochromatized by Ge (331) is used and the data are collected for 1 day in the 2θ range from 20 to 155° with step angle 0.05°. Clear temperature dependence of diffuse scattering intensities is obtained in diffraction patterns of crystals. Sharp Bragg lines and strong oscillatory diffuse scatterings are observed at room temperature. The oscillatory characteristic in the diffuse scattering from ordered crystals at low temperature (~10 K) is not clear. The difference of the diffuse scattering intensities at low and room temperature is from the thermal contribution of atoms. Usually broad peaks of the neutron diffuse scattering from crystals are observed around 2θ~65 and 100°.

In case of angular-dispersive diffractometer HRPD, the observed diffuse scattering data are restricted to a relatively low Q region, where scattering vector Q is equal to $4\pi\sin\theta/\lambda$. The use of TOF spectrometer would be helpful to observe scattering intensity over wide range of Q. TOF neutron scattering measurement was performed for some materials at room temperature on the Sirius spectrometer at Neutron Science Laboratory of High Energy Accelerator Research Organizations, Japan. The sample is placed in a cylindrical vanadium container of 8.0 mm in diameter perpendicular to the incident neutron beam. The scattering data were collected at the back-scattering bank detectors. The results are compared with those by angular-dispersive type diffractometer.

3. NEUTRON DIFFUSE SCATTERING INTENSITY AND LATTICE DYNAMICS

3.1. Correlation Effects

Diffraction intensity $I(Q)$ includes elastic scattering intensity $I_{el}(Q, \omega=0)$ and inelastic scattering intensity $I_{inel}(Q, \omega\neq0)$;

$$I(Q)= I_{el}(Q, \omega=0)+ I_{inel}(Q, \omega\neq0). \tag{1}$$

It is found that information of thermal vibration that corresponds to inelastic scattering would be derived from the analysis of diffraction intensity. The elastic scattering and inelastic scattering intensities are composed of coherent scattering and incoherent scattering. The structure factor F that includes types of atom and atomic positions in a crystal is expressed as:

$$F = \sum_j b_j \exp\left(-i\mathbf{Q}.\mathbf{r}_j\right), \tag{2}$$

where b is a nuclear scattering length in neutron diffraction measurement and r an atomic position [9-10]. The coherent scattering intensity which includes Bragg line intensity I_B and diffuse scattering intensity I_d can be written as:

$$I_B = k\left|\langle F\rangle\right|^2 G \tag{3}$$

$$I_d = k\sum_n \sum_{n'} \exp\{i\mathbf{Q}.(\mathbf{R}_n - \mathbf{R}_{n'})\}\langle \Delta F_n \Delta F_{n'}^*\rangle, \tag{4}$$

where k is a constant depending on the experimental conditions. G is Laue function and ΔF_n defined as the deviation of the structure factor at the nth site from the mean structure factor $<F>$,

$$F_n = \langle F\rangle + \Delta F_n. \tag{5}$$

The deviation of the structure factor ΔF is basically originated due to static disordered arrangements of atoms and the thermal vibration of atoms in the lattice. $\Delta F_{s(i)}$ at $s(i)$ site is defined as:

$$\Delta F_{s(i)} = b_{s(i)}\exp\{i\mathbf{Q}.\Delta\mathbf{r}_{s(i)}\} - \overline{b_{s(i)}\exp\{i\mathbf{Q}.\Delta\mathbf{r}_{s(i)}\}}$$
$$= b_{s(i)}\exp\{i\mathbf{Q}.\Delta\mathbf{r}_{s(i)}\} - p_i b_i\exp\{-M_i\}. \tag{6}$$

$b_{s(i)}$ is either b_i or 0 in the case of disordered arrangement of atoms in the case of averaged structure. The deviation from an equilibrium position is shown by Δr. The probability of finding the atom in any site p_i is equal to the ratio of the number of the i atoms to the number of the i sites in the crystal. The Debye-Waller factor $\exp\{-M_i\}$ is also expressed as $\exp\{-B_i(\sin\theta/\lambda)^2\}$. Isotropic temperature parameters are used in the analysis of diffuse scattering intensities. Anharmonic anisotropic thermal parameters have been applied in the analysis of Bragg lines intensity. However, the theoretical study is insufficient for the treatment of diffuse scattering intensity by anharmonic anisotropic thermal vibrations. From the values of $\langle \Delta b_s \Delta b_{s'} \rangle$ and $\langle \Delta r_s \Delta r_{s'} \rangle$, the static disordered arrangements among atoms (short-range order) and the thermal dynamical disordered arrangements among atoms (thermal correlation effect) are obtained, respectively, where Δb_s is b_s - $<b_s>$. In this study it is assumed that the correlation among Δb and Δr is neglected; $<\Delta b_s \Delta r_s>=0$ and $<\Delta r_s \Delta b_s>=0$. We will study the diffuse scattering analysis in the case of $<\Delta b_s \Delta r_s> \neq 0$, for example, correlation effects among atomic defect and the atomic displacements near the defect.

The average of the thermal displacements of atoms is obtained by cumulant expansion. The thermal average of $\exp\{-i\mathbf{Q}.(\Delta r_j - \Delta r_k)\}$ is written as:

$$\left\langle \exp\{-i\mathbf{Q}.(\Delta r_j - \Delta r_k)\} \right\rangle \cong \exp\left[-\frac{\mathbf{Q}^2}{2}\left(\langle \Delta r_j^2 \rangle + \langle \Delta r_k^2 \rangle\right)\left(1 - 2\frac{\langle \Delta r_j \Delta r_k \rangle}{\langle \Delta r_j^2 \rangle + \langle \Delta r_k^2 \rangle}\right)\right]$$

$$= \exp\left\{-\left(M_j + M_k\right)\left(1 - \mu_{jk}\right)\right\}. \tag{7}$$

The correlation effects among the thermal displacements of atoms μ_{jk} are defined as follows;

$$\mu_{jk} = 2\langle \Delta r_j \Delta r_k \rangle \Big/ \left(\langle (\Delta r_j)^2 \rangle + \langle (\Delta r_k)^2 \rangle\right) \tag{8}$$

It is observed that the information of averaged structure factor $<F>$ can be obtained from the analysis of Bragg line intensities. The space group of the crystal and the structural parameters including lattice constants, Debye-Waller temperature parameters B and atomic positions are determined from the Rietveld refinement analysis of Bragg line intensities. However, we could not derive the information of short-range order and correlation effects among thermal vibrations of atoms in crystals from the analysis of Bragg lines. Short-range order and correlation effects in crystals are obtained from the analysis of diffuse scattering intensities.

3.2. Disordered Crystals

Crystalline superionic conductors at high temperature have a disordered structure in which the number of available atomic sites is greater than that of atoms [11-20]. Room temperature phase of Ag_3SI shows disordered arrangements of Ag atoms at room temperature

(Figure1). Metallic alloys also show disordered arrangements of atoms at high temperature [21-25]. The diffuse scattering intensity I_d from a powder samples including the static disordered arrangements and the correlation effects among the thermal displacements of atoms is expressed as follows;

$$I_d = kN_0 \sum_i u_i b_i b_i \left\{ 1 - \exp\left(-2M_i\right) \right\}$$

$$+kN_0 \sum_i \sum_j \sum_{s'(j)}{}' u_i b_i b_j \left[\alpha_{r_{s(i)s'(j)}} \left[\exp\left\{ -\left(M_i + M_j\right)\left(1 - \mu_{r_{s(i)s'(j)}}\right) \right\} - \exp\left\{ -\left(M_i + M_j\right) \right\} \right] \right.$$

$$\left. + \left(1 - p_i\right)\left(\alpha_{r_{s(i)s'(j)}} - \beta_{r_{s(i)s'(j)}} \right) \exp\left\{ -\left(M_i + M_j\right) \right\} \right] Z_{r_{s(i)s'(j)}} S_{r_{s(i)s'(j)}} + kN_0 \sum_i u_i \sigma_i^{inc}, \qquad (9)$$

where $Z_{r_{s(i)s(j)}}$ is the number of sites belonging to the s'th j type neighbor around a sth i type site. S_r is equal to $\sin(Qr)/Qr$. This term gives oscillatory form to the diffuse scattering intensity pattern. Two sites $s(i)$ and $s'(j)$ are apart by the distance r. The probability function α_r gives the probability of finding an atom at a site of distance r from a site occupied by an atom, and β_r the probability of finding an atom at a site apart by r from a vacant site. u_i corresponds to the number of i atoms per unit cell. The prime added to the summation symbol is to omit the term of $r_{s(i)s'(j)} = 0$. σ_i^{inc} is incoherent scattering cross section of atom i. Eq. (9) can be applied as a shape function of a background in Rietveld analysis.

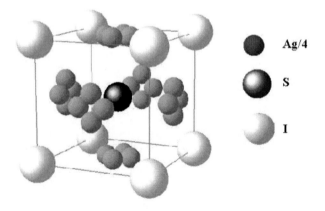

Figure 1. Disordered arrangement of Ag atoms in β-Ag$_3$SI.

3.3. Ordered Crystals

At low temperature many superionic crystals show ordered arrangement of atoms (Figure 2) [26]. The diffuse scattering intensity from ordered crystals including the correlation effects among thermal displacements of atoms is expressed as;

$$I_{\mathrm{d}} = kN_{\mathrm{o}}\left[\sum_i n_i b_i^2 \left(1 - \exp(-2M_i)\right) + \sum_i \sum_j \sum_{s'} {}' n_i b_i b_j \left[\exp(-(M_i + M_j)(1 - \mu_{r_{s(i)s'(j)}})) \right.\right.$$

$$\left.\left. - \exp(-(M_i + M_j)) \right] Z_{s(i)s'(j)} \sin(Qr_{s(i)s'(j)}) / (Qr_{s(i)s'(j)}) + kN_0 \sum_i u_i \sigma_i^{\mathrm{inc}} . \right. \tag{10}$$

At low temperature (~10 K) the values of thermal vibration are very small. The diffuse scattering intensity from ordered crystals is almost from the incoherent scattering cross section. Near room temperature oscillatory diffuse scattering that has a form $\sin(Qr)$ appears in *all* crystals. The peaks of diffuse scattering intensity appear near Q positions $Q=(\pi/2+2n\pi)/r$, where r is inter-atomic distance of first nearest neighboring atoms and n integer.

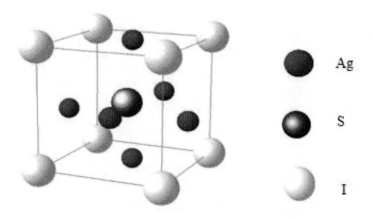

Figure 2. Ordered arrangement of Ag atoms in γ-Ag$_3$SI.

3.4. Force Constants

The information on inter-atomic distances, coordination numbers around atoms, parameters of Debye-Waller temperature factor and correlation effects among thermal displacements of atoms can be obtained from the analysis of diffuse scattering intensity using Eq. (9). Literature also reveals the use of other techniques like EXAFS measurements to obtain information on inter-atomic distances, coordination numbers around atoms, parameters of Debye-Waller temperature factor and correlation effects among thermal displacements of atoms [27]. The MSD (mean square displacements) and DCF (displacements correlation function) of cubic $A^N B^{8-N}$ crystals had been studied by EXAFS measurement. The ratio of DCF to MSD is similar to the definition of correlation effects μ in the analysis of diffuse scattering measurements. Most of the values of correlation effects by EXAFS measurement agree with that by diffuse scattering measurements. The correlation effects among first nearest neighboring atoms had been reported by Beni and Platzman by theoretical approach in the EXAFS analysis [28]. Using a simple Debye model, they pointed out that the correlation effects of bcc are similar to that of fcc. The correlation effects would not depend much on the type of crystal structure near room temperature.

The value of mean square displacements of atoms $<\Delta u_{ss'}^{2}>$, that is usually expressed by δu^{2} in EXAFS measurements [9], is written as

$$< \Delta u_{s\,s'}^{2} >= \left(\langle \Delta r_{s}^{2}\rangle + \langle \Delta r_{s'}^{2}\rangle\right)\left(1-2\frac{\langle \Delta r_{s}\,\Delta r_{s'}\rangle}{\langle \Delta r_{s}^{2}\rangle + \langle \Delta r_{s'}^{2}\rangle}\right)$$

$$= \frac{1}{8\pi^{2}}\left(B_{s} + B_{s'}\right)\left(1-\mu_{ss'}\right). \tag{11}$$

The definition of $\Delta \boldsymbol{u}$ is shown in Figure 3. Debye-Waller temperature parameters B is defined by $B=8\pi^{2}<\Delta r^{2}>$. Using the equation

$$<(\Delta \boldsymbol{u})^{2}> = \frac{k_{B}T}{\alpha} \tag{12}$$

the force constants α among nearest neighboring atoms are determined from Eq. (11);

$$k_{B}T/\alpha = \frac{1}{8\pi^{2}}\left(B_{s} + B_{s'}\right)\left(1-\mu_{ss'}\right). \tag{13}$$

From the values of correlation effects μ among first, second and third nearest neighboring atoms and the values of Debye-Waller temperature factor B, we could estimate the values of force constants among first, second and third nearest neighboring atoms.

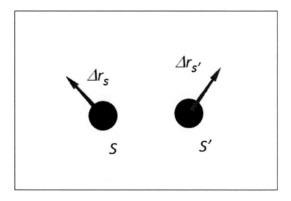

Figure 3. Definition of $\Delta u_{ss'}$; $\Delta u_{ss'}=\Delta r_{s}-\Delta r_{s'}$.

The phonon dispersion relation could be roughly calculated by the force constants among atoms by computer simulation. As the phonons are direction-dependent atomic displacements, we need to include the information of crystal structure in simulation program.

4. DIFFUSE SCATTERING INTENSITY FROM CRYSTALS

Non-linear Rietveld refinement method is one of the most widely used technique of characterizing structure parameters of polycrystalline materials. The structural parameters including the atomic positions, lattice constants and Debye-Waller temperature parameters of crystals are determined from the analysis of Bragg line intensities. Coordination numbers Z and inter-atomic distances r are calculated from the crystal structure after iterative refinements. The profile-shape function for Bragg reflections in the Rietveld method was extensively studied to obtain better fits between observed and calculated diffraction patterns. Background function for Rietveld analysis has been approximated by a finite sum of Legendre polynomials though it has no physical correlation [29]. The background function can be written in terms of the Legendre polynomials as:

$$I_d = \sum_j c_j F_j ,$$ (14)

where F_j is Legendre polynomial and c_j is coefficient that is optimized in the refinement calculations. The observed background intensity is well fitted at low temperature using Legendre polynomial. However, as the temperature increases, it is difficult to explain the oscillatory intensity of diffuse background. Diffuse scattering has information about crystal structure and thermal vibration of atoms. With an aim of obtaining such information, we replace the Legendre polynomials by incorporating correlation effects among thermal displacements of atoms in the background function of Rietveld analysis. Rietveld refinements of neutron powder diffraction patterns in ionic crystals are performed with the two background functions separately; the correlation effect among the thermal displacements of atoms and the Legendre polynomials. The background function including the correlation among thermal displacements of atoms gives somewhat lower R factors at 290 K than that with Legendre polynomials.

4.1 Ordered Crystals

4.1.1. Ionic Crystals

Crystals of copper halides CuCl, CuBr, CuI belong to zincblende-type with the space group $F\bar{4}3m$ below room temperature [30-35]. Copper and halogen atoms occupy 4c and 4a sites, respectively. The observed diffraction intensities are explained by the background function that includes correlation effects among thermal displacements of first, second and third nearest neighboring atoms. Least squares fitting of the neutron diffraction pattern at 280 K was performed with the background function. The obtained values of correlation effects among thermal displacements of first nearest neighboring atoms in CuCl, CuBr and CuI are 0.72, 0.70 and 0.69, respectively [36-38]. The correlation effects among thermal displacements of atoms in typical ionic solids are shown in Figure 4. If there are no correlation effects among thermal displacements of atoms, it corresponds to classical Einstein vibration. The value is equal to zero in the case of Einstein vibration. Although the observed diffuse scattering intensities could be described by the correlation effects of first nearest neighboring atoms, the oscillatory schemes of the observed diffuse scattering intensities are sufficiently explained by the analysis which includes the correlation effects among thermal displacements of first, second and third nearest neighboring atoms. The Debye-Waller temperature parameters in CuBr are B_{Cu}=2.550 Å^2 and B_{Br}=1.693 Å^2 at 280 K which are much greater than those of B_{Ga}=0.45 Å^2 and B_{As}=0.67 Å^2 in GaAs at 290 K [39]. The Debye-Waller temperature parameter is observed to be proportional to the mean-square amplitude of vibration of atoms around their equilibrium positions. The interatomic distance dependence of the highest bonding orbital energy in CuBr and GaAs has been discussed by Aniya [40]. The highest occupied bonding orbital energy of GaAs is very sensitive to the interatomic distance when compared with CuBr; the inter-atomic distance can be varied to large extent without altering so much the bonding orbital energy in CuBr.

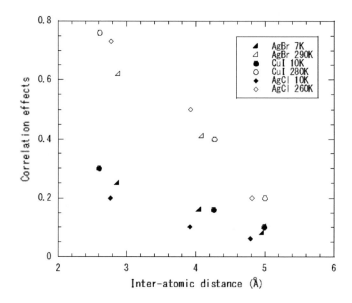

Figure 4. Inter-atomic distance dependence of correlation effects in ionic crystals.

The observed neutron diffuse scattering intensity of AgCl at 260 K shows an oscillatory form and the peaks of diffuse scattering intensity appear in the positions around $2\theta\sim35$, 65 and 100° (Figure 5). Diffuse scattering intensity is calculated with the correlation effect among the thermal displacements of atoms. The crystal of AgCl belongs to rock-salt type with the space group $Fm\overline{3}m$ [41]. The values of parameter of Debye-Waller temperature factor are relatively large at room temperature (Figure 6). The peaks at $2\theta\sim65$ and 100° are expected to appear on the basis of thermal correlation effects among the first nearest neighboring atoms (Ag-Cl). The weak peak of $2\theta\sim35$° is the influence of the correlation effects among the second nearest neighboring atoms (Ag-Ag, Cl-Cl). The three peaks in the diffuse scattering are produced by the superposition of these correlation effects. The value of correlation effects between the first nearest neighboring atoms is about 0.7 and that between the second nearest neighboring atoms is ~ 0.5. The values of correlation effects decrease rapidly with the inter-atomic distance. The values of the correlation effects do not depend much on the composition of copper halides and silver halides [42].

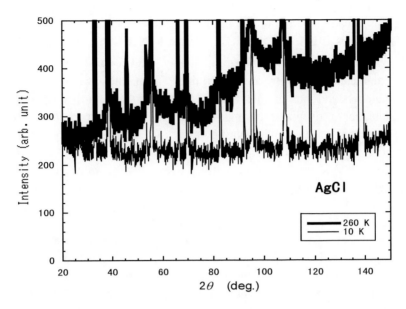

Figure 5. Observed neutron diffuse scattering intensity of AgCl.

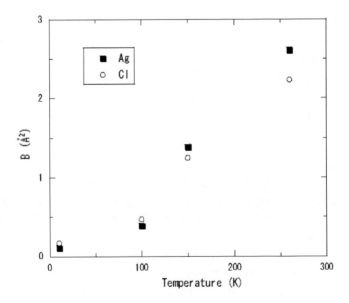

Figure 6. Parameters of Debye-Waller temperature factor of AgCl.

Most of the values of correlation effects in ionic crystals studied by EXAFS measurement agreed with those by diffuse scattering measurements. However, reported value of the correlation effects of first nearest neighboring atoms in KBr by EXAFS measurement was very small compared to other ionic crystals at room temperature. The inter-atomic distance and temperature dependence of correlation effects in KBr is studied by neutron diffraction method. The oscillatory feature of the diffuse scattering intensity of KBr at 7 K is very weak by neutron diffraction measurement with HRPD. Several sharp Bragg lines and clear oscillatory diffuse scattering are observed near room temperature (Figure 7). From the analysis of Bragg lines it is found that KBr belongs to NaCl type structure with the space group $Fm\overline{3}m$. The structural parameters including lattice constants and Debye-Waller temperature parameters for KBr are determined from the analysis of Bragg lines. Coordination numbers and inter-atomic distances are obtained from the crystal structure and used in the calculation of diffuse scattering intensity. The values of correlation effects among first, second and third nearest neighboring atoms at 298 K are 0.68, 0.30 and 0.10, respectively. The values of correlation effects among first, second and third nearest neighboring atoms at 7 K are 0.24, 0.12 and 0.06, respectively. Value of correlation effects among first nearest neighboring atoms in KBr near room temperature is almost the same as that of other reported ionic crystals ($\sim$ 0.7) [43]. The value of correlation effects decreases with the decrease of temperature. In general, the value of the correlation effects is $\sim$ 0.7 at the inter-atomic distance $r \sim$ 2.5 Å and $\sim$ 0.5 for $r \sim$ 4 Å in silver and copper based halides near room temperature as obtained from diffuse scattering measurements. The suggested value of correlation effects among first nearest neighboring atoms ($\sim$0.4) in KBr by EXAFS measurement is less than that of other ionic crystals by EXAFS measurement [44]. When we calculate the diffuse scattering intensity with the value 0.4 in KBr, the oscillatory feature of the calculated intensity is very weak comparing with that of the observed diffuse scattering intensity.

The values of correlation effects at 90 K are estimated from the values at 7 K and 298 K of KBr using the results of correlation effects in other ionic crystals by diffuse neutron scattering measurements. The estimated correlation effects among first, second and third nearest neighboring atoms at 90 K are 0.54, 0.24 and 0.08, respectively. The lattice constant and Debye-Waller temperature parameters of K and Br atoms at 90 K are calculated as 6.564 Å, 0.87 Å^2 and 0.78 Å^2, respectively. From Eq. (13) the force constants among first, second and third nearest neighboring atoms at 90 K are obtained as 0.83 eV/ Å^2, 0.50 eV/ Å^2 and 0.41 eV/ Å^2, respectively. The phonon dispersion relation of KBr at 90 K is roughly calculated using the values of lattice constant and force constants by computer simulation. As the phonons are direction-dependent atomic displacements, the information of crystal structure is needed in simulation program [45]. The obtained dispersion curves of KBr for phonons at 90 K by computer simulation (Figure 8) agree with the earlier report of observed phonons by inelastic neutron diffraction measurement at 90 K qualitatively [46,47]. As we did not include the values of the transverse charge in computer simulation, we could not produce the split of optical phonon into two branches at Γ point.

The theoretical treatment including correlation effects among thermal displacements of atoms has been applied to explain the oscillatory profile of diffuse scattering intensity observed by angular-dispersive X-ray and neutron diffraction experiments. From these measurements, the diffuse scattering intensity in relatively low Q region was discussed. The use of time-of-flight spectrometer could lead us to observe scattering intensity over wide range of Q. The theoretical treatment including the correlation effects is applied to explain the diffuse scattering profile of ordered crystals over wide range of Q. The theoretical expression including the correlation effects among thermal displacements of atoms is effective for the analysis of the diffuse scattering data by TOF measurement.

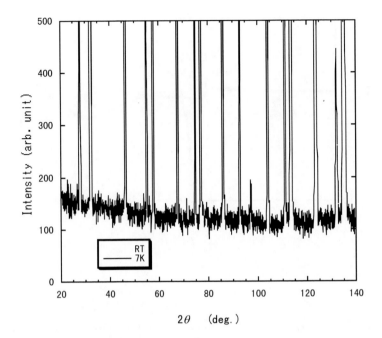

Figure 7. Neutron diffuse scattering intensity of KBr.

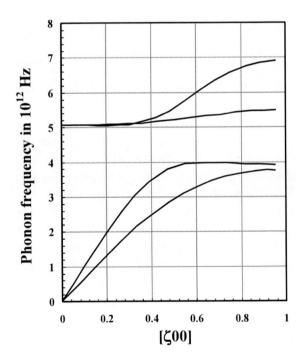

Figure 8. Calculated dispersion curves of KBr for phonon propagating in the [100] direction.

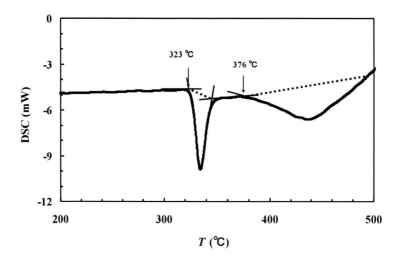

Figure 9. DSC measurement of PbF_2.

PbF$_2$ has three solid phases α, β and γ (Figure 9) [48,49]. The crystal structure of the low temperature α-phase belongs to orthorhombic space group *Pnma* (Figure 10). It changes to a cubic β-phase with ordered arrangements of F atoms at 316°C. The high temperature γ-phase appears at 450°C and shows high ionic conductivity. TOF neutron diffraction measurement of powder PbF$_2$ was performed at room temperature on the Sirius spectrometer at Neutron Science Laboratory of High Energy Accelerator Research Organizations, Japan [50]. The scattering data have been collected at the back-scattering bank detectors. The observed scattering intensity of PbF$_2$ by TOF neutron measurement is shown after the correction for the spectrum $I(\lambda)$ of incident neutron and the absorption of the sample container (Figure 11). The horizontal axis is changed to Q ($Q = 2\pi/d$) in order to compare with the angular-dispersive neutron diffraction data. Rietveld refinement analysis of the scattering intensity is performed using RIETAN-2001T [51]. The crystal structure of PbF$_2$ has ordered arrangements where the Pb and F atoms occupy 4c sites at room temperature. The diffuse scattering intensity increases with the increase of Q value. The observed intensity by angular-dispersive method with HRPD shows an oscillatory diffuse scattering form. A similar oscillatory scheme is also observed in the intensity by TOF scattering measurement. As the crystal structure of PbF$_2$ shows ordered arrangement of atoms at room temperature in the Rietveld analysis, the diffuse scattering intensity of PbF$_2$ by TOF scattering measurement is analyzed using Eq. (10). In simulating the diffuse scattering intensity of PbF$_2$, the following values of correlation effects used are: $\mu_{Pb-F} = 0.70$, $\mu_{F-F} = 0.67$ and $\mu_{Pb-Pb} = 0.40$. The values of correlation effects decrease with inter-atomic distances. The tendency of the oscillatory diffuse scattering intensity is explained by the model. It is found that Eq. (10) could be used to analyze the profile of diffuse scattering intensity from energy-dispersive as well as angular-dispersive neutron experiment. The oscillatory profile of diffuse scattering comes from the second term of Eq. (10) which includes the function sin(Qr). The profiles of diffuse scattering intensity of PbF$_2$ in the range $3 \leq Q$ (Å^{-1}) ≤ 7 by TOF scattering could be explained by including the correlation effects among thermal displacements of atoms. However, it is difficult to separate the observed intensity by TOF measurement into Bragg lines and diffuse scattering at high Q

region. To perform the detail investigation of diffuse scattering at high Q region, a measurement with high Q resolution is necessary.

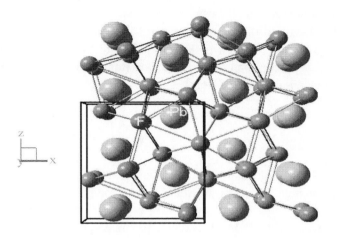

Figure 10. Crystal structure of α-PbF$_2$.

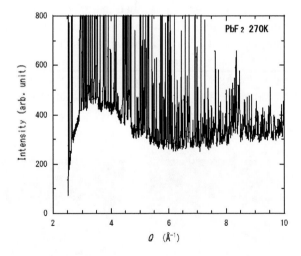

Figure 11. Observed intensity of PbF$_2$ by TOF neutron diffraction experiment.

4.1.2. Semiconductors

The correlation effects among the thermal displacements of atoms are not specific only in ionic crystals. With an interest to explore the correlation effects in semiconductors, diffuse neutron scattering measurement has been performed on VSe, Ge, ZnSe. The Rietveld analysis with the background function including correlation effects is applied to explain the neutron scattering intensity.

To prepare VSe, vanadium and selenium are weighed and enclosed in a quartz tube in vacuum condition and reacted at 1500 K for 3 days. It is annealed at 500 K for 2 days and gradually cooled to room temperature. Neutron diffraction experiment has been performed on powder VSe at 294 K using HRPD [52]. The nuclear scattering length of V (b_V= -0.0382× 10^{-12} cm) is much less than that of Se (b_{Se}= 0.7970× 10^{-12} cm). In this case, the contribution from far neighboring Se-Se pair to the diffuse scattering intensity is very large. From the measurements of neutron scattering intensity the contribution from far neighboring pair could be observed directly. X-ray diffraction measurements were also carried out to determine the atomic position and thermal vibration of V atoms. The diffuse scattering intensity shows oscillatory profile. The space group of VSe is $P6_3/mm$ from the analysis of Bragg intensities. The crystal has ordered arrangement of atoms;

2V in 2a 0, 0, 0,
2Se in 2c 1/3, 2/3, 1/4.

Using the results of Rietveld analysis of X-ray and neutron diffraction intensities electron density distribution and nuclear density distribution of VSe at (100) plane are shown in Figure 12. Nuclear density distribution on the (100) plane of VSe is obtained by MEM (Maximum Entropy Method) analysis with PRIMA of VENUS program system.

The observed oscillatory scheme in the diffuse neutron scattering intensity of VSe is explained by the contribution of Se-Se pair. The peak positions Q of the oscillatory diffuse scattering intensities coincide with the condition $\sin(Qr_{Se-Se})$=1, where r_{Se-Se} is the inter-atomic distance between neighboring Se-Se atoms. Although the value of the correlation effects of far neighboring Se-Se atoms is less than that of nearest neighboring V-Se pair, the large contribution to diffuse scattering from Se-Se atoms is occurred by the value of nuclear scattering length of atoms. The values of correlation effects decrease with the increase of inter-atomic distances. The value of correlation effects is ~ 0.7 for r ~ 2.5 Å and ~ 0.5 for r ~ 4 Å in silver and copper halides [53]. The values of correlation effects in hexagonal VSe show almost the same tendency. The values of the correlation effects do not depend much on the type of the crystal binding and the crystal structure.

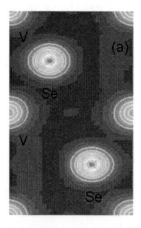 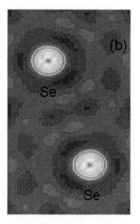

Figure 12. (a) Electron density distribution by X-ray diffraction measurement and (b) nuclear density distribution by neutron diffraction measurement for VSe at (100) plane.

To investigate the correlation effects in semiconductor Ge, diffuse neutron scattering measurement of powder Ge is performed at 6, 150 and 300 K. The Rietveld analysis with the background function including correlation effects is applied to explain the neutron scattering intensity of Ge. Oscillatory diffuse scattering intensity is clearly observed at 150 and 300 K. The diffuse scattering intensity increases with the increase of temperature. Rietveld refinement analysis has been performed on the observed diffraction intensity using RIETAN-94 [54], where the background intensity function of Eq. (10) is used.

The crystal structure of Ge [55,56] is cubic with the space group $Fd\bar{3}m$. The obtained values of correlation effects among thermal displacements of atoms decrease rapidly with increase of inter-atomic distance (Figure 13). The obtained values of the correlation effects of Ge near room temperature are slightly smaller than those of superionic conductors. The obtained value of correlation effects for first nearest neighboring atoms in Ge is almost the same with that obtained by Yoshiasa et al. by EXAFS analysis [44]. In addition to the neutron scattering measurements of crystals, low-frequency excitations in amorphous germanium was also reported [57,58].

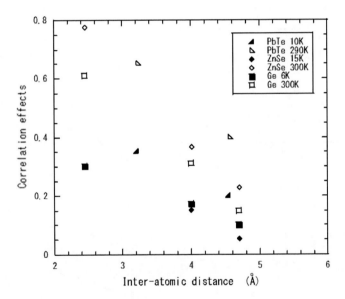

Figure 13. Inter-atomic distance dependence of correlation effects in semiconductors.

Neutron scattering measurements have been performed on powder ZnSe at 15, 150 and 300 K by HRPD [59]. Oscillatory diffuse scattering intensity is clearly observed at 150 and 300 K (Figure 14). The diffuse scattering intensity of Eq. (10) including correlation effects among thermal displacements of atoms is used as background function in Rietveld analysis. The Bragg lines and oscillatory diffuse neutron scattering intensity are explained by including the correlation effects among thermal displacements of atoms into the shape function of background (Figure 15). The oscillatory profile of diffuse background intensity consists of three parts: first (Zn-Se), second (Zn-Zn and Se-Se) and third (Zn-Se) nearest neighboring contribution. The diffuse scattering peak around $Q\sim3.5$ Å^{-1} is mainly from contribution of first and second nearest neighboring atoms. However, the diffuse scattering peak around $Q\sim2$

$Å^{-1}$ would have come from the contribution of second and third nearest neighboring atoms. The values of correlation effects decrease by increasing inter-atomic distances. The values of the correlation effects decrease also with the decrease of temperature.

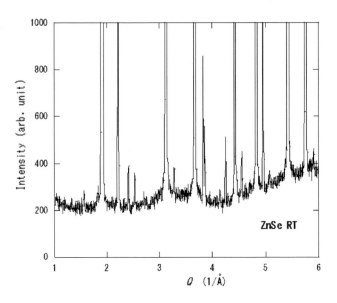

Figure 14. Observed diffraction intensity of ZnSe at room temperature.

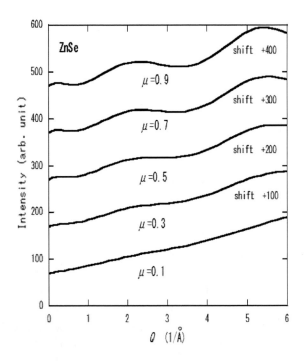

Figure 15. Calculated diffuse scattering intensity of ZnSe at room temperature with correlation effects μ among first nearest neighboring atoms.

4.1.3. Metals

The diffuse neutron scattering measurement of metals has also been performed. The values of correlation effects in metals are obtained near room temperature and compared to those of semiconductors and ionic crystals. The inter-atomic force constants are calculated in these three types of materials. The observed diffuse neutron scattering intensities of Al near room temperature is relatively weak compared to semiconductors and ionic crystals (Figure 16). The difference between the observed scattering intensity at 300 K and that at 10 K of Al is small. Weak peaks of oscillatory diffuse scattering are observed near room temperature at $2\theta \sim 50$ and $100°$. Rietveld analysis has been performed on the neutron scattering intensities of Al using RIETAN-2000 [60]. The crystal of Al has fcc type structure with the space group $Fm\overline{3}m$. The calculated values of Debye-Waller temperature parameter of Al are $B = 0.344$ Å^2 at 10 K and $B = 0.540$ Å^2 at 300 K. The difference of B at these temperatures is very small. The value of thermal parameter B at room temperature is less in comparison to the typical value of ionic crystals and semiconductors. The diffuse scattering intensity of Al has been calculated at 10 K and 300 K. Oscillatory form of the diffuse neutron scattering intensity for Al can be explained by the theory including the thermal correlation effects. The values of correlation effects among first, second and third nearest neighboring atoms in Al at 300 K are 0.67, 0.40 and 0.12, respectively. The values of correlation effects in metals at 300 K are almost the same as those in ionic and covalently-bonded crystals (Figure 17). The values of correlation effects decrease with the decrease of temperature and with the increase of the inter-atomic distance.

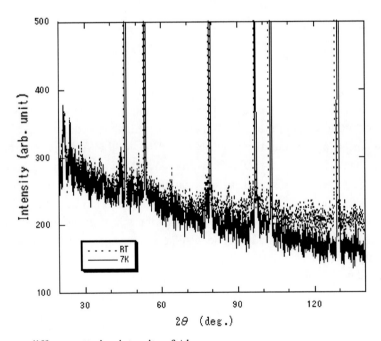

Figure 16. Neutron diffuse scattering intensity of Al.

The values of force constants among first, second and third nearest neighboring atoms of Al, ZnSe and CuCl near room temperature are calculated using Eq. (13). The inter-atomic force constants varied widely among these materials depending on the Debye-Waller temperature factor of the corresponding material. The calculated force constant values for first nearest neighboring atoms in case of ZnSe (6.16 eV/ $Å^2$) and CuCl (1.51 eV/ $Å^2$) are matching well with the earlier reports [40,61]. However, for Al, the calculated value of the force constant for Al (5.72 eV/ Å2) is greater than of the former report [62].

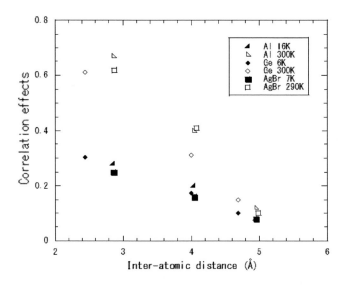

Figure 17. Inter-atomic distance dependence of correlation effects in crystals.

4.2. Disordered Crystals

AgI has been considered as a prototype of superionic conductors [63-68]. Two phases, namely, the β-phase with the wurtzite structure and the γ-phase with the zinc-blend structure of AgI exist at room temperature. The β-phase is more stable than the γ-phase. By increasing temperature, the β-phase and the γ-phase transform to the superionic α-phase at 147°C (Figure 18). The β-phase and α-phase show ordered and disordered arrangement of Ag atoms, respectively. The oscillatory strong diffuse scattering is observed in β-AgI where silver and iodine atoms show ordered arrangements. The diffuse scattering intensity of β-AgI has been calculated by considering hexagonal system with the space group $P6_3mc$. [69]. The inter-atomic distances of the first and second diffuse scattering peaks of hexagonal β-AgI are almost the same as those of cubic α-AgI.

In the α-phase, I atoms form a bodycentered cubic lattice, whereas the highly mobile Ag ions are randomly distributed throughout the equivalent atomic positions. Anomalously strong and oscillatory diffuse scattering from AgI has been observed in X-ray and neutron diffraction measurements [70]. From the measurements of the neutron diffuse scattering intensities in β-AgI and α-AgI near the phase transition temperature 147°C, oscillatory forms of the diffuse scattering intensity appear in both phases of AgI and the difference of the scattering intensities seems significant in low Q region (Figure 19). Neutron diffuse scattering intensity of α-AgI powder crystal was recorded by the $\Delta E{\sim}0$ scan method [71] that collected the scattered intensity from static disordered arrangements of atoms and a part of thermal vibration of atoms within the limit of energy resolution. The contribution from the disordered arrangements of silver atoms in α-AgI having averaged structure is observed in low Q region (Figure 20). The most of the oscillatory scheme of the diffuse scattering of β-AgI and α-AgI are coming from the thermal contribution as observed by double axis measurement (diffraction measurement).

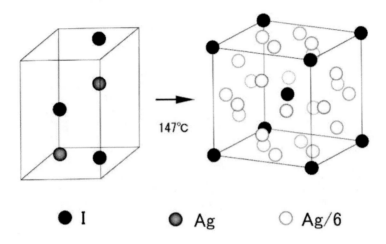

Figure 18. Phase transition of AgI at 147°C.

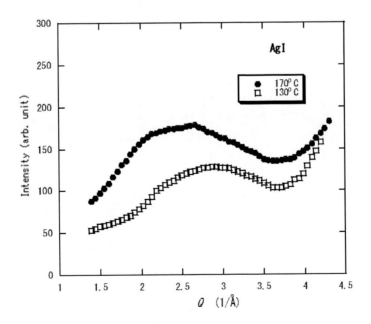

Figure 19. Neutron diffuse scattering curves of β–AgI and α-AgI at 130 and 170°C, respectively, without the crystal analyzer (diffraction measurement).

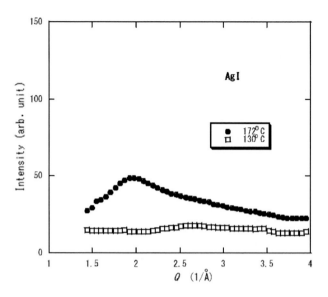

Figure 20. Neutron diffuse scattering curves of β-AgI and α-AgI at 130 and 172°C, respectively, with the crystal analyzer at elastic position ($\Delta E\sim 0$ scan method).

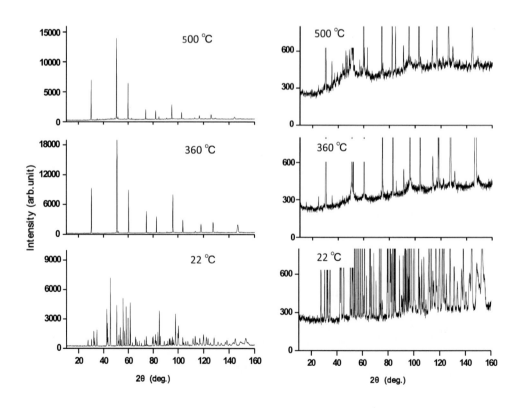

Figure 21. Observed diffraction intensity (left) and diffuse scattering intensity (right) of PbF_2 by HRPD.

The contribution from the disordered arrangements of silver atoms in α-AgI single crystal is also observed in low Q region. The strong diffuse scattering around the 110 reciprocal lattice point is being noticed. The observed diffuse scattering intensity of α-AgI single crystal with $\Delta E \sim 0$ scan method can be explained on the basis of disordered arrangement model where two Ag atoms in the unit cell are distributed into 48j sites.

Another disordered material studied is PbF_2, a typical superionic conductor. Its high conductivity has been linked to strong disordering in the fluorine sub-lattice [72-75]. Neutron diffuse scattering measurements of α-PbF_2 at 22°C, β-PbF_2 at 360°C and γ-PbF_2 at 500°C are observed by HRPD (Figure 21). As the temperature increases from 22°C to 360°C, the diffuse scattering intensity increases slightly. On further increase in temperature from 360°C to 500°C, the obvious increase of the diffuse scattering intensity near $2\theta \sim 50°$ is observed [76]. The structure analysis of γ-PbF_2 at 500°C has been performed with a disordered arrangement model of F atoms using RIETAN-2000 (Figure 22). Eight F atoms are statistically distributed on two 32f sites (Figure 23). The diffuse scattering intensity from γ-PbF_2 is calculated from Eq. (9). In the equation, $(\alpha - \beta)_r$ corresponds to short-range order parameter. The probability functions α_r and β_r satisfy the following relations;

$$\alpha_{r_{F(I)F(I)}} - \beta_{r_{F(I)F(I)}} = \frac{\alpha_{r_{F(I)F(I)}} - W_{F(I)}}{1 - W_{F(I)}} \tag{15}$$

$$\alpha_{r_{F(I)F(II)}} - \beta_{r_{F(I)F(II)}} = \frac{\alpha_{r_{F(I)F(II)}} - W_{F(II)}}{1 - W_{F(I)}} \tag{16}$$

$$\alpha_{r_{F(II)F(II)}} - \beta_{r_{F(II)F(II)}} = \frac{\alpha_{r_{F(II)F(II)}} - W_{F(II)}}{1 - W_{F(II)}} \tag{17}$$

$$\alpha_{r_{F(II)F(I)}} - \beta_{r_{F(II)F(I)}} = \left| \left(\frac{W_{F(I)}}{W_{F(II)}} \right) \alpha_{r_{F(I)F(II)}} - W_{F(I)} \right| / (1 - W_{F(II)}) . \tag{18}$$

The probability functions α_r and β_r have to satisfy the restriction conditions;

$$\sum_r Z_{rF(I)F(I)}(\alpha_{rF(I)F(I)} - \beta_{rF(I)F(I)}) + \sum_r Z_{rF(I)F(II)}(\alpha_{rF(I)F(II)} - \beta_{rF(I)F(II)}) = -1 \tag{19}$$

$$\sum_r Z_{rF(II)F(II)}(\alpha_{rF(II)F(II)} - \beta_{rF(II)F(II)}) + \sum_r Z_{rF(II)F(I)}(\alpha_{rF(II)F(I)} - \beta_{rF(II)F(I)}) = -1 . \tag{20}$$

The diffuse scattering intensity is being calculated with these probability functions and the values of correlation effects among thermal displacements of atoms. The result of the calculated diffuse scattering intensity can qualitatively explain the observed peaks of the diffuse scattering intensity around $2\theta \sim 50°$ and $100°$. The oscillatory profile of the diffuse scattering intensity is mainly explained from the correlation effects among thermal displacements of Pb-F atoms and short-range order of F atoms.

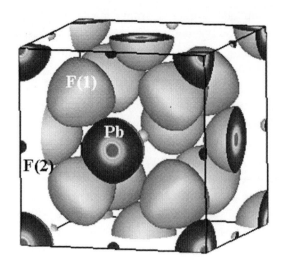

Figure 22. MEM analysis of γ-PbF$_2$.

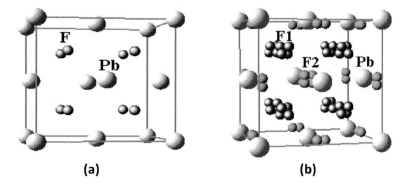

(a)   **(b)**

Figure 23. Crystal structure of (a) β-PbF$_2$ and (b) γ-PbF$_2$.

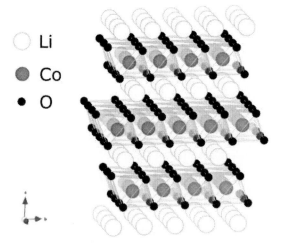

Figure 24. Crystal structure of LiCoO$_2$.

We also investigated the diffuse scattering behavior in a mixed ionic-electronic superionic conductor like copper selenide. $Cu_{2-\delta}Se$ is a mixed ionic-electronic superionic conductor with homogeneity range of $\delta=0$-0.25 [77-80]. The α-phase is stable in the range $\delta=0.15$-0.25 at room temperature. The β-α phase transition temperature is 414 K for Cu_2Se. Crystal structures of $Cu_{2-\delta}Se$ have been studied in the superionic phase by X-ray and neutron powder diffraction measurements [81]. Neutron diffuse scattering from a powder sample of α-Cu_2Se is measured using double axis diffractometer (diffraction experiment) and triple-axis spctrometer with a fixed energy analyzer to measure the elastic scattering (elastic scattering measurement) [82-84]. A part of the damped phonon component is observed in the elastic scattering measurement. The crystal belongs to space group $Fm\bar{3}m$. Se atoms occupy 4a site. Cu atoms occupy mainly in the tetrahedral 8c sites (or 32f sites around 8c) and a small fraction in trigonal 32f sites. Oscillatory strong diffuse scattering intensities are observed. The diffuse scattering intensities were analyzed under the assumption that the Cu atoms were distributed over two 32f crystallographic sites. The diffuse scattering intensities by diffraction and elastic scattering measurements are explained by correlation effects among thermal displacements of atoms [85,86]. The dynamic structure factor including elastic peak, the broad quasi-elastic component and low energy excitations with frequencies of 3.5–4 meV of $Cu_{1.78}Se$ and Cu_2Se are studied [87]. The broad component could be connected with fast-localized diffusion of Cu ions between split 32f sites.

Superionic properties appear in crystalline and glassy materials [88,89]. Many superionic glasses have also been studied by neutron inelastic scattering and diffraction measurements [90-95].

5. SPECIAL APPLICATION OF DIFFUSE NEUTRON SCATTERING IN OPTIMIZING ANNEALING PROCESS

Lithium-based metal oxides $LiCoO_2$, $LiMn_2O_4$ and $LiFePO_4$ are important materials that can be used as cathode materials of rechargeable lithium-ion batteries because of their high energy density [96-105]. At present, $LiCoO_2$ is widely used as commercial product for cathode in lithium-ion batteries (Figure 24). However, the cobalt in $LiCoO_2$ is a relatively scarce, expensive and toxic. Nowadays, $LiNiO_2$ and $LiMn_2O_4$ are studied as alternative materials. $LiMn_2O_4$ (Figure 25) is electrochemically more stable than $LiNiO_2$ composites. Also, $LiNiO_2$ is more difficult to prepare compared to $LiCoO_2$. $LiNiO_2$ exists in two crystal structures, cubic and layered hexagonal. Only the layered hexagonal structure has the potential to be applied as cathode material. Different methods and starting materials have been reported in literature to prepare these materials. In a solid state reaction route, one may get the final product as composite of $LiNiO_2$ and its precursors to some extent. Neutron scattering experiments have been performed on $LiNiO_2$ composites and $LiMn_2O_4$ composites in the process of optimization of annealing parameters. The analysis of crystallite size of $LiNiO_2$ and $LiMn_2O_4$ is performed from the FWHM of the Bragg lines using Scherrer equation. The intensities of the Bragg lines and the diffuse background scattering of $LiNiO_2$ composites and $LiMn_2O_4$ composites have been used to estimate the crystallite size and the amount of crystalline $LiNiO_2$ and $LiMn_2O_4$ produced in the annealing process.

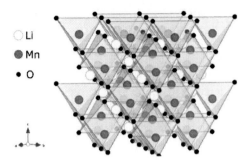

Figure 25. Crystal structure of LiMn$_2$O$_4$.

The sample of LiNiO$_2$ are prepared by dry mixing of starting materials LiOH and NiO and then milled by high-energy planetary milling for 3 hours at frequency 3 Hz to perform a ground product [106]. The mixture is heated at temperature 400°C for 5 hours. The final product has been treated in three ways to get three kinds of samples, (A) which is not annealed further, (B) annealed at 500°C for 5 hours and (C) annealed at 650°C for 5 hours. The reaction that leads to make LiNiO$_2$ from the ground product of LiOH and NiO at high temperature is assumed as follows

$$2LiOH + 2Ni^{(II)}O \xrightarrow{O_2} 2LiNi^{(III)}O_2 + H_2O \qquad (21)$$

Neutron diffraction measurements at room temperature have been performed for LiNiO$_2$ composites (sample A, B and C) by HRPD. Figure 26 shows the neutron scattering intensity from powder LiNiO$_2$ composites for sample A, B and C. Bragg lines of LiNiO$_2$ from sample A are observed at 2θ around 52 and 77°. In further annealing, the Bragg lines become more distinct in sample B. The most clear and sharp Bragg lines of LiNiO$_2$ appear in sample C which is annealed at 650°C.

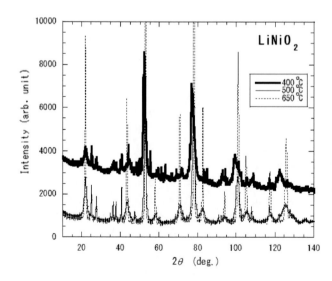

Figure 26. Annealing temperature dependence of neutron diffraction intensities of LiNiO$_2$ composites at room temperature.

Very strong intensity of background scattering from sample A was also found in Figure 26. The background intensities which include the contribution from incoherent scattering of hydrogen atoms and other atoms become smaller in the sample B and C. It indicates although the crystalline $LiNiO_2$ has been formed, a large fraction of starting precursors LiOH could still exist in the sample A and B. The differences of the background intensities are proportional to the number of hydrogen atoms (Δn_H) as they have large incoherent scattering cross section as:

$$\Delta I_{background} \propto \Delta n_H . \tag{22}$$

The amount of hydrogen atoms in the ground product decreases by annealing process and hence the background intensity in sample C.

The crystal $LiNiO_2$ belongs to hexagonal symmetry with space group $R\bar{3}m$. Lithium, nickel and oxygen atoms occupy 3a, 3b and 6c sites, respectively. The width of Bragg lines of $LiNiO_2$ in the sample A, B and C shows a tendency to decrease by increasing the annealing temperature. The average crystallite size of $LiNiO_2$ formed in the annealing process from ground product has been analyzed from peak width of Bragg line using Scherrer equation:

$$t = \frac{K\lambda}{D\cos\theta}, \tag{23}$$

where t is the average crystallite size in angstrom and K a constant and equal to 0.9. λ is the wavelength of incident beam and D is FWHM (full width at half maximum) of Bragg line. The average sizes of $LiNiO_2$ are determined to be about 30, 40 and 120 Å for sample A, B and C, respectively. The average crystallite size of $LiNiO_2$ increases with the increase of annealing temperature. The size of these materials is of nano-dimension. Assuming that 100% of the starting materials changes to crystalline $LiNiO_2$ in the sample C, the percentage of crystalline $LiNiO_2$ formation in sample A and B are about 61 and 87%, respectively as observed from the comparison of area of the Bragg line of $LiNiO_2$. The fraction of crystalline $LiNiO_2$ as formed by annealing process increases with annealing temperature.

In a similar line, powder sample of $LiMn_2O_4$ has been prepared from starting materials LiOH and MnO_2. The mixture is milled by high-energy planetary milling for 3 hours at frequency 2 Hz to perform a ground product. Afterward, it is heated at temperature 200°C for 16 hours. In order to analyze the anneal temperature effect to the crystallite size of $LiMn_2O_4$, the sample is divided into three parts: sample A which is annealed at 400°C for 3 hours, sample B annealed at 500°C for 3 hours and sample C annealed at 800°C for 3 hours. Neutron scattering experiments have been performed on $LiMn_2O_4$ (sample A, B and C) at room temperature using HRPD [107]. Figure 27 shows neutron scattering intensity from sample A, B and C. Neutron scattering intensity of sample A shows very weak and broad Bragg peaks of $LiMn_2O_4$ around $2\theta \sim 45°$, 95° and 118°. These three peaks become stronger and narrower in the sample B. The strongest and narrowest Bragg peaks are observed in the sample C. In the case of $LiMn_2O_4$ the diffuse scattering intensity does not change so much in sample A, B and C. The amount of starting material LiOH is very small even in sample A. This means that the solid state reaction to perform $LiMn_2O_4$ is almost completed by annealing the sample at

$400^{\circ}C$. Crystal of $LiMn_2O_4$ belongs to cubic type structure with the space group $F d\bar{3} m$. The crystallite size of $LiMn_2O_4$ in samples A, B and C are calculated from three Bragg peaks (400), (331) and (622) using Scherrer equation. The average crystallite sizes of $LiMn_2O_4$ are ~50, 140 and 370 Å for sample A, B and C, respectively. The average size increases almost linearly with annealing temperature.

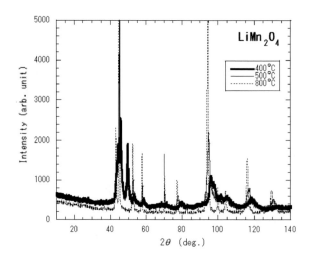

Figure 27. Annealing temperature dependence of neutron scattering intensities of $LiMn_2O_4$ composites at room temperature.

Recently, new neutron sources have been constructed in many places around the world. Diffuse neutron scattering studies of various kinds of materials including ionic crystals have been performed by using research reactor OPAL, ANSTO (Australian Nuclear Science & Technology Organization). New TOF neutron scattering instruments have also started operating in J-PARC, Japan. Neutron diffraction patterns from crystals can be obtained within 40 min. by iMATERIA diffractometer [108-110]. If the power of accelerator could be increased in coming years, we would obtain diffraction data in 5 min. Neutron diffraction measurement techniques and theory are making rapid progress for studying static and dynamic characters of atoms in solids. Neutron diffuse scattering analysis is going to be a very useful tool for exploring fundamental lattice dynamics parameters like inter-atomic forces constants, phonon dispersion, elastic constants and density of states.

CONCLUSIONS

The diffuse scattering intensities of ionic crystals, semiconductors and metals are studied by neutron diffraction measurement. Oscillatory diffuse scattering patterns are commonly observed from ordered and disordered crystals near room temperature. The diffuse scattering theory including the correlation effects among thermal displacements of atoms has been applied to the analysis of diffuse scattering intensity. The values of correlation effects among first, second and third nearest neighboring atoms are determined from the analysis. The values of the correlation effects decrease rapidly with the increase of the inter-atomic distance. The values of the correlation effects also decrease with the decrease of temperature. The tendency of temperature dependence of the values of correlation effects agrees with the result of EXAFS analysis using simple Debye model. The values of force constants are derived from the correlation effects. Phonon dispersion relation has been estimated using the force constants. As the phonons are direction-dependent atomic displacements, the information of crystal structure is used in the simulation program.

The diffuse scattering from disordered crystals has also been investigated with probability functions and the values of correlation effects among thermal displacements of atoms. The result of the calculated diffuse scattering intensity can qualitatively explain the observed pattern. The oscillatory profile of the diffuse scattering intensity is mainly determined from the correlation effects among thermal displacements of first nearest neighboring atoms and short-range order among disordered atoms.

Diffuse scattering analysis has been used as a tool to optimize the annealing process of nano-sized lithium cathode materials prepared by solid state reaction. The average crystallite sizes of lithium metal oxides increases with increasing annealing temperature. The large background intensities of some samples indicate the presence of some fraction of starting materials in the samples.

ACKNOWLEDGMENTS

Authors would like to thank Prof. H. Takahashi, Prof. O. Kamishima, Dr. M. Arai, Dr. N. Igawa, Dr. K. Basar, Dr. E. Kartini, and Dr. S. Danilkin for their useful discussions. Authors wish to express their thanks to Ibaraki Prefecture for financial support. This research was partially supported by the Ministry of Education, Science Sports, and Culture, Japan, Grant-in-Aid for Scientific Research (C), 21540314, 2010. One of the authors (X.) wishes to express her thanks for financial support to the Inner Mongolia Autonomous Region University, China (Grant No. NJ10114) and the Doctor-Master research fund of Inner Mongolia University for the Nationalities, China (Grant No. BS239).

REFERENCES

[1] Warren, B. E. *X-ray Diffraction*, Addison-Wesley, London. 1969.
[2] Miyake, S. *Diffraction of X-rays*, Kyoritsu Publ., Tokyo. 1969.
[3] Sakuma, T. *B. Electrochem.* 1995, 11, 57-80.
[4] Sakuma, T.; Basar, K.; Shimoyama, T.; Hosaka, D.; Xianglian; Arai, M. In *Physics of Solid State Ionics*; Sakuma; Ed.; Research Signpost Pub., Kerala, 2006, 323-346.
[5] Beeken, R. B.; Sakuma, T. In *Modern Topics in Chemical Physics*; George; Ed.; Research Signpost, Trivandrum, 2002, 353-372.
[6] Ortiz, A. L.; Sánchez-Bajo, F.; Padture, N. P.; Cumbrera, F. L.; Guiberteau, F. *J. Euro. Ceramic Soc.* 2001, 21, 1237-1248.
[7] Sakuma, T.; Thomas, J. O. *J. Phys. Soc. Jpn.* 1993, 62, 3127-3134.
[8] Sakuma, T.; Thazin, A.; Arai, M.; Takahashi H. In *Recent Research Developments in Solid State Ionics*; Pandalai; Ed.; Transworld Research Network, Kerala, 2004, 2, 207-224.

[9] Sakuma, T. *J. Phys. Soc. Jpn.* 1992, 61, 4041-4048.

[10] Sakuma, T. *J. Phys. Soc. Jpn.* 1993, 62, 4150-4151.

[11] Geller, S. *Solid Electrolytes*; Springer-Verlag, New York. 1997.

[12] Salamon, M. B. *Physics of Superionic Conductors*; Springer-Verlag, New York. 1978.

[13] Nilsson, L.; Thomas, O, J.; Tofield, B, C. *J. Phys.* 1980, *C*13, 6441-6451.

[14] Kaber, R.; Nilsson, L.; Andersen, N. H.; Lundén, A.; Thomas, O. J. *J. Phys. Condens. Matter* 1992, 4, 1925-1933.

[15] Chandra, S. *Superionic Solids*; North-Holland, Amsterdam. 1995.

[16] Inoue, N.; Zou, Y. *Solid State Ionics* 2005, 176, 2341-2344.

[17] Wakamura, K. *Solid State Ionics* 2009, 180, 1343-1349.

[18] Schneider, J.; Schulz, H. *Zeit. Kristall.* 1993, 203, 1-15.

[19] Nishida, K.; Asai, T.; Kawai, S. *Solid State Commun.* 1983, 48, 701-704.

[20] Kataoka, K.; Takahashi, Y.; Kijima, N.; Hayakawa, H.; Akimoto, J.; and Ohshima, K. *Solid State Ionics* 2009, 180, 631-635.

[21] Kohda, T.; Ono, S.; Kobayashi, M.; Iyetomi, H.; Kashida, S. *J. Phys. Soc. Jpn.* 2001, 70, 2689-2693.

[22] Hoshino, S.; Sakuma, T.; Fujii, Y. *J. Phys. Soc. Jpn.* 1979, 47, 1252-1259.

[23] Hoshino, S.; Shapiro, S. M.; Fujishita, H.; Sakuma, T. *J. Phys. Soc. Jpn.* 1988, 57, 4199-4205.

[24] Sakuma, T.; Saitoh, S. *J. Phys. Soc. Jpn.* 1985, 54, 3647-3648.

[25] Aniya, M.; Wakamura, K. *Solid State Ionics* 1996, 86-88, 183-186.

[26] Azzoni, C. B.; Camagni, P.; Samoggia, G.; Paleari, A. *Solid State Ionics* 1993, 60, 233-226.

[27] Stern, E. A.; Livins, P.; Zhang, Z. *Phys. Rev.* 1990, *B*43, 8850-8860.

[28] Beni, G.; Platzman, P. M. *Phys. Rev.* 1976, *B*14, 1514-1518.

[29] Sakuma , T.; Nakamura , Y.; Hirota , M.; Murakami , A.; Ishii, Y. *Solid State Ionics* 2000, 127, 295-300.

[30] Kamishima, O.; Ishii, T.; Maeda, H.; Kashino, S. *Solid State Commun.* 1977, 103, 141-144.

[31] Hoshino, S.; Fujii, Y.; Harada, J.; Axe, J. D. *J. Phys. Soc. Jpn.* 1976, 41, 965-973.

[32] Ishii, T.; Kamishima, O. *J. Phys. Soc. Jpn.* 1999, 68, 3580-3584.

[33] Yokoyama, T.; Satsukawa, T.; Ohta, T. *Jap. J. Appl. Phys.* 1989, 28, 1905-1908.

[34] Keen, D. A.; Hull, S.; *J. Phys. Condens. Matter* 1995, 7, 5793-5804.

[35] Vardeny, Z.; Gilat, G.; Moses, D. *Phys. Rev.* 1978, *B*18, 4487-4496.

[36] Sakuma, T.; Shimoyama, T.; Basar, K.; Xianglian.; Takahashi, H.; Arai, M.; Ishii, Y. *Solid State Ionics* 2005, 176, 2689-2693.

[37] Xianglian.; Siagian, S.; Basar, K.; Sakuma, T.; Takahashi, H.; Igawa, N.; Ishii, Y.; *Solid State Ionics* 2009, 180, 480-482.

[38] Basar, K.; Xianglian.; Sakuma, T. *Indonesian J. Physics* 2006, 17, 95-98.

[39] Arai, M.; Ohki, K.; Mutou, M.; Sakuma, T.; Takahashi, H.; Ishii, Y. *J. Phys. Soc. Jpn.* 2001, Suppl. A. 70, 250-252.

[40] Aniya, M. *Solid State Ionics* 1999, 121, 281-284.

[41] Srinivas, K.; Sirdeshmukh, D. B. *Pramana* 1984, 23, 595-597.

[42] Arai, M.; Shimoyama, T.; Sakuma, T.; Takahashi, H.; Ishii, Y. *Solid State Ionics* 2005, 176, 2477-2480.

[43] Basar, K.; Xianglian.; Sakuma, T.; Takahashi, H.; Igawa, N. *ITB J. Sci.* 2009, 41, 50-58.

[44] Yoshiasa, A.; Koto, K.; Maeda, H.; Ishii, T. *Jpn. J. App. Phys.* 1997, 36, 781-784.

[45] Kamishima, O.; Ishii, T.; Maeda, H.; Kashino, S. *Jpn. J. Appl. Phys.* 1997, 36, 247-253.

[46] Kittel, C. *Introduction to Solid State Physics*, 7th edn; John Wiley & Sons, Inc., New York, 1996.

[47] Woods, A. D. B.; Brockhouse, B. N.; Cowley, R. A.; Cochran, W. *Phys. Rev.* 1963, 131, 1025-1029.

[48] Boldrini, P.; Loopstra, B. O. *Acta Cryst.* 1967, 22, 744-745.

[49] Kamijo, N.; Koto, K.; Ito, Y.; Maeda, H.; Tanabe, K.; Hida, M.; Terauchi, H. *J. Phys. Soc. Jpn.* 1984, 53, 4210-4220.

[50] Basar, K.; Xianglian.; Siagian, S.; Sakuma, T.; Takahashi, H.; Yonemura, M.; Kamiyama, T.; Ishigaki, T.; Igawa, N. *Physica* 2008, *B403*, 2557-2560.

[51] Ohta, T.; Izumi, F.; Oikawa, K.; Kamiyama, T. *Physica* 1997, *B234-236*, 1093-1095.

[52] Sakuma, T.; Xianglian.; Siagian, S.; Basar, K.; Takahashi, H.; Igawa, N.; Kamishima, O. *J. Thermal Anal. Cal.* 2010, 99, 173-176.

[53] Sakuma, T.; Nakamura, Y.; Hirota, M.; Arai, M.; Ishii, Y. *J. Phys. Chem. Solids* 1999, 62 1503-1506.

[54] Izumi, F. In *The Rietveld Method;* Young; Ed; Oxford Univ. Press, Oxford, 1993, Chapter 13.

[55] Nisson, G.; Nelin, G. *Phys. Rev.* 1971, *B3*, 364-369.

[56] Wei. S.; Chou, M. Y. *Phys. Rev.* 1994, *B50*, 2221-2226.

[57] Buchenau, U.; Prager, M.; Kmitakahara, W. A.; Shanks, H. R.; Nucker, N. *Europhys. Lett.* 1988, 6, 695-700.

[58] Sakuma, T.; Aoyama, T.; Tsuchiya, Y. *J. Phys. Soc. Jpn.* 1995, 64, 3770-3774.

[59] Basar, K.; Siagian, S.; Xianglian.; Sakuma, T.; Takahashi H. Igawa, N. *Nuclear Instruments and Methods in Physics Research* 2009, *A600*, 237-239.

[60] Izumi, F.; Ikeda, T. *Mater. Sci. Forum* 2000, 198-203, 321-324.

[61] Kushwaha, A. K. *Physica* 2010, *B405*, 1638-1642.

[62] Walker, C. B. *Phys. Rev.* 1956, 103, 547-557.

[63] Hoshino, S. *Solid State Ionics* 1991, 48, 179-201.

[64] Jagodzinski, H. *Zeit. Kristall.*1959, 112, 80-87.

[65] Hoshino, S. *J. Phys. Soc. Jpn.* 1957, 12 315-326.

[66] Hull, S.; Keen, D. A.; Sivia, D. S.; Madden, P. A.; Wilson, M. *J. Phys. Condens. Matter* 2002, 14, 9-17.

[67] Hoshino, S.; Sakuma T.; Fujii, Y. *Solid State Commun.*1977, 22, 763-765.

[68] Dalba, G.; Fornasini, P.; Gotter, R.; Rocca, R. *Phys. Rev.* 1995, *B52*, 149-157.

[69] Thazin, A.; Arai, M.; Sakuma, T.; *Solid State Ionics*; Chowdari; Ed; World Scientific Pub., Singapore, 2002, 777-784.

[70] Hoshino, S.; Fujishita, H.; Sakuma, T. *Phys. Rev.* 1982, *B25*, 2010- 2011.

[71] Sakuma, T.; Hoshino, S. *J. Phys. Soc. Jpn.* 1993, 62, 2048-2050.

[72] Hull, S.; Berastegui, P.; Eriksson, S. G.; Gardner, N. J. G. *J. Phys: Condens. Matter* 1998, 10, 8429-8446.

[73] Koto, K.; Schulz, H.; Huggins, R. *Solid State Ionics* 1981, 3-4, 381-384.

[74] Kosacki, I. *Solid State Ionics* 1988, 28-30, 449-451.

[75] Merlino, S.; Pasero, M.; Perchiazzi N.; Kampf, A. R. *American Mineralogist* 1966, 81, 1277-1281.

[76] Xianglian.; Sakuma, T.; Basar, K.; Takahashi, H.; Igawa, N. *J. Phys. Soc. Jpn.* 2010, 79

Supplement A, 29-32.

[77] Ishikawa, T.; Miyatani, S. *J. Phys. Soc. Jpn.* 1977, 42, 159-167.

[78] Korzhuev, M. A.; Bankina, V. F.; Gruzinov, B. F.; Bushmarina, G. S. *Sov. Phys. Semicond.* 1989, 23, 959-962.

[79] Ohtani, T.; Tachibana, Y.; Ogura, J.; Miyake, T.; Okada Y.; Yokota, Y.; *J. Alloys and Compounds* 1998, 279, 136-141.

[80] Ohtani, T.; Shohno, M. *J. Solid State Chemistry* 2004, 177, 3886-3890.

[81] Sakuma, T.; Sugiyama, K.; Matsubara, E.; Waseda, Y. *Materials Transactions, JIM* 1989, 30, 365-369.

[82] Sakuma, T.; Aoyama, T.; Takahashi, H.; Shimojo, Y.; Morii, Y. *Physica* 1995, *B*213&214, 399-401.

[83] Danilkin, S. A. *Solid State Ionics* 2009, 180, 483-487.

[84] Danilkin, S. A. *J. Alloys and Compounds* 2009, 467, 509-513.

[85] Sakuma, T.; Aoyama, T.; Takahashi, H.; Shimojo, Y.; Morii, Y.; *Physica* 2005, *B*213&214, 399-401.

[86] Basar, K.; Shimoyama, T.; Hosaka, D.; Xianglian.; Sakuma, T.; Arai, M. *J. Therm. Anal. Cal.* 2005, 81, 507-510.

[87] Danilkin, S. A.; Skomorokhov, A. N.; Hoser, A.; Fuess, H.; Rajevac V.; Bickulova, N. N. *J. Alloys and Compounds* 2003, 361, 57-61.

[88] Aniya, M. *Solid State Ionics* 2000, 136-137, 1085-1089.

[89] Wasiucionek, M.; Garbarczyk, J. E.; Wngtrzewski, B.; Machowski, P.; Jakubowski, W. *Solid State Ionics* 1996, 92, 155-160.

[90] Kartini, E.; Nakamura, M.; Arai, M.; Inamura, Y.; Taylor, J. W.; Russina, M. *Solid State Ionics* 2009, 180, 506-509.

[91] Kartini, E.; Arai, M.; Mezei, F.; Nakamura, M.; Russina, M. *Physica* 2006, *B*385-386, 236-239.

[92] Kartini, E.; Collins, M. F.; Lovekin, C. C.; Svensson, E. C.; Musyafaah, E. *Journal of Non-Crystalline Solids* 2002, 312-314, 633-636.

[93] Takahashi, H.; Sakuma, T.; Ishii, Y. *Solid State Ionics* 2003, 160, 103-107.

[94] Takahashi, H.; Rikitake, N.; Sakuma, T.; Ishii, Y. *Solid State Ionics* 2004, 168, 93-98.

[95] Takahashi, H.; Hiki, Y.; Sakuma, T.; Funahashi, S. *Solid State Ionics* 1992, 53-56, 1164-1167.

[96] Dahn, J. R.; von Sacken, U.; Michal, C. A. *Solid State Ionics* 1990, 44, 87-97.

[97] Kobayashi, H.; Arachi, Y.; Kageyama, H.; Sakaebe, H.; Tatsumi, K.; Mori, D.; Kanno, R .; Kamiyama, T. *Solid State Ionics* 2004, 175, 221-224.

[98] Marzec, J.; Świerczek, K.; Przewoźnik, J.; Molenda, J.; Simon, D. R.; Kelder, E. M.; Schoonman, J. *Solid State Ionics* 2002, 146, 225-237.

[99] Tsang, C.; Manthiram, A. *Solid State Ionics* 1996, 89, 305-312.

[100] Tsumura, T.; Shimuzu, A.; Inagaki, M. *Solid State Ionics* 1996, 90, 197-200.

[101] Arai, H.; Okada, S.; Ohtsuka, H.; Ichimura, M.; Yamaki, J. *Solid State Ionics* 1995, 80, 261-269.

[102] Molenda, J.; Wilk, P.; Marzec, J. *Solid State Ionics* 2002, 146, 73-79.

[103] Li, W.; Reimers, J. N.; Dahn, J. R. *Phys. Rev.*1992, *B*46, 3236-3246.

[104] Takahashi, Y.; Kimima, N.; Dokko, K.; Nishizawa, M.; Uchida, I.; Akimoto, J. *Solid State Chemistry* 2007, 180, 313-321.

[105] Piszora, P. *J. Alloys Compounds* 2004, 382, 112-118.

[106] Basar, K.; Xianglian.; Honda, H.; Sakuma, T.; Takahashi, H.; Abe, O.; Igawa, N.; Ishii, Y. In *Solid State Ionics* ; Chowdari; Ed.; World Scientific Pub., Singapore, 2006, 121-128.

[107] Basar, K.; Xianglian.; Siagian, S.; Ohara, K.; Sakuma, T.; Takahashi, H.; Abe, O.; Igawa, N.; Ishii, Y. *Indonesian Journal of Physics.* 2009, 20, 9-12.

[108] Ishigaki, T.; Harjo, S.; Yonemura, M.; Kamiyama, T.; Aizawa, K.; Oikawa, K.; Sakuma, T.; Morii, Y.; Arai, M.; Ebata, K.; Takano, Y.; Kasao, T. *Physica* 2006, *B*385-386, 1022-1024.

[109] Hoshikawa, A.; Ishigaki, T.; Nagai, M.; Kobayashi, Y.; Sagehashi, H.; Kamiyama, T.; Yonemura, M.; Aizawa, K.; Sakuma, T.; Tomota, Y.; Arai, M.; Hayashi, M.; Ebata, K.; Takano, Y.; Kasao, T. *Nuclear Instruments and Methods in Physics Research* 2009, *A*600, 203-206.

[110] Ishigaki, T.; Hoshikawa, A.; Yonemura, M.; Morishima, T.; Kamiyama, T.; Oishi, R.; Aizawa, K.; Sakuma, T.; Tomota, Y.; Arai, M.; Hayashi, M.; Ebata, K.; Komatsuzaki, K.; Asano, H.; Kasao, T. *Nuclear Instruments and Methods in Physics Research* 2009, *A* 600, 189-191.

In: Neutron Scattering Methods and Studies
Editor: Michael J. Lyons
ISBN: 978-1-61122-521-1
© 2011 Nova Science Publishers, Inc.

Chapter 2

SMALL ANGLE SCATTERING ANALYSIS OF VIRUS-LIKE PARTICLES FOR BIOMEDICAL DIAGNOSTIC ASSAYS

Susan Krueger[1], Janet L. Huie[2] and Deborah A. Kuzmanovic[3]

[1]NIST Center for Neutron Research, NIST, Gaithersburg, MD, USA
[2]Sciencenter, Ithaca, NY, USA
[3]University of Michigan, Biophysics and the Department of Molecular, Cellular, and Developmental Biology, Ann Arbor, MI, USA

ABSTRACT

Technical advances in biotechnology have made the production of genetically engineered biomaterials commonplace. Many artificially produced biomolecules retain native binding to protein or nucleic acid targets, assemble into higher order complexes, and perform native catalytic functions.

Virus-based tools for biomedical purposes are of particular interest. Virus coat proteins can spontaneously assemble into intact virus-like particles (VLPs), with many structural features of the native virus. Specifically, the VLP displays surface antigens for host recognition, and its protective shell prevents damage to and acts as a delivery agent for its contents. Unlike native viruses, VLPs are noninfectious and usually lack nucleic acid content. These features have been used to advantage to produce vaccines such as for human papillomavirus (HPV). However, VLP vaccine development would be greatly served by a facile, high resolution method for physical characterization of VLPs in the form to be administered (i.e., in solution).

VLPs are also ideally suited as standards for corresponding infectious viral particles in clinical diagnostic assays. In particular, VLP carriers of short nucleic acid sequences show great promise as quantitative reference standards in DNA and RNA diagnostic assays. However, as for VLP vaccine development, the use of nucleic acid-filled VLPs in diagnostic testing has been hampered by a lack of physical techniques to easily measure VLP physical properties in solution.

*Corresponding Author. University of Michigan, Biophysics and the Department of Molecular, Cellular, and Developmental Biology, 930 N. University, 3039 Chemistry Bldg, Ann Arbor, MI 48109, Tel: (734)-647-9121, Fax: (734)-764-3323, Email: kuzman@umich.edu.

Small-angle neutron scattering (SANS) and small-angle x-ray scattering (SAXS) are novel approaches to physical characterization of VLPs for vaccine and diagnostic standard development. SANS offers the unique capability of studying both the coat protein and the RNA or DNA of interest as separate entities, within a VLP containing both components. Both SAXS and SANS measurements are performed using VLPs in solution. This chapter examines the unique power and utility of scattering techniques for structural characterization of native viral particles, nucleic acid-free VLPs (reconstituted capsids), and recombinant VLPs carrying a foreign RNA of interest. In addition, the analysis of SANS and SAXS results will be demonstrated for modeling the structure of the coat protein and encapsulated RNA (required for vaccine development) in solution, as well as for quantifying the amount of RNA in a VLP. Quantification of nucleic acid content is critical for assessment of whether recombinant VLPs meet the requirements for a standard biomedical assay.

INTRODUCTION

For biomedical purposes, the use of genetically-modified, noninfectious versions of viruses is under exploration for a variety of infectious disease prevention (Jennings and Bachmann 2008. The coming of age of virus-like particle vaccines and anti-cancer strategies (Ramqvist *et al.*, 2007) (Ludwig and Wagner, 2007). One of the more successful such systems involves the use of Virus-Like Particles (VLPs). VLPs are genetically engineered versions of natural viruses that are noninfectious. Like natural viruses, VLPs utilize the coat as a packaging system to encapsulate, protect, and deliver agents (drugs, nucleic acids, and proteins) to target cells or to display proteins/epitopes (Ludwig and Wagner, 2007). Among the most successful are the Human papillomavirus (HPV)-VLP vaccines, which have been shown to stimulate high serum antibodies against the most common oncogenic HPV genotypes associated with cervical cancer (GlaxoSmithKline [Cervarix]), as well as to provide immunity against HPV genotypes associated with laryngeal papillomas and most general warts (Merck [Gardasil]) (Cutts *et al.*, 2007). Also commercially available is a VLP vaccine protective against Hepatitis B virus (HBV)(Ludwig and Wagner, 2007).

For therapeutic purposes, designer oncolytic viruses have been engineered to infect specific tissues, which are known to be cancerous, to display tumor associated antigens on their surface for the induction of (or to boost) tumor specific immune responses and lysis (Schiller and Lowy, 2010). Although oncolytic virus technology is still in early experimental stages, a number of systems have shown great promise in animal model experiments and pilot testing in humans (Chalikonda *et al.*, 2008)(Breckpot *et al.*, 2007)(Lundstrom, 2005)(Bauerschmitz *et al.*, 2002)(Peng *et al.*, 2002) (Stojdl *et al.*, 2003) Mahoney and Stoidl, 2010; Morse *et al.*, 2010). Engineered viruses have been increasingly employed as internal standard materials in DNA-based diagnostic testing (Cartwright, 1999; Cheng *et al.*, 2006 ;Zhan *et al.*, 2009; Meng and Li, 2010; Reid et al., 2010). In the case of Armored RNAs, MS2 virus is used as a packaging system to encase short standard sequences of nucleic acid to provide a relative measure of nucleic acid quantity for comparison to challenge samples and/or as positive or negative controls in diagnostic assays, including for HCV (Crowther, 2004; Stevenson *et al.*,2008; Walkerpeach et al., 1999; Zhao et al., 2006; Meng and Li, 2010; Reid et al., 2010).

Despite these advances, virus structural design faces a number of recurring challenges. In particular, engineered viral coats often lack the stability of natural viruses due to modifications employed to regulate artificial expression, control infection, and/or display proteins or epitopes of interest (Walkerpeach *et al.*, 1999). Furthermore, virus vector systems often package nucleic acids less faithfully and in smaller quantities than their counterparts found in nature, although recent reports show expansion of the RNA capacity of the VLP capsids as better positive controls for an HIV RT-PCR detection assay (Walkerpeach et al., 1999; Wei et al., 2008; Zhan et al., 2009).

In general, the virus coat can be thought of as both a primary drug target and as a potential tool in the quest to detect and eradicate cancer and to prevent or treat infectious diseases. The greatest challenge in both areas of research is insight into the structural and molecular interactions necessary for the assembly and large scale organization of the viral coat and its structural rearrangements during viral infection. The MS2-VLP system is an ideal model system both directly for biomedical diagnostics and more broadly for the elucidation of mechanisms of infection critical for the design of virus-based tools.

MS2 as a Genetic and Biochemical Model Organism

MS2 is a 24 nm RNA virus that infects the male form of the bacterium *Escherichia coli* (*E. coli*) and is an ideal model system to quantify the structural dynamics of the coat. The MS2 coat is comprised of 180 copies of the coat protein and a single copy of a maturation protein called A protein. The organization of the MS2 icosahedral coat and its infectious cycle share many conserved features with animal viruses associated with disease. The empty virus coat, similar to the icosahedral RNA influenza and polio viruses, does not break down during infection. Rather, binding of the MS2 coat to its bacterial host results in cleavage of the structural A protein and subsequent ejection of both the virus genome and the cleaved A protein, which enter the host cell.

The completely sequenced MS2 genome encodes four proteins, only two of which comprise the virus coat: coat protein (Mr=13.7KDa) and maturase A protein (Mr=44KDa) (Fiers et al., 1976; Fiers et al., 1975; Min Jou *et al.*, 1972; Min Jou *et al.*, 1979). The coat protein, in addition to its role as the main structural component of the virus coat, binds a specific 19 nucleotide (nt) sequence called the MS2 RNA operator site, which acts as a transcriptional repressor of the replicase gene (Carey *et al.*, 1983; Stockley *et al.*, 1994) and also catalyzes virus coat assembly by nucleation of the initial viral coat dimer complex (Carey *et al.*, 1983; Stockley *et al.*, 1994).

Additionally, MS2, in both infectious and noninfectious forms, is a biochemical model system for the assembly of intact viruses and packaged RNA, using a variety of simple methods (DuBois *et al.*, 1997). The fully infectious virus can be reconstituted *in vitro* by combining purified coat protein dimers, RNA genome, and the maturase A protein, or by *in vivo* transcription-translation of the complete genomic cDNA. RNA-free MS2 capsids with or without the A protein can also be assembled *in vitro* from purified coat protein dimers and A protein, or through transcription-translation of a plasmid containing the coat protein cDNA and a gene coding for heterologous RNA of interest, to create MS2-VLPs for biomedical and research purposes (DuBois *et al.*, 1997; Pasloske *et al.*, 1998; Zhan *et al.*, 2009; Meng and Li, 2010).

MS2-VLPs and Biomedical Diagnostic Assays

MS2-VLPs for biomedical diagnostic use have a number of unusual features, including: (a) the virus coat can spontaneously assemble to form a complete virus coat and incorporate any RNA of interest; (b) each virion appears to encapsulate a single RNA particle (provided the RNA is less than 1000 nt); and (c) those VLPs formed without expression of the A protein are noninfectious (DuBois et al., 1997; Heal et al., 2000; Mastico et al., 1993; Pasloske et al., 1998; Pickett and Peabody, 1993; Van Meerton et al., 2001). For biomedical use, the MS2 RNA packaging system is typically used in assays to detect RNA associated with diseases detectable in the bloodstream, such as HCV and HIV (Cheng et al., 2006; Das et al., 2006; Drosten et al., 2001; Eisler *et al.*, 2004; Legendre and Fastrez, 2005; Walkerpeach et al., 1999; Zhan et al., 2009; Meng and Li, 2010)

For clinical diagnostic purposes, the MS2-VLP (RNA packaging system) has a number of possible uses as (a) a standard diagnostic reference material of the type described above in RNA-based diagnostic assays for infection, as reported for HCV, HIV, SARS, and Rubella, to name a few (Pasloske *et al.*, 1998; Zhan *et al.*, 2009;Cheng *et al.*, 2006; Wei *et al.*, 2008), (b) a quantitative standard to gauge the efficiency of new drug regimens by measuring relative viral load, and (c) a quality assurance standard tool for teaching students and/or testing clinical laboratories either to measure the efficiency of a particular diagnostic laboratory or for technician training in proper preparation of blood samples (Bressler and Nolte, 2004 ; Zhao *et al.*, 2006, 2007).

Although the structural characteristics of the MS2-VLP have not been rigorously characterized, it has achieved international use for detection of RNA associated with disease. The widespread use of MS2-VLPs in the United States, however, is limited to research purposes because fundamental questions remain about its physiological structure and function, the full characterization of which is required prior to human and animal clinical use.

The nature of the structure of the virus-like complexes, nucleic acid packaging characteristics, and overall structural integrity have not been fully characterized using intact particles under physiological conditions analogous to the conditions of its designed use (in solution).

Structure of the MS2 Virus Coat

The structure of the wild type MS2 virus, its RNA-free capsid, and a variety of coat protein mutants have been examined by solid and/or crystalline methods such as X-ray crystallography, transmission electron microscopy (EM), and cryo-EM (Golmohammadi *et al.*, 1993; Konig et al., 2003; Ni et al., 1995; Stonehouse and Stockley, 1993; Stonehouse *et al.*, 1996; Valegard *et al.*, 1990, 1991; Valegard *et al.*, 1997; Valegard *et al.*, 1986; van den Worm *et al.*, 2006; Toropova *et al.*, 2008; Elsawy *et al.*, 2010; Rolfsson *et al.*, 2010; Dykeman *et al.*, 2010; Morton *et al.*, 2010) The impact of viral RNA on assembly pathway selection. The MS2 crystal structure combined with genetic and biochemical studies have revealed important structural features of the native coat and its components: (i) The coat protein dimer is the major subunit of virus assembly; ninety coat protein dimers and a single A protein comprise the MS2 coat. (ii) The coat protein dimers can adopt three possible quasi-equivalent conformers (A, B, C) in the formation of the MS2 icosahedral coat. These conformers, although very similar, vary primarily at the F-G b-strand loop (FG loop). In the A and C subunits, which interact at the quasi 6-fold axis, the FG loop is extended, while in the B subunit it is folded back in the direction of the protein at the 5-fold axis. (iii) A consequence of the interactions of the conformer FG loops is that a hole or channel is created at both the 5-fold and quasi 6-fold axis; one of these channels is postulated to be a site of RNA extrusion during infection or RNA leakage and subsequent instability in the absence of the A protein (Golmohammadi et al., 1993; Ni et al., 1995; Stonehouse and Stockley, 1993; Stonehouse et al., 1996; Valegard et al., 1990, 1991; Valegard et al., 1997; Valegard et al., 1986).

Cryo-EM has revealed a layer of RNA attached to the inner surface of the virus coat in an organized network (Konig et al., 2003; van den Worm et al., 2006; Toropova et al., 2008). A comparison of the cryo-EM images to the crystal structure of MS2 bound to an RNA fragment containing the coat protein operator stem-loop site confirmed that 10-20% of the genomic RNA is bound to the coat protein dimers in the same manner, which may impart structural stability to the coat and/or RNA organization and sequestration (and stability) within the coat (Konig et al., 2003; Lima et al., 2006; van den Worm et al., 2006). Interestingly, despite the fact that high resolution crystal structures of MS2 wild type and both mutant and wild type RNA-free capsids have been described, the positioning of the A protein within the MS2 coat is still unknown (Golmohammadi et al., 1993).

The A protein

The A protein is a coat structural protein with a well-established role in host recognition (Kozak and Nathans, 1971). Upon binding of MS2 to its *Escherichia coli* host pilus, the A protein is cleaved by an unidentified host protease (or auto-cleaved), and the A protein and genomic RNA are transferred into the bacterial host (Figure 1). The empty virus coat is left intact outside of the host cell (Curtiss and Krueger, 1974; O'Callaghan et al., 1973).

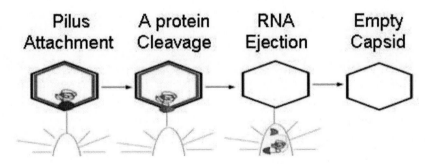

Figure 1. Model of the Bacteriophage MS2 Infectious Cycle. The black hexagons, green lines, and red circles represent the MS2 coat, genomic RNA, and A protein, respectively. The virus particles are attached to the bacterial pilus. Figure not to scale.

Collectively, biochemical and solid/crystalline biophysical MS2 analyses have led to a well-developed understanding of the global organization of the MS2 virus coat and the coat protein in association with itself and with genomic RNA, but fail to address the conformational state/s of MS2-VLP, the mechanistic interactions that may occur during packaging of non-native RNA, natural infection, and the specific role of the A protein in these processes. Therefore, our preliminary experiments using small-angle scattering (SAS) techniques sought to analyze the structure of the MS2-VLP and the relationship between the coat and heterologous RNA under the native solution conditions, as required for the approval of biomedical virus tools for clinical use.

Structural Analysis using Small Angle Scattering (SAS) Techniques

Small-angle scattering (SAS) techniques are powerful tools to study the architecture of complex macromolecular structures such as viruses, nucleic acid-protein, and homo- and hetero-macromolecular protein complexes in solution (Fitter *et al.*, 2006). Particles or photons (light, X-rays, or neutrons) that pass though a sample in solution are scattered in a pattern that can be analyzed to establish the physical properties of the sample, such as average size, shape, molecular weight, and oligomerization state. Of particular advantage for complex macromolecular structures such as viruses, small-angle neutron scattering (SANS) can uniquely be combined with a contrast variation technique to gain structural information for the individual protein, nucleic acid, and lipid components (Fitter *et al.*, 2006; Jacrot and Zaccai, 1981; Struhrmann and Miller, 1978). Since the contrast variation technique simply involves varying the solvent to deuterated water ratio, biological samples are analyzed intact, under physiological conditions, and the data obtained reveal *in situ* structural information (Fitter *et al.*, 2006; Koch *et al.*, 2003; Krueger, 1998).

Similarly, small-angle X-ray scattering (SAXS) provides in solution structure data, but more specifically for nucleic acids (Graille *et al.*, 2005; Olah *et al.*, 1995; Thuman-Commike *et al.*, 1999; Mylon *et al.*, 2010). In the case of complex macromolecular structures, such as viruses, that contain both nucleic acid and protein components, the nucleic acids are uniquely resolved by SAXS due to the fact that X-rays interact more strongly with nucleic acids than with proteins (Jacrot and Zaccai, 1981; Koch *et al.*, 2003). Therefore, SAXS provides unique insight into the physical characteristics of RNA or DNA in complex biological systems. The scattering intensity from SAXS provides quantitative and qualitative measures of the *in situ* physical characteristics of the viral RNA in its native environment (surrounded by its virus coat), in solution.

Together, solution based SANS and SAXS methods both complement solid/crystalline techniques such as X-ray crystallography, transmission- and cryo-EM studies, and, uniquely, are ideal for the analysis of VLP and intact viruses. The samples can be quantitatively measured intact and under conditions that most closely mimic those used in typical diagnostic assays and under physiologically natural conditions.

METHODS

Sample Preparation

MS2 bacteriophage strain 15597-B1 and its *Escherichia coli (E. coli)* host 15597 were purchased from the American Type Culture Center (Manassas, Va.). MS2 phage were grown and purified by ultra centrifugation as described in (Kuzmanovic *et al.*, 2003). RNA-free MS2 capsids were produced as described in (Kuzmanovic *et al.*, 2006b) using purified WT MS2 virus as starting material.

In general, to produce MS2-HCV and MS2-λ virus particles, an expression vector was genetically engineered to express only the MS2 coat protein and either the HCV or λ RNA sequences. Expression of the MS2 coat protein is followed by the spontaneous assembly and incorporation of the HCV or λ RNA. Note that these virus particles do not contain the genetic information to encode the A protein. Therefore, there are no pleiotropic effects associated with chemical removal of the A protein because the A protein was not present at any stage of the virus production or assembly process. Specifically, MS2-HCV virus particles were produced from the *E.coli* expression vector pAR-HCV-2b, which contains a 412-nucleotide sequence from the 5' noncoding core region of HCV subtype2b (Genbank Accession No. M62321). MS2-λ virus particles were produced from the *E. coli* expression vector pAR-l-1.0, which contains 908 nucleotides (1329 to 421) of l sequence (Genbank Accession No. M17233) (Ambion Diagnostics, personal communication). Both vectors were generous gifts from Ambion Diagnostics, Inc. (Austin, Texas) and were grown as described in (DuBois *et al.*, 1997) and purified by cesium gradient ultra centrifugation as described in (Kuzmanovic et al., 2003). SDS/ polyacrylamide gel electrophoresis was performed as described in (Kuzmanovic *et al.*, 2003).

SANS Measurements and Data Analysis

Details of the SANS and SAXS measurements of MS2 particles and the data analyses were described previously (Kuzmanovic *et al.*, 2003)(Kuzmanovic *et al.*, 2006a)(Kuzmanovic *et al.*, 2006b). In both cases, scattered neutrons were detected using a two-dimensional position-sensitive detector. After making necessary instrument-dependent corrections, the data were radially-averaged to produce one-dimensional scattering intensity, I(Q), vs. Q curves, where $Q = 4\pi\sin(\theta)/\lambda$, λ is the neutron wavelength and 2θ is the scattering angle. A typical Q range covered for the MS2 samples is $0.005 \text{ Å}^{-1} \leq Q \leq 0.17 \text{ Å}^{-1}$. The scattering intensities from the samples were then further corrected for buffer scattering and, in the case of SANS, for incoherent scattering from hydrogen in the samples. SANS data were also placed on an absolute scale by normalizing the scattering intensity to the incident beam flux.

The Guinier approximation, $I(Q) = I(0)\exp(-Q^2 R_g^2/3)$, was used on the low-Q portions of the data to obtain initial values for the radius of gyration, Rg, and the forward scattering intensity, I(0), of the samples. This analysis is valid only in the region where QRg ~ 1. Rg and I(0) were also found from the distance distribution function, P(r) vs r, where $4\pi P(r)$ represents the number of distances within the scattering particle (Glatter and Kratky, 1982). Since P(r) is related to I(Q) by a Fourier transform, it gives a real space view of the shape of the scattering particles.

Since the WT MS2, MS2-HCV and MS2-λ samples contain both a protein and a RNA component, the scattering intensities from the coat protein shells were obtained by decomposition of the scattering intensities from the protein/RNA complexes as described in (Kuzmanovic *et al.*, 2003). Specifically, the scattering intensities from the MS2 protein/RNA complexes were decomposed into the scattering from their components, $I_{PROT}(Q)$ and $I_{RNA}(Q)$ using the equation:

$$I(Q) = \Delta\rho_{PROT}^2 I_{PROT}(Q) + \Delta\rho_{PROT}\Delta\rho_{RNA} I_{PROTRNA}(Q) + \Delta\rho_{RNA}^2 I_{RNA}(Q), \tag{1}$$

where $\Delta\rho = (\rho-\rho_s)$ is the contrast, or the difference between the scattering length density of the molecule (ρ) and the solvent (ρ_s). The cross-term, $I_{PROTRNA}(Q)$, represents the interference function between the protein and RNA components. The known quantities in Eq. 1 are $\Delta\rho_{PROT}$ and $\Delta\rho_{RNA}$ and the unknowns are $I_{PROT}(Q)$, $I_{RNA}(Q)$ and $I_{PROTRNA}(Q)$. Since measurements were made at four different contrasts, or D_2O/H_2O buffer conditions, there is sufficient information to solve for the three unknown component intensities from the set of simultaneous equations for $I(Q)$ at each contrast.

The Mw values of the protein and RNA components of the MS2 complexes were calculated using the relation (Kuzmanovic *et al.*, 2003):

$$I(0) = n(\Delta\rho_{PROT}V_{PROT} + \Delta\rho_{RNA}V_{RNA})^2, \tag{2}$$

where n is the total number density of scattering particles and V particle volume. Using the relations $V_{PROT} = M_{WPROT}/(N_A d_{PROT})$ and $V_{RNA} = M_{WRNA}/(N_A d_{RNA})$, where N_A is Avogadro's number, M_W is the molecular weight and d is the mass density, Eq. 2 can be rewritten as (Kuzmanovic et al., 2003):

$$\left[\frac{I(0)}{n}\right]^{\frac{1}{2}} = \left(\frac{|\Delta\rho_{PROT}|}{N_A d_{PROT}}\right)M_{WPROT} + \left(\frac{|\Delta\rho_{RNA}|}{N_A d_{RNA}}\right)M_{WRNA}, \tag{3}$$

Now, there are only 2 unknowns, M_{WPROT} and M_{WRNA}, i.e., the molecular weights of the two components in the complex. It is important to note that $I(0)$ must be on an absolute scale, usually i.e. in cm^{-1}, in order to obtain accurate Mw values from Eq. 3. Eq. 3 also requires that the number density of particles at each contrast be known independent of the measured sample concentration. This is because the number density itself is a function of concentration, c, and Mw since

$$n = c N_A / M_W . \tag{4}$$

If such independent number densities cannot be measured, then the total Mw cannot be separated into the M_{WPROT} and M_{WRNA} components. Independent measurements of number density at each contrast were obtained for the WT MS2 samples (Kuzmanovic et al., 2003) using the Integrated Virus Detection System, a specialized instrument developed for counting virus particles (Kuzmanovic et al., 2003; Wick, 2002a, b; Wick and McCubbin, 1999).

Since MS2 can be approximated very well by a spherical shell at the resolution level of the SANS measurements, the data were also fit to a core-shell sphere model in order to obtain the radii of the protein shell and the solvent core (Hayter, 1983). The neutron scattering length density of the core was treated as an additional fitting parameter that allowed the amount of water versus RNA in the core to be calculated using the relation:

$$\rho_{CORE} = X\rho_{RNA} + (1-X)\rho_{SOLVENT}, \tag{5}$$

where X is the mass fraction of RNA in the core, ρ_{CORE} is the fitted scattering length density of the core portion of the core-shell model and ρ_{RNA} and $\rho_{SOLVENT}$ are the known scattering length densities of the RNA and the solvent, respectively. Thus, if there is no RNA in the core, $\rho_{CORE} = \rho_{SOLVENT}$.

PRELIMINARY STUDIES

Analysis of the MS2 Coat in WT MS2 and MS2-VLP: Conformational changes in the MS2 coat are uniquely A protein dependent

Four types of MS2 based particles were characterized structurally using SANS and SAXS: wild type (WT) MS2 and its RNA-free capsid (MS2-capsid), both of which contain A protein, as well as two MS2-VLPs containing non-native viral RNA fragments typically used in diagnostic assays. These MS2-VLPs lack the A protein but contain 500nt of Human cytomegalovirus (HCV) or 800nt of bacteriophage lambda RNA, and are referred to as MS2-HCV and MS2-λ, respectively.

Our initial analysis focused on solving the solution structure of WT MS2 using SANS (Kuzmanovic *et al.*, 2003). From this analysis, we were able to describe a physical model of the shape of the MS2 coat in solution and to determine the dimensions and molecular weight of the individual coat and RNA components, under *in vivo* conditions (Kuzmanovic *et al.*, 2003). As diagrammed in Figure 2, the shape of an MS2 particle can be approximated very well by a spherical shell, with inner radius, R1, outer radius, R2, and shell thickness, t = R2 – R1. This shape is generally known as a core-shell model (Hayter, 1983).

As summarized in Table 1, the experimental WT MS2 coat has an (inner radius) R1 = 115 ± 1 Å, (outer radius) R2 = 136 ± 1 Å, and (thickness) t = 21 ± 1 Å in solution. Recombinant forms of the bacteriophage MS2 (MS2-VLP) and MS2-capsid were later analyzed similarly to determine if RNA content and/or the A protein play a role in the global arrangement of the virus coat (Kuzmanovic et al., 2006b).

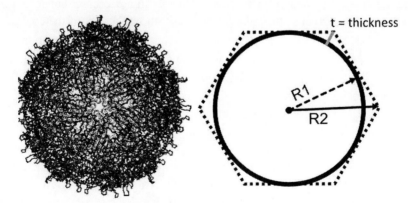

Figure 2. The shape of MS2 fits core-shell model. R1, R2, and t represent the inner radius, outer radius and thickness, respectively. The MS2 virus was rendered using the VMD program (Humphrey *et. al.*, 1996) and the wild type crystal structure (2MS2) (Golmohammadi *et al.*, 1996; Shepherd *et al.*, 2006). The solution structure was solved by Kuzmanovic *et al.*, 2003.

Table 1. MS2 Protein Shell Radii from SANS Contrast Variation Studies[*]

	R1 (Å)	R2 (Å)	t (Å)
Measured Solution Data			
WT MS2 (A-plus)	115 ± 1	136 ± 1	21 ± 1
MS2 Capsid$_{100\% \, D2O}$ (A-plus)	115 ± 1	140 ± 2	25 ± 1
MS2-HCV (A-minus)	105 ± 2	135 ± 1	31 ± 1
MS2-λ (A-minus)	101 ± 1	136 ± 1	37 ± 1
Crystal Derived Model Data[+]			
[2]MS2 (A-plus)	111 ± 1	130 ± 1	19 ± 1
[1]ZDI (A-minus)	111 ± 1	130 ± 1	19 ± 1
[1]ZDH (A-minus)	111 ± 1	131 ± 1	20 ± 1

[*]From Kuzmanovic et al, 2006 capsid
[+] 2MS2 refers to wild type MS2, and 1ZDI and 1ZDH refer to MS2-VLPs analogous to MS2-HCV and MS2-λ (Stonehouse et al., 1996; Valegard et al., 1997).

From these studies we demonstrated that the coat of MS2-capsid (lacking RNA) that contain A protein do not significantly differ in size or shape when compared to WT MS2. The MS2-capsid coat has R1 = 115 ± 1 Å, R2 = 140 ± 2 Å, and t = 25 ± 1Å. This suggests that RNA content does not influence the structure of the MS2 coat. However, analysis of the coat of MS2-VLPs (MS2-HCV, 500nt, and MS2-λ, 800nt), which lack the A protein, revealed dramatic differences compared to WT MS2 in solution. MS2-VLP coats are of approximately the same size or outer radius (R2 = between 135 and 136 ± 1 Å) as compared to the size of WT MS2 and the MS2-capsid particles, in which R2 = 136 ± 1 and 140 ± 2 Å, respectively. However, the inner radius (R1) is significantly smaller than WT MS2 or MS2-capsid particles. Since the thickness (t) of the coat is defined as the size or outer radius (R2) minus the inner radius (R1), the maintenance of overall size with a decreased inner radius results in particles with a coat of greater thickness. In this case, the thickness of the coat of MS2-VLP particles is t = 31 ± 1 Å and 37 ± 1 Å, respectively, considerably thicker than the coat formed by either the WT MS2 or the MS2-capsid, with coats of thickness of t = 21 ± 1 Å and 25 ± 1 Å, respectively. The MS2-VLPs (MS2-HCV and MS2-λ) differ from the WT MS2 and MS2-capsid samples in one significant respect: these recombinant viruses lack the A protein while the WT MS2 and MS2-capsid viruses contain the A protein. Therefore, these results are consistent with the idea that the A protein, not the RNA, is uniquely responsible for structural rearrangements in the MS2 coat. Furthermore, this A protein-dependent difference in virus coat structure represents the first description of a conformational specific phenotype in MS2. Finally, modeling of the SANS data from the crystal structure of a variety of MS2 mutants and related viruses (fr, Qb, and GA), which are analogous to the recombinant particles used in this study, was used to generate simulated SANS data, revealed only the thin (t = 21 ± 1 Å) conformation of the MS2 coat (Kuzmanovic et al., 2006b).

To better understand the influence of RNA and the A protein on protein shell arrangement in vivo, this paper presents the use of SANS and the contrast variation techniques to examine the amount and spatial distribution of RNA of two recombinant forms of MS2 virus lacking the A protein but containing varying amounts of RNA, (MS2-HCV and MS2-λ) as compared to WT MS2. In addition, empty capsids (MS2-capsid), containing coat protein and A protein, but lacking RNA, were measured for comparison to the protein shell structures of the above-mentioned samples.

RESULTS AND DISCUSSION

SANS data for MS2-VLPs

The contrast variation SANS data, I(Q) vs Q, for MS2-HCV and MS2-λ are shown in Figure 3. The corresponding distance distribution functions, P(r) vs r, are shown in Figure 4, along with that obtained from SAXS (Kuzmanovic *et al.*, 2006a) for comparison. Table 2 lists the Rg and I(0) values that were obtained from the P(r) analysis from both SANS and SAXS (Kuzmanovic *et al*, 2006), along with those values for WT MS2 (Kuzmanovic *et al.*, 2003) and (Kuzmanovic *et al.*, 2006b). is at lower r values, as compared to the contrast conditions where the RNA scattering is weaker, i.e., 85% D2O and 100% D2O. However, the difference in the peak position is not as large as it is for WT MS2, as shown in Figure 5 (Kuzmanovic *et al.*, 2003). This indicates that the protein component surrounds the RNA component, as expected. To quantify this difference, several parameters were studied for both the protein and RNA components as described below.

Figure 4 shows that, under contrast conditions where the RNA scattering is strong, i.e., 0% D_2O, 10% D_2O and SAXS, the peak in the P(r) function

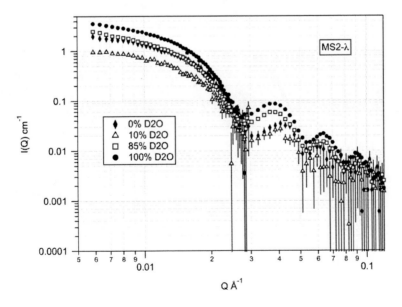

Figure 3 (Continued)

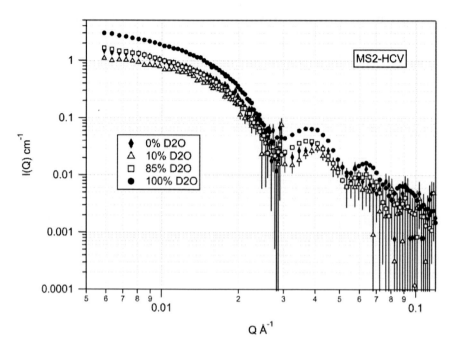

Figure 3. Contrast Variation Data for MS2-VLPs. I(Q) vs Q data are shown for four different contrasts as indicated in the figure legends: MS2-HCV (upper) and MS2-λ (lower).

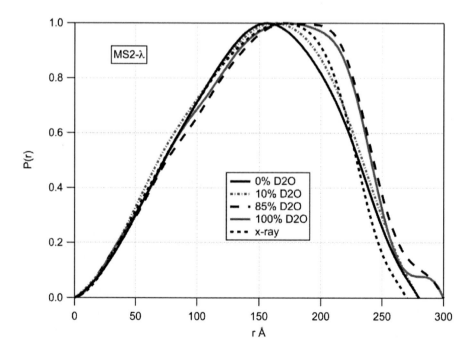

Figure 4 (Continued)

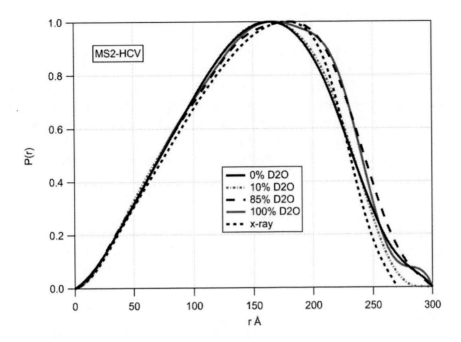

Figure 4. Distance Distribution Functions for MS2-VLPs. P(r) vs r curves from SANS data are shown at four different contrasts as indicated in the figure legends: MS2-HCV (upper) and MS2-λ (lower). P(r) vs r curve from SAXS is also shown for comparison (Kuzmanovic *et al*, 2006). P(r) is normalized such that P(r) = 1 at its maximum value so that the shapes of the curves can be easily compared.

Table 2. MS2 Parameters from Distance Distribution Function Analysis

Sample	SANS Data		SAXS Data	
	Rg (Å)	I(0) cm^{-1}	Rg (Å)	I(0) cm^{-1}
WT MS2$_{0\% D20}$	114 ± 1	2.10 ± 0.02	106.5 ± 0.5	2.6 ± 0.1
WT MS2$_{10\% D20}$	114 ± 1	2.60 ± 0.02	—	—
WT MS2$_{65\% D20}$	130 ± 1	0.46 ± 0.07	—	—
WT MS2$_{85\% D20}$	120.0 ± 0.5	2.10 ± 0.01	—	—
WT MS2$_{100\% D20}$	122.0 ± 0.5	6.70 ± 0.01	—	—
MS2-capsid$_{100\% D20}$	123 ± 2	1.28 ± 0.06	—	—
MS2-capsid$_{0\% D20}$	—	—	115.0 ± 0.5	0.57 ± 0.03
MS2-HCV0% D20	115 ± 1	1.62 ± 0.02	113.5 ± 0.5	0.81 ± 0.04
MS2-HCV10% D20	113 ± 1	1.19 ± 0.02	—	—
MS2-HCV85% D20	118.0 ± 0.5	1.60 ± 0.01	—	—
MS2-HCV$_{100\% D20}$	117.0 ± 0.5	3.00 ± 0.01	—	—
MS2-λ$_{0\% D20}$	113 ± 1	2.07 ± .01	112.0 ± 0.5	1.3 ± 0.1
MS2-λ$_{10\% D20}$	113 ± 1	1.10 ± 0.03	—	—
MS2-λ$_{85\% D20}$	119 ± 1	2.41 ± 0.02	—	—
MS2-λ$_{100\% D20}$	117 ± 1	3.68 ± 0.05	—	—

RNA Packaging Defects in MS2-VLPs

The amount and structure of the RNA packaged into MS2-VLPs containing varying amounts of foreign RNA (MS2-HCV, 500nt, and MS2-λ, 800nt) was initially examined using SANS, and later by SAXS, and compared to WT MS2, which contains the A protein and normal RNA levels. The results are shown in Table 3. This information was obtained from the results of core-shell model fits of the contrast variation data for MS2-HCV and MS2-λ (Figure 3). Eq. 5 was used to determine the fraction of RNA in the core at each contrast measured. When first measured by SANS, MS2-HCV and MS2-λ appeared to package RNA levels greater than or equal to WT MS2 RNA, containing 23 ± 1% and 22 ± 1% RNA, respectively, compared to 19 ± 4% RNA in the WT MS2 (Kuzmanovic *et al.*, 2003). The same samples were reexamined more than six months later by SAXS. At that time, the MS2-HCV, MS2-λ, and WT MS2 contained 8% ± 2%, 12% ± 2%, and 27% ± 2% of RNA, respectively (Kuzmanovic *et al.*, 2006a). These later SAXS results, paradoxically, were in good agreement with the minimum amount of RNA expected based on the sequence

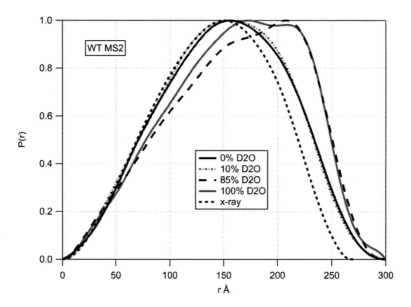

Figure 5. Distance Distribution Functions for WT MS2. P(r) vs r curves from SANS data are shown at four different contrasts (Kuzmanovic *et al.*, 2003) as indicated in the figure legend. P(r) vs r curve from SAXS is also shown for comparison (Kuzmanovic *et al.*, 2006). P(r) is normalized such that P(r) = 1 at its maximum value so that the shapes of the curves can be easily compared.

contained in the expression vectors that were used to generate these recombinant particles *(Walkerpeach et al.*, 1999) and on the published sequence information for the wild type MS2 virus (Fiers *et al.*, 1976). However, compared to the SANS results, the SAXS results also suggest that versions of MS2 particles that lack the A protein but contain foreign RNA initially package RNA of equal or greater amount compared to wild type. However, over time, the foreign RNA has likely degraded (Kuzmanovic *et al.*, 2006a). The molecular weight, size, and shape of the coat of all of the samples remained constant over time, indicating that although the RNA seems to degrade over time in the absence of A protein, the coat remains stable and intact (Kuzmanovic *et al.*, 2006a). These data support observations (Heisenberg, 1966) indicating that the genomic RNA was degraded in the absence of the A protein. The MS2-VLP samples were also compared to MS2-capsid that contain the A protein but lack RNA. The MS2-capsid samples were found to be essentially RNA free, with a RNA content of 4% ± 1% by SANS and 2% ± 2% by SAXS.

Table 3. MS2 RNA Content and Spatial Distribution

	RNA Mass Fraction[+] in Core (%)	RNA Radius[*] (Å)
Measured SANS Data		
WT MS2 (A-plus)	19 ± 4	84 ± 1
MS2 Capsid$_{100\% \, D2O}$ (A-plus)	4 ± 1	—
MS2-HCV (A-minus)	23 ± 1	97 ± 1
MS2-λ (A-minus)	22 ± 1	96 ± 1
Measured SAXS Data		
WT MS2 (A-plus)	27 ± 2	85 ± 2
MS2 Capsid$_{0\% \, D2O}$ (A-plus)	2 ± 2	90 ± 2
MS2-HCV (A-minus)	8 ± 2	94 ± 2
MS2-λ (A-minus)	12 ± 2	96 ± 3

+SANS values were averaged after applying Eq. 5 to the best-fit ρCORE values at each contrast measured.
*SANS values listed are for data obtained in buffer with 10% D2O, where the scattering from the RNA dominates the total scattering. Since the MS2-capsid was not measured by SANS in a 10% D2O solution, no value is listed. The RNA contrast for the SAXS measurements is approximately equal to the 10% D2O SANS measurements.

The conformation and spatial distribution of the RNA were also measured by SAS. Our group has previously shown that the wild type MS2 genomic RNA appeared to be tightly packed within the MS2 virus core with a radius of 84 ± 1 Å by SANS (Kuzmanovic *et al*, 2003) and 85± 2 Å by SAXS (Kuzmanovic *et al.*, 2003). This was determined from the inner radius (R1) from the core-shell model fit to the data, under conditions where the scattering from the RNA is strong compared to that of the protein. This is automatically the case for SAXS data, as well as for SANS data taken in 10% D_2O. On the other hand, the radii of the RNA cores contained in the MS2-VLP (A protein minus) MS2-HCV and MS2-λ virus particles are 97 ± 1 Å and 96 ± 1 Å, respectively, by SANS and 94 ± 2 Å and 96 ± 3 Å, respectively, by SAXS. This indicates that the RNA in the MS2-HCV and MS2-λ particles is not packed as tightly in the core of the recombinant viruses as compared to the wild type MS2.

Protein Shell Stability of MS2-VLPs

While the RNA packaging in MS2-VLPs was found to vary from the expected values for MS2-VLPs, SAS analysis shows that this is not true for the protein shell. This can be seen by an analysis of the Mw of the WT MS2 and MS2 VLP complexes and their individual components. Since independent number densities were available for WT MS2 and MS2-HCV, the Mw of the protein and RNA components were determined for these samples. The results are shown in Table 4. The Mw for MS2-capsid was obtained in a similar manner. However, since a contrast variation series of experiments was not performed on this sample, the total Mw was found using the relation

$$\frac{I(0)}{c} = \frac{(\Delta\rho)^2}{N_A\, d^2}\, M_W \tag{6}$$

by combining Eqs. 3 and 4 for the protein component alone. The results show that, while the Mw of the RNA component did not agree initially with the calculated value for MS2-HCV, the Mw of the protein shell was in agreement with expected values. For WT MS2, by comparison, both the RNA and protein Mw results were in agreement with expected values.

Another measure of protein shell stability can be obtained from the outer shell radius value, R2, obtained from the core-shell model fits to the SANS and SAXS data for WT MS2, MS2-HCV and MS2-λ. The average R2 values obtained from the best core-shell model fits to the SANS data at each contrast are shown in Table 5, along with the R2 values obtained from the best core-shell model fits to the SAXS data. While the protein shell outer radii for the WT MS2 and MS2 VLPs are systematically smaller for the SAXS data, they agree well with the values obtained for MS2-capsid in each case. Furthermore, unlike the RNA radii, the protein shell outer radii are the same for WT MS2, MS2-HCV and MS2-λ. This result, the lack of significant change in the size of the protein shell overtime for the latter samples as measured by SAXS, and SANS has been described in detail previously by our group (Kuzmanovic *et al.*, 2006a).

Table 4. MS2 Molecular Weight Calculations from SANS Data

	Measured Values		
Sample	Mw_{RNA} x 10^6 g/mol	Mw_{PROT} x 10^6 g/mol	Mw_{TOTAL} x 10^6 g/mol
WT MS2	1.0 ± 0.2	2.5 ± 0.3	3.5 ± 0.5
MS2-capsid$_{100\% D2O}$	N/A	2.7 ± 0.4	2.7 ± 0.4
MS2-HCV	1.6 ± 0.1	2.1 ± 0.2	3.7 ± 0.1
	Calculated Values		
Sample	Mw_{RNA} x 10^6 g/mol	Mw_{PROT} x 10^6 g/mol	Mw_{TOTAL} x 10^6 g/mol
WT MS2	1.2	2.1	3.3
MS2-capsid	N/A	2.1	2.1
MS2-HCV	0.2	2.1	2.3

The M_w value for the MS2-capsid sample was obtained from Eq. 6 using $\Delta\rho = 3.0$ x 10^{10} cm^{-2}. The M_w values for WT MS2 and MS2-HCV are the M_{PROT} and M_{RNA} values from Eq. 3, using independent number densities measured at each contrast by the Integrated Virus Detection System (Kuzmanovic *et. al.,* 2003) Values are not listed for MS2-λ, since independent number density measurements weren't available at each contrast for this sample. The calculated Mw values were determined assuming 90 coat protein dimers and 1 genomic RNA, where appropriate, per sample.

Table 5. MS2 Protein Spatial Distribution

Sample	R2 (Å) from SANS[*]	R2 (Å) from SAXS
WT MS2	136 ± 1	134 ± 2
MS2-capsid$_{100\% \text{ D2O}}$	140 ± 2	—
MS2-capsid$_{0\% \text{ D2O}}$	—	130 ± 2
MS2-HCV	136 ± 1	130 ± 2
MS2-λ	137 ± 1	131 ± 2

[*]The R2 values for WT MS2, MS2-HCV and MS2-λ were averaged from the best core-shell model fits to the SANS data at each contrast measured

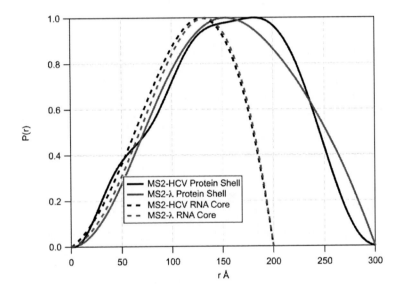

Figure 6. Distance Distribution Functions for protein shell and RNA core of MS2-HCV and MS2-λ. P(r) vs r curves for the RNA core and protein shell components were obtained from I(Q)$_{\text{PROT}}$ and I(Q)$_{\text{DNA}}$, derived from the SANS contrast variation data using Eq. 1. P(r) is normalized such that P(r) = 1 at its maximum value so that the shapes of the curves can be easily compared.

Protein and RNA components of MS2-VLPs

I(Q)$_{\text{PROT}}$ vs Q and I(Q)$_{\text{RNA}}$ vs Q for the protein shell and RNA core components of MS2-HCV and MS2-λ were obtained from the SANS contrast variation data using Eq. 1. The corresponding P(r) vs r curves were subsequently calculated and are plotted in Figure 6. The curves illustrate that the RNA core and protein shell of both MS2 VLPs are similar in structure. The same P(r) vs r functions are shown in Figure 7 for the protein and RNA components separately, along with those from WT MS2 (Kuzmanovic *et al.*, 2003; Kuzmanovic *et al.*, 2006b) for comparison. The differences between the WT MS2 (A-plus) and MS2 VLPs (A-minus) particles are once again illustrated, i.e., the protein shell is thicker for the MS2-VLPs compared to WT MS2 and the RNA is more loosely packed in the MS2-VLPs compared to WT MS2.

Implications of SAS Analysis of MS2-VLPs used in Biomedical Assays

Our analysis of VLP forms of MS2 indicates that they do not package one RNA molecule per coat. Rather, they initially package a wild type or greater genomic equivalent of RNA in the absence of the A protein. Secondly, the foreign RNA that is packaged is not compacted within the virion as is the wild type genomic RNA. Instead, the RNA is diffusely distributed about the virus core and likely protrudes through the holes of the virus coat (Kuzmanovic et al., 2006b). Over time this foreign RNA is degraded, presumably by nucleases, but not in WT MS2 samples containing A protein. The coat, however, remains intact (Kuzmanovic et al., 2006a,b). Since the genomic RNA is normally tethered to the A protein at its 5' and 3' ends, a role for the A protein in regulating the amount and conformation of the RNA within the coat is a distinct possibility.

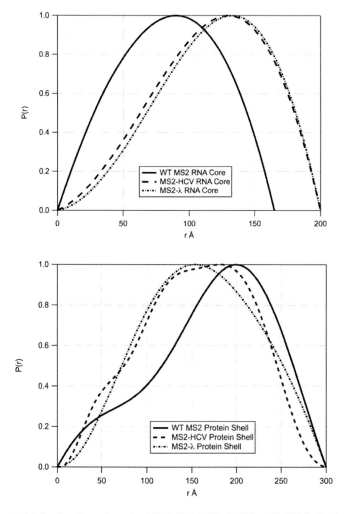

Figure 7. Distance Distribution Functions for WT MS2, MS2-HCV and MS2-λ. P(r) vs r curves were obtained from $I(Q)_{PROT}$ and $I(Q)_{DNA}$, derived from the SANS contrast variation data using Eq. 1. The P(r) vs r curves for WT MS2 (Kuzmanovic *et. al* 2003) are shown for comparision: a) RNA core, b) protein shell. P(r) is normalized such that P(r) = 1 at its maximum value so that the shapes of the curves can be easily compared.

In general, approximately one-fifth of all cancers worldwide are caused by chronic infections of viral origin (WHO, 2008). The majority of cases are attributable to infections by three viruses, Hepatitis B and C viruses (HBV and HCV) and Human papillomavirus (HPV), corresponding to liver (HBV and HCV) and cervical cancers (HPV) (Cutts *et al.*, 2007) (Tan *et al.*, 2008). Three recent anti-cancer strategies have involved the use of genetically modified viruses: (i) oncolytic viruses are designed to infect and mediate lysis of cancerous tissues; (ii) noninfectious, virus-like particles (VLPs) are used to display antigens necessary to stimulate immune response in cancer prevention and therapeutic schemes; and (iii) VLPs have been further developed as standards in cancer and infectious disease diagnosis assays.

The development of these new strategies for virus based disease detection, vaccination, and drug delivery methods require new approaches for analysis of both the nucleic acid and protein components of these biomolecules. Our characterization of MS2 and its virus-like particles offers a strategy for such an detailed biophysical analysis. Using in solution quantitative methods such as SAXS and SANS, virus based technologies can be more fully appraised with the long term goal of improving their efficacy for biomedical use.

ACKNOWLEDGMENTS

We acknowledge the support of the the the University of Michigan, National Institute of Standards and Technology (NIST), U. S. Department of Commerce, and the U.S. Army Aberdeen Proving Grounds in providing facilities used in this work. This material is based upon activities supported by the National Science Foundation under Agreement No. DMR-9986442. DAK was supported in part by the NIST-National Research Council Fellowship at the NIST Biotechnology Division and the University of Michigan.

&Certain commercial materials, instruments, and equipment are identified in this manuscript in order to specify the experimental procedure as completely as possible. In no case does such identification imply a recommendation or endorsement by the National Institute of Standards and Technology nor does it imply that the materials, instruments, or equipment identified is necessarily the best available for the purpose.

REFERENCES

Bauerschmitz, G.J., Barker, S.D., and Hemminki, A. 2002. Adenoviral gene therapy for cancer: from vectors to targeted and replication competent agents (review). Int J Oncol *21*, 1161-74.

Breckpot, K., Aerts, J.L., and Thielemans, K. 2007. Lentiviral vectors for cancer immunotherapy: transforming infectious particles into therapeutics. Gene Ther *14*, 847-62.

Bressler, A.M., and Nolte, F.S. 2004. Preclinical evaluation of two real-time, reverse transcription-PCR assays for detection of the severe acute respiratory syndrome coronavirus. J Clin Microbiol *42*, 987-91.

Carey, J., Cameron, V., de Haseth, P.L., and Uhlenbeck, O.C. (1983). Sequence-specific interaction of R17 coat protein with its ribonucleic acid binding site. *Biochem 22*, 2601-2610.

Cartwright, C. (1999). Synthetic viral particles promise to be valuable in the standardization of molecular diagnostic assays for hepatitus c virus. *Clin Chem 45*, 2057-2059.

Chalikonda, S., Kivlen, M.H., O'Malley, M., Dong, X.D., McCart, J.A., Gorry, M.C., Tin, X.-Y., Brown, C.K., Zeh, H.J., Guo, Z.S., *et al.* (2008). Oncolytic virotherapy for ovarian carcinomatosis using a replication-selective vaccinia virus armed with yeast cytosine gene. *Cancer Gene Ther 15*, 115-125.

Cheng, Y., Niu, J., Zhang, Y., Huang, J., and Li, Q. (2006). Preparation of His-tagged armored RNA phage particles as a control for real-time reverse transcription-pcr detection of severe acute respiratory syndrome coronavirus. *Journal of Clin Microbiol 44*, 3557-3561.

Crowther, R.A. (2004). Viruses and the development of quantitative biological electron microscopy. *IUBMB LIFE 56*, 239-248.

Curtiss, L.K., and Krueger, R.G. (1974). Localization of Coliphage MS2 A-Protein. *J Virol 14*, 503-508.

Cutts, F., Franceschi, S., Goldie, S., Castellsague, X., de Sanjose, S., Garnett, G., Edmunds, W., Claeys, P., Goldenthal, K., Harper, D., *et al.* (2007). Human papillomavirus and HPV vaccines: a review. *Bull WHO 85*, 719-726.

Das, A., Spackman, E., Senne, D., Pedersen, J., and Suarez, D.L. (2006). Development of an internal positive control for rapid diagnosis of avian influenza infections by rea-time reverse transcription-PCR with lyophilized reagents. *J Clin Microbiol 44*, 3065-3077.

Donia, D., Divizia, M., and Pana, A. (2005). Use of armored RNA as a standard to construct a calibration curve for real-time RT-PCR. *J Virol Meth 126*, 157-163.

Drosten, C., Seifried, E., and Roth, W. (2001). TaqMan 5'-nuclease human immunodeficency virus type 1 PCR assay with phage-packaged competitive internal control for high-throughput blood donor screening. *J Clin Microbiol 39*, 4302-4308.

DuBois, D.B., Walkerpeach, C., Pasloske, B.L., and Winkler, M. 1997. Universal ribonuclease resistant RNA standards (Armored RNA) for RT-PCR and bDNA-based hepatitis virus RNA assays. Clin Chem *43*, 2218.

DuBois, D.B., Winkler, M.M., and Pasloske, B.L. (1997). U. S. patent 5,677,124.

Dykeman, E.C., Stockley, P.G., and Twarock, R. (2010). Dynamic allostery controls coat protein conformer switching during MS2 phage assembly. *J Mol Biol 395*, 916-923.

Eisler, D., McNabb, A., Jorgensen, D., and Issac-Renton, J. (2004). Use of an internal positive control in a multiplex reverse transcription-PCR to detect West Nile virus RNA in mosquito pools. *J Clin Microbiol 42*, 841-843.

Elsawy, K.M., Caves, L.S., and Twarock, R. (2010). The impact of viral RNA on the association rates of capsid protein assembly: bacteriophage MS2 as a case study. *J Mol Biol 400*, 935-947.

Fiers, W., Contreras, R., Duerinck, F., Haegman, G., Iserentant, D., Merregeart, J., Min Jou, W., Molemans, F., Raeymaeker, A., Van den Berge, A., *et al.* (1976). Complete Nucleotide Sequence of Bacteriophage MS2 RNA: Primary and Secondary Structure of the Replicase Gene. *Nature 260*, 500-507.

Fiers, W., Contreras, R., Duerinck, F., Haegman, G., Merregeart, J., Jou, W.M., Raeymaeker, A., Volckaert, G., and Ysebaert, M. (1975). A-Protein gene of bacteriophage MS2. *Nature 256*, 273-278.

Fitter, J., Gutberlat, T., and Katsaras, J. (2006). *Neutron scattering in biology: Techniques and Applications* (Berlin, Springer-Verlag).

Glatter, O., and Kratky, O., eds. (1982). *Small angle scattering* (London, Academic Press, Inc).

Golmohammadi, R., Fridborg, K., Bundle, M., Valegard, K., and Liljas, L. (1996). Crystal Structure of Bacteriophage QB at 3.5 *A Resolution. Structure 4*, 543-554.

Golmohammadi, R., Valegard, K., Fridborg, K., and Liljas, L. (1993). Refined Structure of Bacteriophage MS2 at 2.8A Resolution. *J Mol Biol 234*, 620-639.

Graille, M., Zhou, C.-Z., Receveur-Brechot, V., Collinet, B., Declerck, N., and van Tilbeurgh, H. (2005). Activation of the LicT transcriptional antiterminator involves a domain swing/lock mechanism provoking massive structural changes. J *Biol Chem* in press.

Hayter, J.B. (1983). *Physics of Amphiphiles--Micelles, Vesicles, and Microemulsions. In Proceedings of the International School of Physics "Enrico Fermi"*, Course XC, V. DeGiorgio, and M. Corti, eds. (North Holland, Elsevier Science Publishing Co), pp. 59-93.

Heal, K.G., Hill, H.R., Stockely, P.G., Hollongdale, M.R., and Taylor-Robinson, A.W. (2000). Expression and Immunogenicity of a Liver Stage Malaria Epitope Presented as a Foreign Peptide on the Surface of RNA-free MS2 Bacteriophage Capsids. *Vaccine 18*, 251-258.

Heisenberg, M. (1966). Formation of Defective Bacteriophage Particles by Fr. Amber Mutants. *J Mol Biol 17*, 136-144.

Humphrey W., Dalke, A. and Schulten, K. (1996). VMD - Visual Molecular Dynamics. J Mol Graph *14.1*, 33-38.

Jacrot, B., and Zaccai, G. (1981). Determination of Molecular Weight by Neutron Scattering. *Biopolym 20*, 2413.

Jennings, G.T. and Bachmann, M.F. (2008). The Coming of Age of Virus-like Particle Vaccines. Biol Chem *389(5)*, 521-536.)

Koch, M.H.J., Vachette, P., and Svergun, D., I. (2003). Small-angle scattering: a view on the properties, structures and structural changes of biological macromolecules in solution. *Quart Rev Biophys 36*, 147-227.

Konig, R., van den Worm, S., Plaiser, J., Van Duin, J., Abrahams, J.P., and Koerten, H. (2003). Visualization by Cryo-electron Microscopy of Genomic RNA that Binds to the Protein Capsid inside Bacteriophage MS2. *J Mol Biol 332*, 415-422.

Kozak, M., and Nathans, D. (1971). Fate of maturation protein during infection by coliphage MS2. *Nature New Biol 234*, 209-211.

Krueger, S. (1998). SANS Provides Unique Information on the Structure and Function of Biological Macromolecules in Solution. *Physica* B *241*, 1131-1137.

Kuzmanovic, D.A., Elashivili, I., Wick, C.H., O'Connell, C., and Krueger, S. (2003). Bacteriophage MS2: Molecular Weight and Spatial Distrbution of the Protein and RNA Components by Small-Angle Neutron Scattering and Virus Counting. *Structure 11*, 1339-1348.

Kuzmanovic, D.A., Elashivili, I., Wick, C.H., O'Connell, C., and Krueger, S. (2006a). Quantification of RNA in Bacteriophage MS2-like Viruses in Solution by Small Angle X-ray Scattering. *Rad Phys Chem J 75*, 359-3

Kuzmanovic, D.A., Elashivili, I., Wick, C.H., O'Connell, C., and Krueger, S. (2006b). The MS2 Protein Shell is Assembled under Tension: A Novel Role for the MS2 Bacteriophage A Protein as Revealed by Small Angle Neutron Scattering. *J Mol Biol 355*, 1095-1111.

Legendre, D., and Fastrez, J. (2005). Production in Saccharomyces cerevisiae of MS2 virus-like particles packing functional heterologous mRNAs. *J Biotechnol 117*, 183-194.

Lima, S., Vaz, A., Souza, T., Peabody, D.S., Silva, J., and Oliveira, A. (2006). Dissecting the role of protein-protein and protein-nucleic acid interactions in MS2 bacteriophage stability. *FEBS J 273*, 1463-1475.

Ludwig, C., and Wagner, R. (2007). Virus-like particles-universal molecular toolboxes. *Curr Opin Biotechnol 18*, 537-545.

Lundstrom, K. 2005. Biology and application of alphaviruses in gene therapy Gene Ther *12(S1)*, S92-7.

Mahoney, D.J., and Stojdl, D.F. 2010. Potentiating oncolytic viruses by targeted drug intervention. Curr Opin Mol Ther *12*, 394-402.

Mastico, R.A., Talbot, S.J., and Stockley, P.G. (1993). Multiple Presentation of Foreign Peptides on the Surface of an RNA-Free Spherical Bacteriophage Capsid. *J Gen Virol 74*, 541-548.

Meng, S., and Li, J. (2010). A novel duplex real-time reverse transcriptase-polymerase chain reaction assay for the detection of hepatitis C viral RNA with armored RNA as internal control. *Virol J 7, 717*.

Min Jou, W., Haegeman, G., Ysebaert, M., and Fiers, W. (1972). Nucleotide Sequence of the Gene Coding for the Bacteriophage MS2 Coat Protein. *Nature 237*, 82-88.

Min Jou, W., Raeymaeker, A., and Fiers, W. (1979). Crystallization of Bacteriophage MS2. Eur *J of Biochem 102*, 589-594.

Morse, M.A., Hobeika, A.C., Osada, T., Berglund, P., Hubby, B., Negri, S., Niedzwiecki, D., Devi, G.R., Burnett, B.K., Clay, T.M., *et al.* (2010). An alphavirus vector overcomes the presence of neutralizing antibodies and elevated numbers of Tregs to induce immune responses in humans with advanced cancer. *J Clin Invest* Epub ahead of print.

Morton, V.L., Dykeman, E.C., Stonehouse, N.J., Ashcroft, A.E., Twarock, R., and Stockley, P.G. (2010). The impact of viral RNA on assembly pathway selection. *J Mol Biol 401*, 298-308.

Mylon, S.E., ., Rinciog, C.I., Schmidt, N., Gutierrez, L., Wong, G.C., and Nguyen, T.H. (2010). Influence of salts and natural organic matter on the stability of bacteriophage MS2. *Langmuir 26*, 1035-1042.

Ni, C.-Z., Syed, R., Kodandapani, R., Wickersham, J., Peabody, D.S., and Ely, K.R. (1995). Crystal Structure of the MS2 Coat Protein Dimer: Implications for RNA Binding and Virus Assembly. *Structure 3*, 255-263.

O'Callaghan, R., Bradley, R., and Paranchych, W. (1973). Controlled Alterations in the Physical and Biochemical Properties of R17 Bacteriophage Induced by Guanidine Hydrochloride. *Virol 54*, 476-494.

Olah, G.A., Gray, D.M., Gray, C.W., Kergil, D.L., Sosnick, T.R., Mark, B.L., Vaughan, M.R., and Trewhella, J. (1995). Structures of fd Gene 5 Protein-Nucleic acid

Complexes: A Combined Solution Scattering and Electron Microscopy Study. *J Mol Biol 249*, 576-594.

Pasloske, B.L., Walkerpeach, C.R., Obermoeller, R.D., Winkler, M., and DuBois, D.B. (1998). Armored RNA Technology for Production of Ribonuclease-Resistant Viral RNA Controls and Standards. *J Clin Microbiol 36*, 3590-3594.

Peng, K.W., TenEyck, C.J., Galanis, E., Kalli, K.R., Hartmann, L.C., and Russell, S.J. 2002. Intraperitoneal therapy of ovarian cancer using an engineered measles virus. Cancer Res *62*, 4656-62.

Pickett, G.G., and Peabody, D.S. (1993). Encapsidation of Heteologous RNAs by Bacteriophage MS2 Coat Protein. *Nucleic Acids Research 21*, 4621-4626.

Ramqvist, T., Andreasson, K., and Dalianis, T. (2007). Vaccination, immune and gene therapy based on virus-like particles against viral infections and cancer. E*xp Opin Biol Ther 7*, 997-1007.

Reid, S.M., Pierce, K.E., Mistry, R., Bharya, S., Dukes, J.P., Volpe, C., Wangh, L.J., and King, D.P. (2010). Pan-serotypic detection of foot-and-mouth disease virus by RT linear-after-the-exponential *PCR. Mol Cell Probes* Epub ahead of print.

Rolfsson, O., Toropova, K., Ranson, N.A., and Stockley, P.G. (2010). Mutually-induced conformational switching of RNA and coat protein underpins efficient assembly of a viral capsid. *J Mol Biol 401*, 309-322.

Schiller , J.T., and Lowy, D.R. (2010). Vaccines To Prevent Infections by Oncoviruses.. *Annu Rev Microbiol Epub ahead of print.*

Shepherd, C., Borelli, I., Lander, G., Natarian, P., Siddavanahali, V., Bajaj, C., Johnson, J., Brooks, C., and Reddy, V. (2006). Viperdb: A relational database for structural biology. *Nucleic Acids Research 1*, D386-389.

Stevenson, J., Hymas, W., and Hillyard, D. 2008. The use of Armored RNA as a multi-purpose internal control for RT-PCR. J Virol Meth *150*, 73-6.

Stockley, P.G., Stonehouse, N.J., and Valegard, K. (1994). Molecular Mechanism of RNA Phage Morphogenesis. *Int J Biochem 26*, 1249-1260.

Stojdl, D.F., Lichty, B.D., tenOever, B.R., Paterson, J.M., Power, A.T., Knowles, S., Marius, R., Reynard, J., Poliquin, L., Atkins, H., Brown, E.G., Durbin, R.K., Durbin, J.E., Hiscott, J., and Bell, J.C. 2003. VSV strains with defects in their ability to shutdown innate immunity are potent systemic anti-cancer agents. Cancer Cell *4*, 263-75.

Stonehouse, N.J., and Stockley, P.G. (1993). *Effects of amino acid substitutions on the thermal stability of MS2 capsids lacking genomic RNA.* FEBS Letter *334*, 355-389.

Stonehouse, N.J., Valegard, K., Golmohammadi, R., van den Worm, S., Walton, C., Stockley, P.G., and Liljas, L. (1996). Crystal Structures of MS2 Capsid with Mutations in the Subunits FG Loop. *J Mol Biol 256*, 330-339.

Struhrmann, H.B., and Miller, A. (1978). Small-angle neuton scattering of biological structures. *J Appl Crystrallog 11*, 325-345.

Tan, A., Yeh, S., C, L., Cheung, C., and Chen, P. (2008). Viral hepatocarcinogenesis:from infection to cancer. *Liver Int* 28. 175-188.

Thuman-Commike, P.A., Tsuruta, H., Greene, B., Prevelige, P.E., King, J., and Chiu, W. (1999). Solution X-ray Scattering-Based estimation of Electron Cryomicroscopy Imaging Parameters for Reconstruction of Virus Particles. *Biophys J 76*, 2249-2261.

Toropova, K., Basnak, G., Twarock, R., Stockely, P.G., and Ranson, N.A. (2008). Three-dimensional structure of genomic RNA in bacteriophage MS2: Implications for assembly. *J Mol Biol 375*, 824-836.

Valegard, K., Liljas, L., Fridborg, K., and Unge, T. (1990). Three Dimensional Structure of the Bacterial Virus MS2. *Nature 345*, 36-41.

Valegard, K., Liljas, L., Fridborg, K., and Unge, T. (1991). Structure Determination of the Bacteriophage MS2. *Acta Cryst B47*, 949-960.

Valegard, K., Murrary, J., Stonehouse, N.J., Van den Worm, S., Stockely, P.G., and Liljas, L. (1997). Three-dimensional Structures of Two Complexes between Recombinant MS2 Capsids and RNA Operator Fragments Reveal Sequence-specific Protein-RNA interactions. *J Mol Biol 270*, 724-738.

Valegard, K., Unge, T., Montelius, I., Strandberg, B., and Fiers, W. (1986). Purification, Crystallization and Preliminary X-ray Data of the Bacteriophage MS2. *J Mol Biol 190*, 587-591.

van den Worm, S., Konig, R., Warmehoven, H., Koerten, H., and van Duin, J. (2006). Cryo electron microscopy reconstructions of the Leviviridae unveil the denest icoshedral RNA packing possible. *J Mol Biol 363*, 858-865.

Van Meerton, D., Olsthoorn, R.C.L., Van Duin, J., and Verhaert, R.M.D. (2001). Peptide Display on Live MS2 Phage: Restrictions at the RNA Genome Level. *J Gen Virol 82*, 1797-1805.

Walkerpeach, C.R., Winkler, M., DuBois, D.B., and Pasloske, B.L. (1999). Ribonuclease-resistant RNA Controls (Armored RNA) for Reverse Transcription-PCR, Branched DNA, and Genotyping Assays for Hepatitis C Virus. *Clin Virol 45*, 2079-2085.

Wei, Y., Yang, C., Wei, B., Huang, J., Wang, L., Meng, S., Zhang, R., and Li, J. (2008). RNase-resistant virus-like particles containing long chimeric RNA sequences produced by two-plasmid coexpression system. *J Clin Microbiol 46*.

WHO, W.H.O. (2008). *Ten facts about cancer*.

Wick, C.H. (2002a). Method and Apparatus for Counting Submicron Sized Particles (U.S.A., US. patent 6,485,686).

Wick, C.H. (2002b). Method and System for Detecting and Recording Submicron Sized Particles (U.S.A., US patent 6,491,872).

Wick, C.H., and McCubbin, P.E. (1999). Characterization of Purified MS2 Bacteriophage by the Physical Counting Methodology used in the Integrated Virus Detection System (IVDS). *Toxicol Meth 9*, 245-252.

Zhan, S., Li, J., Xu, R., Wang, L., Zhang, K., and Zhang, R. (2009). Armored long RNA controls or standards for branched DNA assay for detection of human immunodeficiency virus type 1. *J Clin Microbiol 47*, 2571-2576.

Zhao, L., Ma, Y., Zhao, S., and Yang, N. (2006). Armored RNA as positive control and standard for quantitative reverse transcription-polymerase chain reaction assay for rubella virus. *Arch Virol* Epub ahead of print.

Zhao, L., Ma, Y., Zhao, S., and Yang, N. (2007). Armored RNA as a positive control and standard for quantitative reverse transcription-polymerase chain reaction assay for rubella virus. *Arch Virol 152*, 219-224.

In: Neutron Scattering Methods and Studies
Editor: Michael J. Lyons

ISBN: 978-1-61122-521-1
© 2011 Nova Science Publishers, Inc.

Chapter 3

SMALL ANGLES NEUTRONS SCATTERING APPLICATION IN BIOLOGICAL MATERIALS

Adel Aschi and Abdelhafidh Gharbi

Lab. De Physique de la Matiere Molle, Department de Physique Faculte des Sciences
de Tunis, Campus Universitaire, Tunisia

ABSTRACT

In this part of book, we presented the different steps to study the structure and properties of different proteins by small angles neutrons scattering (SANS). We have shown in this case, how to treat the neutron scattering spectra i.e the corrections of spectra. The form and structure factors used to determine respectively the structure and interactions effects. The figures presented are good witnesses and examples for these.

INTRODUCTION

Small-angle neutrons scattering (SANS) is a routine technique available at neutron-scattering facilities associated with research nuclear reactors. For over 50 years, SANS has been used to study the microstructure of alloys, ceramics, polymers, colloids, and other materials. Neutron scattering technique plays an important role to determine the structural and properties of colloidal medium.

In fact, the understanding of protein folding remains one of the major goals of contemporary structural biology. This requires a detailed characterization of both the folded and the unfolded states. The unfolded state is frequently viewed as unstructured and featureless, and thus described by a random coil [1]. While X-ray crystallography and nuclear magnetic resonance yield exponentially growing data on native proteins, only few biophysical techniques can provide structural information about denatured state considered as unstructured. Among these techniques, small-angle scattering (SAS) of either neutrons (SANS) or X-rays (SAXS) is a very powerful tool that yields low and medium-resolution information about the structure of macromolecules in solution [2]. Recently it has been used increasingly in the field of protein folding [3-10]. This technique provides a direct measurement of the radius of gyration of a molecule, and therefore very sensitive to the molecule's compactness which is a key parameter characterizing the degree of denaturation of a protein. Furthermore, SAS can also give description of overall shape of a macromolecule and distribution of configuration of an unfolded chain [11,12]. Still, In the literature relatively few articles focus on present with these techniques and methods used to analyze resultants [14-16]. In this part of book, we will present the different steps to study the structure and properties of different proteins by small angles neutrons scattering (SANS).

1. SMALL ANGLES NEUTRONS SCATTERING

The elastic small angle neutron scattering (SANS) probes the matter at size scales ranging from a few angstroms, to several hundred of angstroms. This technique is complementary techniques of X-ray scattering and light. In addition, thanks to the interaction of neutrons with matter, the isotopic labeling allows, in a mixture, to distinguish a specific scatterer among the others. In any scatterer medium, the variations and fluctuations density of matter cause diffusion. These variations are located at interfaces then the fluctuations occur in the bulk. Neutron scattering is essential for the study of medium with interfaces, such as colloidal solutions or porous medium. By choosing an appropriate contrast and range of scattering vectors, this technique allows for different types of information as the size of scatterers and their internal structure, the material quantity at the surface or present in a micelle, the concentration profile of an interface or the amplitude and the extent of correlations of density within a material, etc…

1.2. Measurement of neutron scattering

1.2.1. Equipments

The small angles neutron scattering experiments can be achieved with different spectrometer. This technique is used in some laboratory in the world. The most successful time-of-flight SANS instruments are:

North America:

- o   Spallation Neutron Source, Oak Ridge
- o   Los Alamos Neutron Science Center (LANSCE)
- o   University of Missouri Research Reactor Center
- o   High Flux Isotope Reactor, Oak Ridge
- o   Canadian Neutron Beam Centre, Chalk River, Canada
- o   Indiana University Cyclotron Facility

Europe:

- o   ISIS-Rutherford-Appleton Laboratories, United Kingdom
- o   Institut Laue-Langevin, Grenoble, France
- o   Leon Brillouin Laboratory, Saclay, France
- o   Berlin Neutron Scattering Center, Germany
- o   GKSS Geesthacht, Germany
- o   Juelich Center for Neutron Science, Germany
- o   FRM-II, Munich, Germany
- o   Budapest Neutron Centre, Hungary
- o   RID, Delft, The Netherlands
- o   SINQ, Paul Scherrer Institut (PSI), Switzerland
- o   Frank Laboratory of Neutron Physics, Dubna, Russia
- o   St. Petersburg Neutron Physics Institute, Gatchina, Russia

Asia and Australia:

- o   ISSP Neutron Scattering Laboratory, Tokai, Japan
- o   JAEA Research Reactors, Tokai,Japan
- o   KENS Neutron Scattering Facility, Tsukuba, Japan
- o   Hi-Flux Advanced Neutron Application Reactor, Korea
- o   Bragg Institute, ANSTO, Australia.

2. SMALL ANGLES NEUTRONS SCATTERING PRINCIPLE

Consider a parallel beam of neutrons and monochromatic wavelength λ passing through a sample. This beam can be likened to a wave plane of incident wavevector, $\vec{k}_i$. The interaction between neutrons and atomic nuclei of the sample causes the neutron scattering. In one direction at θ angle with the incident beam, the scattering wavevector is $\vec{k}_s$. The wavevector transfer is:

$$\vec{q} = \vec{k}_i - \vec{k}_s \tag{1}$$

and its module:

$$q = \frac{4\pi}{\lambda} \sin \frac{\theta}{2} \tag{2}$$

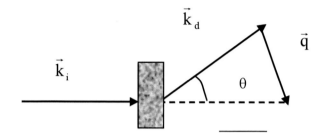

Figure 1. Representation of the neutron scattering by a sample.

2.1. Scattering Cross Section

A particle (neutron or photon) interacts with an atom can be either absorbed or scattered. Noting I_i the number of incident particles per unit of time and area, we can define a scattering intensity I_s and absorption intensity I_a corresponding to the number of neutrons or photons scattered and absorbed per unit of time, respectively. We define also the scattering and absorption cross sections σ_s and σ_a [17].

$$I_s = I_i \, \sigma_s \tag{3}$$

$$I_a = I_i \, \sigma_a \tag{4}$$

σ_s and σ_a have the dimension of a surface. The unit used by neutron physicists is barn: 1 barn = 10^{-24} cm^2.

In fact, there are two types of interaction between a neutron and atom: an interaction between the neutron and the nucleus of the atom, corresponding to nuclear scattering and interaction of the magnetic moment of the neutron with the electron spin of the atom, called magnetic scattering, we do not consider here. The event involving the interaction between a nucleus and a neutron is characterized by a length b_i, called length scattering, which depends on nuclear spin states of the atom and the neutron. Then, we define a coherent scattering length b^{coh} and incoherent scattering length b^{inc} such as:

$$b^{coh} = <b_i> \tag{5}$$

and

$$b^{inc} = \sqrt{<b_i^2> - <b_i>^2} \tag{6}$$

$<b_i>$ is the average over all isotopes i and all the spin states of the scattering object. b^{inc} is a unique term to neutron scattering, which represents the deviations b_i versus the average $<b_i>$.

This difference is actually due to spin fluctuations for each atom. The scattering lengths b^{coh} and b^{inc} are related to scattering cross sections of coherent and incoherent by the following relations:

$$\sigma_{coh} = 4\pi \, (b^{coh})^2 = 4\pi <b_i>^2 \tag{7}$$

and

$$\sigma_{inc} = 4\pi(b^{inc})^2 = 4\pi \, (<b_i^2> - <b_i>^2) \tag{8}$$

with

$$\sigma_s = \sigma_{coh} + \sigma_{inc} \tag{9}$$

Table 1 gives the values of coherent, incoherent scattering cross sections and absorption of neutrons of any isotope. It is important to note the differences between the values of σ_{coh} and σ_{inc} of hydrogen (H) and deuterium (D):

$\sigma_{coh}(H) = 1.76$ barn $\sigma_{coh}(D) = 5.60$ barn
$\sigma_{inc}(H) = 79.91$ barn $\sigma_{inc}(D) = 2.04$ barn

The strong incoherent scattering of hydrogen is an essential property that must be considered and used in neutron scattering studies. The incoherent signals of highly hydrogenated molecule, as is the case for example a protein, are dominated by that of the hydrogen molecule. The H-D isotopic exchange can then vary the contribution of different hydrogenated groups according to what we want to observe.

The values of these cross sections determine the choice of elements used in the experiments of neutron scattering. For example, in order to remove air and especially water vapor, a scavenging with gas is always performed in the chamber where the sample cells are placed. In addition its chemical neutrality towards the samples, the chosen gas must have cross sections of absorption and scattering as low as possible for not modify the samples signal (example of gas: argon or helium frequently used).

2.2. Experimental Conditions

In general, samples are placed in quartz cells. The thickness of cells is: 5mm for colloids (proteins, polymers, …) solutions and buffer, 1mm for H_2O, vacuum and a sample of B_4C. The temperature inside the cell is regulated by a thermostatic bath. In each configuration of the device, the scattering spectrum and transmission of the colloids solution and the corresponding buffer, those of water and vacuum cell and the scattering spectrum of an absorbent consisting of B_4C and cadmium are measured. In addition, the scattering spectra of each sample are stored for 6 hours or more, time depends on the power of nuclear reactor, in order to have adequate statistical accuracy. In contrary, 30 minutes are sufficient for absorbent which only serves to determine the electronic noise of detectors. In principle, the scattering spectra of water and vacuum cell used primarily to perform correction sensitivity detectors. In fact, water is very incoherent light diffuser with the scattering spectrum independent on q in the angular domain covered by the SANS.

Table 1. Values of coherent, incoherent scattering cross sections and absorption of neutrons of any isotope

Isotope[a]	σ_{coh}(barn)	σ_{inc}(barn)	σ_{abs}(barn)
^{1}H	1.7599	79.91	0.3326
^{2}H	5.597	2.04	0.00051
^{3}He	4.42	1.2	5333
^{4}He	1.34	0	0
B	3.54	1.70	767
C	5.554	0.001	0.0035
N	11.01	0.49	1.90
O	4.235	0.000	0.00019
Al	1.495	0.0085	0.231
Si	2.163	0.015	0.171
S	1.0186	0.007	0.53
Cl	11.531	5.2	33.5
Ar	0.22	0.68	0
V	0.0184	5.187	5.08
Cd	3.3	2.4	2520

[a] When the isotope is not specified, the given values correspond to the average of all isotopes of the same species, weighted by the relative abundance in the nature.

2. 3. Spectra treatment

2. 3. 1. Transmissions

The transmission, T, of a sample in a cell, is the ratio of the sample transmitted intensity by that of the vacuum cell:

$$T_{sample} = I_{sample}(0)/I_{vacuum}(0) \tag{10}$$

The detector becomes saturated when it's exposed to direct beam; a fraction of the incident flow is absorbed through attenuators constituted of Plexiglas plates.

The scattered intensity of sample depends on its transmission, its thickness and its scattering cross section. This latter depends on the incident neutrons wavelength because the diffusion is not always perfectly elastic. This is particularly true for water, whose molecular mass is not fully negligible compared to that of neutrons. It is the same for transmission, even for a perfectly elastic scattering. Indeed, the transmission of a sample is:

$$T(\lambda) \exp\left\{ [\sigma_{Scatt}(\lambda) + \sigma_{Abs}(\lambda)] \times 1 \right\} \qquad (11)$$

where l is the optical trajectory of the neutron beam in the sample. $\sigma_{Scatt}(\lambda)$ and $\sigma_{Abs}(\lambda)$ are the total cross sections of scattering and absorption respectively. They depend on the wavelength of neutrons.

2. 3.2. Scattered intensity

The scattered intensity, by unit volume of any sample, composed of its cell and its contents can be written:

$$I'_{Cont+Cell}(\lambda, q) = I_i(\lambda)S\Delta\Omega C(\lambda)T_{Cont+Cell}(\lambda) \times$$
$$\times [e_{cell} \cdot I_{Cell}(\lambda, q) + e_{Cont} I_{Cont}(\lambda, q) + N(q)] + N_{Amb} \qquad (12)$$

where S is the incident beam section of the intensity $I_i(\lambda)$. $\Delta\Omega$ is the collection solid angle of the detector and $C(\lambda)$ its performance for neutron wavelength λ. e_{cell} is the thickness of cell walls, its internal optical path e_{Cont}, i.e. the thickness of its content. $I_{Cell}(\lambda, q)$ and $I_{Cont}(\lambda, q)$ are the scattering cross sections, per unit volume and solid angle, of the vacuum cell and its contents. B(q) is a direct beam noise due to imperfect collimation and scattering by air in the vicinity of the sample. B_{amb} is ambient background noise.

Similarly, the scattered intensity by the vacuum cell is:

$$I'_{Cell}(\lambda, q) = I_i(\lambda)S\Delta\Omega C(\lambda)T_{Cell}(\lambda) \times [e_{cell} I_{Cell}(\lambda, q) + N(q)] + N_{Amb} \qquad (13)$$

It follows that the quantity

$$I'_{Cont}(\lambda, q) = \frac{I'_{Cont+Cell}(\lambda, q) - B_{Amb}}{T_{Cont+Cell}(\lambda)} - \frac{I'_{Cell}(\lambda, q) - B_{Amb}}{T_{Cell}(\lambda)} =$$
$$I_i(\lambda)S\Delta\Omega C(\lambda) e_{Cont} I_{Cont}(\lambda, q) \qquad (14)$$

is independent of background noise and represents the only scattering of the cell contents.

The spectra of water and the vacuum cell can correct the samples spectra and their buffer. Indeed, for small values of q, the scattered intensity of the water (H_2O):

$$I'_{H_2O}(\lambda, q) = \frac{I'_{H_2O+Cell}(\lambda, q) - B_{Amb}}{T_{H_2O+Cell}(\lambda)} - \frac{I'_{Cell}(\lambda, q) - B_{Amb}}{T_{Cell}(\lambda)} =$$
$$I_i(\lambda)S\Delta\Omega C(\lambda) e_{H_2O} I_{H_2O}(\lambda, q) \qquad (15)$$

is principally incoherent origin and therefore independent of q. In general the thickness of the water sample is set at 1 mm in order to have a transmission about 0,5.

2 .3.3. Treatment of raw spectra

The experimental data are processed in the usual manner, as suggested by above considerations and described by different authors [16] and [18]. Here we present an example of simple program called PASIDUR, written by D. Lairez [19] in the Leon Brillouin Laboratory (LLB), France. For each value of q, this program computes the following quantity:

$$I_{sample}(\lambda, q) = \frac{I'_{Solution}(\lambda, q) - I'_{Buffer}(\lambda, q)}{I'_{H_2O}(\lambda, q)} \frac{1}{V} \left. \frac{d\sigma_{H_2O}(\lambda)}{d\Omega} \right|_{Measured} \tag{16}$$

where I'$_{Solution}$(λ, q) and I'$_{Buffer}$ (λ, q) are given by the equation (14) for the sample and its

buffer, respectively, and I'$_{H2O}$(λ, q) by the equation (15). $\left. \dfrac{1}{V} \dfrac{d\sigma_{H_2O}(\lambda)}{d\Omega} \right|_{Measured}$ is the

scattering cross section of the unit volume of water sample measured with neutron wavelength λ.

2.3.4. Absolute calibration

To obtain the absolute value of the incoherent scattering cross section of water sample, the detector is moved so as to measure the direct beam with the same detectors as those used in measurements of scattered intensities. To avoid saturation of these, the direct beam is first attenuated. The attenuation factor is obtained by taking the ratio of the scattered intensity by a strongly diffusing material such as graphite or Teflon, in absence and presence of the attenuator. With this manner, we can estimate the scattering cross section of water in small cell with 1 mm thick at a given wavelength. The forward scattering intensity, I$_1$ (0, c) [20], for a solution of polymer or protein is such that:

$$\lim_{c \to 0} \frac{I_1(0, c)}{c} \tag{17}$$

2.3.5. Incoherent noise

To obtain the spectrum of coherent scattering of a sample, it is necessary to correct the result given by the expression (16) of the incoherent scattering of the solute. When a colloid (protein or polymer) is dissolved in a deuterated buffer, a number, N of hydrogen atoms are not exchanged against the deuterium atoms of the medium and give incoherent contribution. Then:

$$I_{sample}(\lambda, q) = I_{sample}{}^{Coh}(\lambda, q) + \Delta I_{sample}{}^{Inc}(\lambda, q) \tag{18}$$

where the first term of second member denotes the coherent contribution to the diffusion. While the second term denotes the incoherent contribution.

The latter equation can be evaluated as follows:

$$\Delta I_{sample}^{Inc}(\lambda, q) = \frac{N_{sample}}{N_{H_2O}} I_{H_2O}(\lambda, q) \tag{19}$$

where N_{sample} and N_{H_2O} are the molarities of hydrogen atoms in the sample (protein, polymer,…) solution and H_2O, respectively.

The expression (19) gives an estimate of the incoherent contribution due to the hydrogen atoms of the sample, because it assumes that multiple scattering, and thus the transmission, are the same in the sample solution and water of 1 mm thick. On the other hand, even if great care is taken in samples preparation in deuterated medium, they remain a slight accidental contamination. In consequence, $\Delta I_{potein}^{Inc}(\lambda, q)$ will often viewed as a adjustable parameter, especially when interpreting the obtained scattering spectra for $q > 0.1$ Å^{-1}, where the scattered intensity is very low.

3. EXPRESSION OF THE SCATTERED INTENSITY

3.1 General Case

The total scattering amplitude of a sample is:

$$a(\vec{q}) = \sum_i b_i \exp(i\,\vec{q}.\vec{r}_i) \tag{20}$$

where $\vec{r}_i$ is the position of the atomic nucleus i with the scattering length b_i.

The total scattered intensity per unit of solid angle and volume of sample is:

$$I(\vec{q}) = \frac{1}{V} <|a(\vec{q})|^2> = \frac{1}{V} < \sum_{i,j} b_i b_j \exp[i\,\vec{q}.(\vec{r}_i - \vec{r}_j)] > \tag{21}$$

where V is the scattering volume. The brackets indicate the statistical average. I (q) have the dimension of the inverse of product of length by a solid angle (cm-1 sr-1). The shape of the scattering profile, I (q), provides information on a spatial scale of the order q-1 on the structure of the sample. The previous expression can be rewritten as follows :

$$I(q) = \frac{1}{V} \left\langle \left| \int_V \rho(\vec{r}) e^{i\vec{q}.\vec{r}} d^3r \right|^2 \right\rangle \tag{22}$$

where $\rho(\vec{r}) = \sum_i b_i \delta(\vec{r} - \vec{r}_i)$, is a local scattering length density.

Figure 2 shows the scattering length density for water and various biological macro-molecules as a function of the heavy water concentration [2]. The hydrogen/deuterium exchange influences the intensity scattered by solutions.

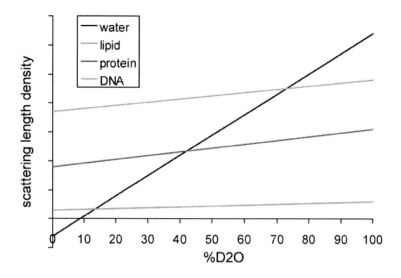

Figure 2. The relationship between the scattering length densities $\rho(\vec{r})$ of various biological macromolecules versus D_2O concentration (SANS by ref. [2]).

The total intensity of neutrons scattered by any sample is the sum of two terms: coherent intensity, I (q), whose general expression has been given and incoherent intensity resulting from the change of the state of neutron spin in their interactions with the nuclear spin of scattering atoms. The largest incoherent scatterer is hydrogen. For small angles, the incoherent scattering is independent of q. It may therefore be regarded as a "mere" background noise level which can be estimated from the atomic composition of the sample.

3.2. Colloidal Solutions Ideal

In a system composed by large objects in very dilute solution where the solvent constituted of small molecules, these objects produce an excess of scattered intensity

$$I(q) = n \left\langle \left| \int_{v} [\rho(\vec{r}) - \rho_s \frac{v'}{v}] e^{i\vec{q}.\vec{r}} d^3 r \right|^2 \right\rangle_{\Omega} \tag{23}$$

compared to the solvent alone. In this expression n and v are the number density and volume of objects, respectively. v' is the partial volume in the solution. $\rho(\vec{r})$ is the local scattering length density of one of the objects and ρ_s is the average scattering length density of the solvent. Here the bracket $<...>_{\Omega}$, indicates the average over all possible orientations of macromolecules for example. In this case, the scattering intensity provides information on the size, shape and structure of the scatterers. When scatterers are sufficiently homogeneous, the previous expression can be rewritten as follows:

$$I(q) = n(<\rho>v - \rho_s v')^2 P(q) \qquad (24)$$

Where $<\rho> = v^{-1}\sum_i b_i$ is the average scattering length density of the objects and

$$P(q) = \left\langle \left| \frac{1}{(<\rho>v - \rho_s v')} \int_v r^2 \left[\rho(r) - \rho_s \frac{v'}{v} \right] e^{i\vec{q}.\vec{r}} d\vec{r} \right|^2 \right\rangle \qquad (25)$$

For q = 0, the form factor, P(0) ≡ 1.

It is very common use to write the expression (24) as follows:

$$I(q) = \frac{cM}{N_A} K^2 P(q) \qquad (26)$$

where, N_A is Avogadro number. $c = nM/N_A$ is the mass concentration of diffusing objects (g cm^{-1}), M their molar weight and $\overline{K} = (<\rho>\overline{v} - \rho_s\overline{v}')$ is the average specific contrast. $\overline{v}$ and $\overline{v}'$ are their specific and partial specific volumes (cm^{3}g^{-1}), respectively.

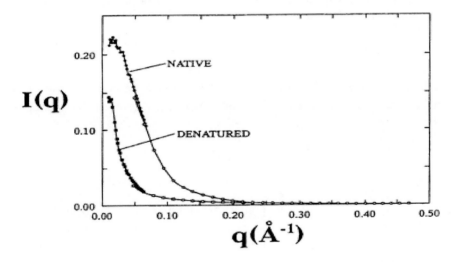

Figure 3. SANS profiles for protein, PGK, in the native form and denatured in 4 M guanidinium chloride solution. The contrast between the solvent and protein scattering has been maximized. Each scattering profile is a merger of two data sets [21].

Example of SANS experiments is shown in Figure 3. The scattering intensities curves are the superposition of two sets of data [21]. The errors are relatively large for the four points at lowest q, mainly because of the decreased detection of the solid angle at the corresponding scattering angles. Here, the scattering intensities were corrected for non-uniformity of the detector response by normalization to the mainly incoherent scattering from a 1.0 mm-thick water sample, whose absolute differential cross section was measured to be 0.81 ± 0.02 cm^{-1} sr^{-1} at $\lambda = 5.0$ Å and 0.98 ± 0.02 cm^{-1} sr^{-1} at $\lambda = 11.0$ Å. The scattering spectrum of the solvent alone was subtracted in each case from the solution spectrum. The result was further corrected for the excess incoherent scattering from the un-exchanged hydrogen atoms in the protein, such that the remaining intensity almost vanishes for values of $q > 0.55$ Å^{-1}.

3.3. Interaction

When the colloidal solution is not infinitely diluted, it must be taken into account the interactions between particles. The general expression of the scattered intensity is written then:

$$I(q) = \frac{cM}{N_A} K^2 P(q) S(q,c) \tag{27}$$

where

$$S(q,c) = 1 + \frac{c N_A}{M} \int_v h(r) e^{i\vec{q}\cdot\vec{r}} d\vec{r} \tag{28}$$

is the structure factor of the solution. It's the Fourier transform of h $(r) = [g (r) - 1]$, where g (r) is the correlation function of gravity centers of scattering objects. The expression (28) is not valid only for spherical scatterers. Otherwise, it will be modified to take account of possible correlations of orientation [22].

In the thermodynamic limit, the interactions between particles in solution can be evaluated by determining the Virial coefficients. These quantities are physically important that can be deduced from the variation of the forward scattering intensity with the concentration c of the sample. The forward scattering intensity can be written as follows:

$$\frac{K^2 Mc}{N_A I(0,c)} = \frac{1}{S(0,c)} = \frac{M}{RT} \frac{\partial \Pi}{\partial c}\bigg|_{T,P} = (1 + 2M\,A_2\,c + 3\,MA_3\,c^2 + ...) \tag{29}$$

where R is the ideal gas constant, T the temperature and Π is the osmotic pressure. A_2, A_3, ... are the second, third,...Virial coefficients of the solution. A_2 describes the interactions between two particles, A_3 those between three particles, etc... The coefficient A_2 is positive when the interactions are repulsive and negative when they are attractive.

In general, the scattering spectrum can be written as follows:

$$\frac{K^2 Mc}{N_A I(q,c)} = \frac{1}{S(q,c)} = \frac{1}{P(q)}\left[1 + 2M\,A_2\,cF_2(q) + 3\,MA_3\,c^2 F_3(q) + ...\right], \tag{30}$$

where $F_2(q)$, $F_3(q)$, ... are the Fourier transforms, normalized to unity for $q = 0$, of the direct correlation functions of 2, 3,... particles. Their expressions can be calculated from the results given in some statistic physics textbooks [23, 24].

4. RADIUS OF GYRATION

4.1. Definition

The size of a diffuser is obtained experimentally by determining the radius of gyration, which is derived directly from the small angle scattering spectrum. The radius of gyration of a particle is defined as follows:

$$R_g^2 = \frac{1}{<\rho>v - \rho v'} \int_v r^2 \left[\rho(r) - \rho_s \frac{v'}{v} \right] d^3r \qquad (31)$$

where r is the measured distance from the gravity center of the diffuser.

Table 2. Radius of gyration of a few homogeneous particles with simple shapes

Radius of sphere R	$Rg^2 = 3/5.\ R^2$
Hollow sphere (inner radius R_1, outer radius R_2)	$Rg^2 = 3/5.(R_2^5 - R_1^5) / (R_2^3 - R_1^3)$
Ellipsoid with a,b,c axes	$Rg^2 = (a^2 + b^2 + c^2) / 5$
Based elliptical cylinder with a, b axes and height h	$Rg^2 = (a^2 + b^2) / 4 + h^2 / 12$
Hollow cylinder (inner radius R_1, outer radius R_2)	$Rg^2 = (R_1^2 + R_2^2) / 2 + h^2 / 12$

The radius of gyration, by definition, contains information on the size and shape of the diffusing entity. However, its value depends to contrast. As shown in the table 2, the radius of gyration is connected, at least more complex, to the geometric parameters of the particle. These results concern only homogeneous particle with constant scattering length density. ρ (r) = $<\rho>$.

4.2. Guinier Approximation

The Guinier approximation is commonly used to determine the value of the apparent radius of gyration, $R_g(c)$, of the scatterers [25]. It is a rather arbitrary but it is very useful and acceptable as long as $qR_g(c) \leq 1$. The scattering intensity is written as follows:

$$I(q,c) \cong I(0,c) \exp\left[-\frac{q^2 R_g^2(q,c)}{3} \right] \qquad (32)$$

A plot of Ln [I(q,c)] as a function of q^2 allows the forwards scattering intensity, I(0,c) and the apparent radius of gyration, R_g(q, c), to be easily determined.

Figure 4 shows an example of Guinier plots of SANS spectra [26]. In a q-domain where the Guinier approximation is valid, the slope of the linear regression is related to the value of the apparent radius of gyration Rg (q, c). Here, the value of the apparent radius of gyration increases with increasing pressure. The estimated error of the apparent radius of gyration obtained from the regression analysis varies from ±1% to ±2% according to the protein concentration.

The previous model was used to infer from each spectrum the value of the forward scattered intensity would have in absence of interaction. The actual value of the forward scattered intensity I(0,c), was deduced from the small q region of the spectra by means of the Guinier approximation. In fact, the forward intensity I(0, c) can be written

$$I(0, c) = KcMw(c) / [1 + Bw(c) Mw(c) \times c] \tag{33}$$

where Mw(c) is the weight average of the solute molecular weight and Bw(c) is a mean interaction parameter related to the virial coefficients of the osmotic pressure.

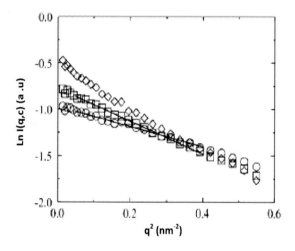

Figure 4. Guinier plot of SANS spectra. Logarithm of the scattered intensity, I(q, c), as a function of the square of the wave-number transfer, q. β-lactoglobulin concentration, c (Pression = 0.1 MPa): c= 14.3 mg cm^{-3}. Pressure, P: (o) 50 MPa, (□) 150 MPa, and (◊) 300 MPa. The full lines are a fit with Eq. (32).Temperature is 20 °C (from ref.[26]).

Using Eq. (33), the quantity Bw(c) ×Mw(c) will be estimated. Examples of results are shown in Figure 5. This figure show that close to the critical micellar concentration "CMC" (~1 mg/ml in this case) the mean interactions are attractive and concentration dependent [27]. When either the protein concentration or the temperature is high enough the mean interactions become weakly repulsive.

From equation (30) it is possible to show that the apparent radius of gyration depends on the concentration and can be expanded as follows:

$$R_g^2(c) = R_g^2(1 + 2B_2Mc + 3B_3Mc^2 + ...)^{-1} \qquad (34)$$

where the coefficients B_2, B_3,... are functions of both the virial coefficients and the spatial extent of the interactions. The actual radius of gyration, $R_g = R_g(0)$, is simply obtained by extrapolation to zero concentration. Therefore it is absolutely necessary to perform scattering measurements at various concentrations when the interactions between the solute molecules are not negligible.

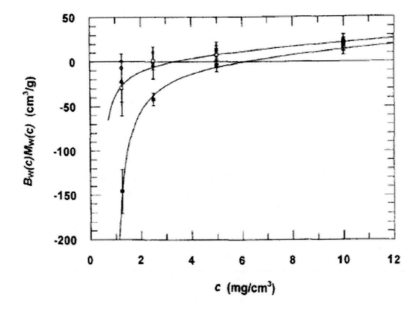

Figure 5. Variation of the interaction coefficient $B_w(c)$ $M_w(c)$ with the protein concentration c. $B_w(c)$ $M_w(c)$ is defined by Eq. (33). The temperatures are 4.5 (■), 9.6 (□), 15 (●) and 23 °C (○). The solid lines are guide for the eye (from ref. [27]).

5. SCATTERING BY POLYMER CHAINS

The fractals objects are irregular or chaotic shape. But, their geometric properties are invariant by scaling change [18]. The trajectory of a Brownian particle is an example of fractal. The compactness of the structure of a fractal object is characterized by its fractal dimension, D. It quantifies how the mass, M, or number of structural units increases with the size R of the object:

$$M \propto R^D \qquad (35)$$

The linear Polymers are examples of fractals objects. Their invariance by change of scale can describe their conformation by "scaling laws" [28-30]. In typical three dimensions space, D = 3 for collapsed chain forming a compact globule, D = 2 for an ideal chain, or Gaussian, and $D = 1.7$ for a chain with excluded volume interactions. Therefore, the determination of D can characterize polymer conformation.

5.1. Flexibles Chains

The flexible chain is a rather simplistic polymer model without rigidity. The expression of the form factor of an ideal flexible chain was established by Debye [31] and writes

$$P_D(x) = \frac{2}{x^2}[e^{-x} - 1 + x] \tag{36}$$

where $x = (q\ R_g)^2$. For large values of x, this expression has the following asymptotic form:

$$\lim_{q \to \infty} P_D(q) = \frac{2}{(qR_g)^2} \tag{37}$$

In the presence of excluded-volume interactions, the Debye approximation (36) remains valid as long as $x \leq 9$ [32]. A very useful approximation of this law is [33]:

$$P_D(x) \cong (1 + 0{,}359x^{1.103})^{-1} \tag{38}$$

where the figures 0.359 and 1.103 gives to this approximation an accuracy better than $\pm$ 0.4% for $x \leq 13$.

Figure 2 shows the reciprocal of the intensity I(q, c) *versus* $q^{2.206}$ for β-casein monomers and the line fits to Eqn (38). This plot shows that Eqn (38) correctly describes the scattering profiles of β-casein monomers. From the fits to these curves, it is possible to get both the apparent radius of gyration and the aggregation number of the micelle [34].

Generally, for values of q larger than $3R_g^{-1}$, the form factor of a flexible random coil behaves as [28, 30]:

$$P(q) \approx \frac{P_\infty}{(qR_g)^{1/v}} \tag{39}$$

where P_∞ is the amplitude and $D = 1/v$ is the fractal dimension. v is the Flory exponent [35].

$P_\infty = 2$ and $v = 0.5$ for an ideal chain whereas $P_\infty = 1.22$ and $v = 0.588$ for a chain with excluded volume interactions [36]. Determination of the two quantities P_∞ and v gives information about the average conformation of the chain and the nature of the interactions.

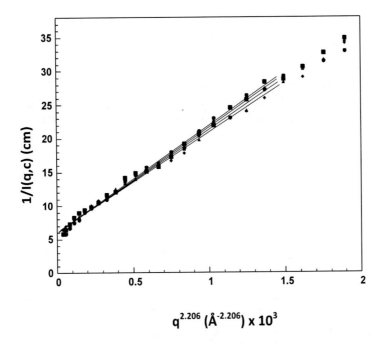

Figure 6. Plots of $[I(0, c)]^{-1}$ versus $q^{2.206}$ for β-casein monomer. The GdmCl concentrations are: (▲) 1.5, (♦) 2.0, (■) 3.0, (●) 4.0 mol L^{-1}. Using the approximation (Eqn (38)) to the Debye function (Eqn (36)) the radius of gyration of β-casein is given by Rg = (s/0.359i)$^{0.453}$, where s is the slope and i the intercept of the straight line (ref. [34]).

5.2. Semi-flexible Chain

A semi-flexible chain corresponds to a more realistic polymer model. It is a random chain the structure of which evolves continuously to that of a rod when the distance between two points of the chain becomes lower than the statistical length. The latter represents the extent of the orientation correlations along the chain. A chain with persistence length thus has certain rigidity. Two parameters define such a system: its contour length, L, and its statistical length, b. Its form factor is still given by the law of Debye (36) up to $qR_g \cong 3$. The radius of gyration of the chain has been calculated by Benoît and Doty [37]. They found

$$R_g^2 = R_g'^2 \left[1 - \frac{3}{2n} + \frac{3}{2n^2} - \frac{3}{4n^3}(1 - e^{-2n}) \right]$$ (40)

where $R_g'^2 = Lb/6$ and $n = L/b$ is the number of statistical links of the chain.

The behavior described by Eq. (39) remains valid in the intermediate range of transfer wave vector $3R_g^{-1} \leq q \leq 2b^{-1}$. For higher values of q the form factor may be approximated by the following approximation [38]:

$$P(q) = \frac{1}{qL}\left(\pi + \frac{4}{3qb}\right) \tag{41}$$

which is only valid for infinite chain length ($L \gg b$). This approximation is not sufficient to describe correctly the scattering spectra of completely unfolded proteins because they are not sufficiently long. For such chains, much better analytical expressions were obtained by Pedersen and Schurtenberger using the results of Monte Carlo simulations [39]. These authors have given three different methods to establish a precise approximation of the form factor for a random chain. In what follows, we will only give and use the results of the third method which allows the form factor of a random chain with or without excluded volume to be obtained in the same manner. Only the case of the relatively long chains ($L > 4b$) will be considered.

5.3. Ideal Worm-like Chain

The form factor is described like that of an ideal chain for $qb < 3.1$ and by that of a rod for $qb \to \infty$. In the crossover region, $qb \cong 3.1$, the form factor is described by an empirical function of L and b. This approximation is based on the result of Burchard and Kajiwara [40] in which the scattering function calculated by Sharp and Bloomfield [41] is used for small q values.

For $qb < 3.1$, the form factor of an ideal chain is represented by the following expression:

$$P_{SB}(q,L,b) = P_D(x) + \frac{b}{15L}\left[4 + \frac{7}{x} - \left(11 + \frac{7}{x}\right)e^{-x}\right] \tag{42}$$

where $x = (qR'_g)^2$ This expression differs slightly from that of Sharp and Bloomfield in which the argument of the function is $x = (qR_g)^2$, where R_g is the actual value of the radius of gyration given by Eq.(40).

For $qb > 3.1$ Pedersen and Schurtenberger [39] represent the form factor by a sum of powers of q:

$$P_{PS}(q,L,b) = \frac{a_1}{(qb)^{p_1}} + \frac{a_2}{(qb)^{p_2}} + \frac{\pi}{qL} \tag{43}$$

which behaves as the scattering function of a rod for $qb \to \infty$. In this expression, $p_1 = 4.95$ and $p_2 = 5.29$. These values were determined by fitting to the results of simulations. The parameters a_1 and a_2 are obtained in such a way that the two functions (42) and (43), and their first derivatives with respect to q are continuous for $qb = 3.1$. Therefore the parameter a_i where $i = 1,2$, is:

$$a_i = [P_{SB}(u_0) + \frac{1}{p_j}P'_{SB}(u_0) - \frac{\pi b}{u_0 L}(1 - \frac{1}{p_i})]p_j \frac{u_0^{p_i}}{p_j - p_i} \tag{44}$$

With

$$P'_{SB}(u) = 2u \frac{dP_{SB}(u)}{du} \qquad (45)$$

where $u = qb$ and $u_0 = 3.1$.

5.4. Excluded Volume Worm-like Chain

At intermediate q values, there is a large difference between the scattering functions of a random chain with and without excluded volume interactions. Consequently, Eq. (42) must be modified to take into account this effect. A new expression for the form factor was derived from the results of Utiyama et al. [42] by Pedersen and Schurtenberger [39]. The following approximation has been obtained:

$$P_{EX}(q,L,b) = W(x)P_D(q,L,b) + [1-W(x)](c_1 x^{-1/\nu} + c_2 x^{-2/\nu} + c_3 x^{-3/\nu}) \quad (46)$$

where $\nu = 0,588$ is the Flory exponent and $x = (qR_g)^2$. Here Rg is radius of gyration of chain with excluded volume:

$$R_g^2 = [\alpha_S(L/b)]^2 R_g'^2 \qquad (47)$$

$\alpha_S(L/b)$ is the expansion factor due to the excluded volume interactions. It is given by the following expression:

$$\alpha_S(y) = \left[1 + \left(\frac{y}{3,12}\right)^2 + \left(\frac{y}{8,67}\right)^3\right]^{\frac{2\nu-1}{6}} \qquad (48)$$

In Eq. (46), $W(x)$ is the following empirical cross-over function

$$W(x) = \frac{1}{2}\left(1 - th\frac{x-c_4}{c_5}\right) \qquad (49)$$

Finally the values of the constants appearing in Eqs. (46) and (49) are: $c_1 = 1,2200$, $c_2 = 0,4288$, $c_3 = -1,651$, $c_4 = 1,523$, and $c_5 = 0,1477$.

The second term in the right hand side of Eq.(42), which takes into account the local rigidity of the chain, must also be slightly modified in order to describe the results of the simulations. When excluded volume interactions are present, Pedersen and Schurtenberger have shown that the form factor of the chain may be written as follows [39]:

$$P_{SB}(q,L,b) = P_{EX}(q,L,b) + C(L/b)\frac{b}{15L}\left[4 + \frac{7}{x} - \left(11 + \frac{7}{x}\right)e^{-x}\right] \qquad (50)$$

where

$$C(L/b) = a_4(L/b)^{-p_3} \tag{51}$$

for $L > 10\ b$. The values of the constants are $a_4 = 3.06$ and $p_3 = 0.44$. For shorter chains $C(L/b) = 1$.

The preceding results describe the form factor up to $qb = 3.1$. At higher values of q equations (44) and (45), established for an ideal chain, remain valid.

5.5. Real chain

A real chain is not infinitely thin. Therefore it is necessary to take into account its finite cross section which may be assumed to be circular with an apparent average radius, R_T. The value of R_T strongly depends on both the local structure of the chain and that of solvent in its close vicinity. The form factor of such a chain may be written as follows [32, 39]:

$$P(q,L,b,R_T) = [P_{SB}(q,L,b) + P_{PS}(q,L,b)]\ P_T(q,R_T) \tag{52}$$

where the relations for $P_{SB}(q,L,b)$ and $P_{PS}(q,L,b)$ are given in the previous sections for chains with and without excluded volume.

$$P_T(q,R_T) = \left(\frac{2J_1(qR_T)}{qR_T}\right)^2 \tag{53}$$

is the form factor a disc of radius R_T. $J_1(x)$ is the first order Bessel function. The value of R_T depends strongly on the local contrast and the solvent structure in the immediate vicinity of the chain.

5.6. Kratky representation

A representative scattering curves used can give qualitative information on the shape of the chain, or quantitative internal structure of molecule and interactions at short distances within the molecule. This representation is so-called Kratky. It consist to plot the function $q^2I(q)$ versus q on an extended domain of q ranging up about 0, 6 Å^{-1}. Indeed, the behavior of this function is directly dependent on the structure of the molecule:

• A chain compact, spherical overall shape (polymer in poor solvent) follows the Porod law [43]: at large q, the scattering intensity follows $1/q^4$ profile. In the Kratky representation, the curve has hyperbolic decay q^{-2} for values of q sufficiently large.
• In the case of a Gaussian chain, the equation (36) shows that when q becomes large, the function q^{2}I (q) tends to a constant value, independent of q. In Kratky representation, this curve has a plateau.

• Finally, Kratky and Porod [44] proposed the model of chain "worm" (Wormlike chain), or chain with persistence length, to describe a chain with some rigidity. This rigidity is represented by the statistic length b, defined as twice of the persistence length, which is the minimum length of a segment when one of extremities can take all the possible orientations relative to the other. This rigidity may be due to the presence residual interaction on the chain segment. The fact that a real chain is not infinitely thin, we must introduce a radius of gyration Transverse R_T, [32], whose form factor is writing, using Guinier approximation, $\exp(-q^2R_T^2 / 2)$. The chain form factor can then write [32, 38]:

Figure 7 and Table 3 summarize the different behaviors $q^2I(q)$ function according to the type of molecular chain. In this representation the shape of scattering curves given us immediate qualitative information on the form of the protein i.e. if it is more compact or extended.

Table 3. Kratky representation for different chain conformation

Kratky representation	
Spherical Molecule	$q^2I(q) \propto q^{-2}$
Gaussian Chain	$q^2I(q) = constant$
Infinitely Chain with persistence length	$q^2I(q) \propto q$
Chan with persistence length and R_T	$q^2I(q) \propto q \times \exp(-q^2R_T^2/2)$

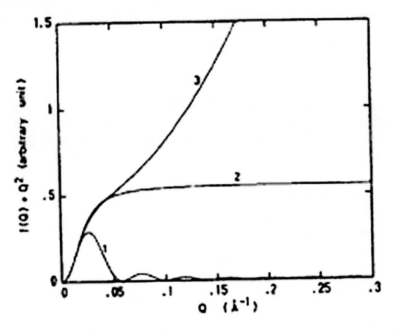

Figure 7. Kratky representation of different types of chains in solution: 1.Spherical molecule, 2. Gaussian chain, 3. Chain infinitely fine with persistence length (from ref. [45]).

The validity of these various approximations have been checked [39] by comparison with the experimental results obtained by Rawiso et al. [32] on polystyrene solutions in carbon disulfide. This is a good solvent of polystyrene and the polymeric chains have excluded volume interactions. Measurements had been carried out with regular, partially and completely deuterated polystyrene.

5.7. Concentration effects

In the case of random chains it is more interesting to consider those existing between the monomers of different chains. Using the method of calculation of Benoit and Benmouna [46], it is easy to show the scattered intensity may be written as

$$I(q) = \frac{Mc}{N_A} K^2 P(q) \left| 1 + \frac{cN_A}{M} P(q) h_T(q,c) \right|$$ (54)

where

$$h_T(q,c) = \int_V [g_T(r,c) - 1] e^{i\vec{q}.\vec{r}} \, d\vec{r}$$ (55)

$g_T(r,c)$ being the correlation function of two monomers belonging to different chains, separated by the distance r. In the thermodynamic limit, $q = 0$, the last equation can be rewritten as follows:

$$\frac{cN_A}{M} h_T(0,c) = \frac{RT}{M} \frac{\partial \Pi}{\partial c} \Big|_{T,P}^{-1} - 1 = -\frac{2A_2 Mc + 3A_3 Mc^2 + ...}{1 + 2A_2 Mc + 3A_3 Mc^2 + ...}$$ (56)

Therefore, to first order in c, Eq.(54) is similar to the result of Zimm [47] if one assumes, like him, point-like interactions between the chains, i.e. that $h_T(r,c) = [g_T(r,c) - 1] = 2A_2 M^2 N_A^{-1} \delta(r)$ where $\delta(r)$ is the Dirac function. Like Eq. (29), Eq. (56) can be generalized in the following form:

$$\frac{cN_A}{M} h_T(q,c) = -\frac{2A_2 McF_{T2}(q) + 3A_3 Mc^2 F_{T3}(q) + ...}{1 + 2A_2 McF_{T2}(q) + 3A_3 Mc^2 F_{T3}(q) + ...}$$ (57)

where $F_{T2}(q)$, $F_{T3}(q)$,... are the Fourier transforms of the direct correlation functions of 2, 3,... chains, normalized to unity for $q = 0$. This general result differs appreciably from that given by Benoît and Benmouna [46].

For the model previously used, in which a polymeric chain is represented by a semi-flexible cylinder of average radius R_T, the Fourier transform of correlation function of 2 chains can be described by the following approximation:

$$F_{T2}(q) \cong \left(\frac{2J_1(qR_C)}{qR_C} \right)^2 \tag{58}$$

where R_C is the correlation length whose value depends on the actual interactions between the two chains. For chains behaving like semi-flexible cylinders of radius R_T, the interactions are only steric and

$$R_C^2 = (2R_T)^2 = \frac{A_2 M^2}{\pi L N_A} \tag{59}$$

For example, to describe the scattering spectra of protein completely unfolded by denaturant, it's commonly to use the following approximation:

$$I(q,c) \cong \frac{Mc}{N_A} K^2 P(q) \left[1 - 2A_2 Mc\, P(q) F_{T2}(q) \right] \tag{60}$$

By expanding this expression to first order in q^2 it can be shown that the square of the apparent radius of gyration of the scatterers is:

$$R_g^2(c) \cong R_g^2(0) \left[1 - 2B_2 Mc \right] \tag{61}$$

with

$$B_2 = A_2 \left(1 + \frac{R_C^2}{4R_g^2(0)} \right) \tag{62}$$

and

$$R_g^2(0) = R_g^2 + \frac{3}{4} R_T^2 \cong R_g^2 \tag{63}$$

where R_g^2 is the radius of gyration of the axis of the chain and R_T^2 the radius of its cross section. As we will see it below $R_T^2/R_g^2 \leq 1\%$. As $R_g^2(0) >> R_C^2$, $F_{T2}(q) \cong 1$ with for small q values. Moreover, if $2A_2 Mc << 1$ Eq. (60) becomes

$$\frac{M K^2 c}{N_A I(q,c)} \cong \frac{1}{P(q)} + 2A_2 Mc \tag{64}$$

The scattering spectrum of the denatured beta-casein is shown in Figure 8 and is plotted in a Kratky representation [48]. One can assume that electrostatic intermolecular interactions are screened. After correction for the concentration effects, no peak observed in the q-range. In fact, the spectra of unfolded b-casein is similar to that of an excluded volume chain [8,9].

Using the third approximation of Pedersen and Schurtenberger for semi-flexible polymer chains [39], important structural parameters can be inferred from the scattering profiles. Values of the radius of gyration Rg(0), of the fractal dimension ν, of the statistical length b, and of the apparent radius of the chain cross-section Rc were obtained at 4M GdmCl. The following values are obtained for the structural parameters: L ≅ 719.5 Å, b ≅31.5 Å and Rc ≅5.5 Å . In this case the value of L is very close to that calculated assuming a distance of 3.45 Å between two successive α-carbon atoms [49]. The statistical length of β-casein is the first correct experimental determination and show that the chain presents a small rigidity. The value of Rc strongly varies with the denaturing conditions because it is very sensitive to the solvent structure near the chain.

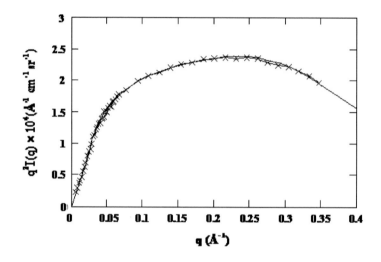

Figure 8. Kratky plot of SANS curve from β-casein in fully unfolded state in 4M GdmCl at 20°C. The protein concentration is ≅ 10 mg/ml (from the ref. [48]). The solid line is the result of the third approximation of Pedersen and Schurtenberger [39] fitting of the experimental data.

The second virial coefficient, A_2, of a polymer solution is given by the following expression [28,50]:

$$A_2 = \Psi\, 4\pi^{3/2} N_A R_g^3 M^{-2} \tag{65}$$

where Ψ is the interpenetration function.

For a chain of infinite length ($L/b \to \infty$) this function has two asymptotic values: $\Psi = 0$ for an ideal chain and $\Psi \cong 0.23$ for an excluded volume one. In fact the value of Ψ depends on both the length of chain and the range of the of excluded volume interactions [51-53]. In fact the value of Ψ depends on the length L of the chain and the interaction of excluded volume, B [53]. In Figure 9 the curves with solid lines show how Ψ varies with L when B remains constant, and those in dotted lines how Ψ varies with B, when L is fixed. The latter case is most interesting because it corresponds to the usually performed experiments. The results of Figure 9 are not general. In fact, they have been calculated for atactic polystyrene using a spiral model that takes into account both, the stiffness and twisting of the chain [53].

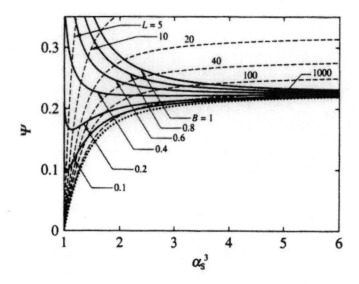

Figure 9. Variation of interpenetration function, ψ, versus to the cube of the expansion factor, α_s, given by the expression (48). B is the interaction range excluded volume. Here L is the length of the chain measured in length statistics units (L / b). From reference [53]).

6. SCATTERING BY THE MICELLES

In absence of strong micelle-micelle correlations, the structure factor S(q,c) is featureless over the whole wavevector range. Thus, in this case, the scattering intensity can be described by I(q) = NP(q) with N is the number of scatters per volume unit. The form factor of protein micelle, P(q), can be written as a sum of three different terms: the contribution of the micelle core, the contribution of the chains in the corona and the cross term between the spherical core and the corona chains, explicitly [54-57]:

$$P(q) = \left[\frac{3\Phi_0}{q^3 R_1^3} F(qR_1) + \frac{3(1-\Phi_0)}{q^3(R_2^3-R_1^3)}[F(qR_2)-F(qR_1)] \right]^2 \qquad (66)$$

where, $F(x) = \sin x - x \cos x$, ϕ_0 is the particles proportion in the core of micelle, compared to the total number and R_1, R_2 are respectively the radius of core and the corona.

This form factor term is established on the basis of lower values of q. The micelle can be considered as a spherical object made up of a core, of radius R_1, surrounded by a corona, of external radius R_2, of which the scattering length densities are different but constant. Thus, the micelle has a density profile corresponds to the distribution function following:

$$\phi(r) = \phi_1 = 3\phi_0 / 4\pi R_1^3 \quad if\ 0 \leq r < R_1,$$

$$\phi(r) = \phi_2 = 3\phi_0 / 4\pi(R_2^3 - R_1^3) \qquad\qquad if\ R_1 \leq r < R_2, \qquad (67)$$

$$\phi(r) = 0 \qquad if\ r > R_2.$$

The radius of gyration of the micelle can be calculated from its definition (32). It is given by the following expression:

$$R_g^2 = \frac{\int_0^\infty r^4 \phi(r)\sin(qr)dr}{\int_0^\infty r^2 \phi(r)\sin(qr)dr} \tag{68}$$

Using the expression (67) for the distribution function φ(r), we obtain

$$R_g^2 = \frac{3}{5}\left| \frac{(\phi_1 - \phi_2)R_1^5 + \phi_2 R_2^5}{(\phi_1 - \phi_2)R_1^3 + \phi_2 R_2^3} \right| \tag{69}$$

The SANS data can be also analyzed with Porod plot (q4I(q, c) vs q) [54]. An example of typical curve shown in Figure 11, for T= 50 °C and denaturant concentration (Guanidine hydrochloride) = 1 M. The scatter of the experimental points and their uncertainty (error bars) increase with the increase of q. This is due to the reduction of SANS with the increase of scattering angle. The solid line was obtained by a fit (Eq. (66)) of the experimental data and so is a reasonable approximation here.

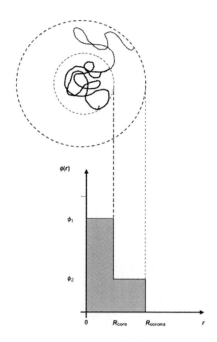

Figure 10. Sketch of spherical micelle made of a hydrophobic core (region I) and a hydrophilic corona (region II) in a selective solvent which is good for (II). We assume constant concentrations in each region, with φ1 > φ2 (from ref. [54]).

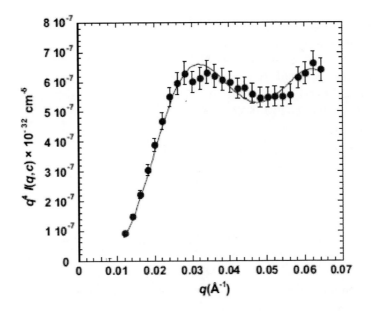

Figure 11. Porod plot, $q^4 I(q, c)$ vs. q, of SANS curve from beta-casein in 1 M GdmCl at 50 °C. The solid line is the result of the Eq. (66). The radius of core, Rcore and corona, Rcorona and the forward intensity I (0,c) are equal to ~ 56.58Å , ~127.05Å and 5.30 cm^{-1} respectively. (from ref. [54]).

CONCLUSION

We have demonstrated that small-angle neutrons scattering measurements can provide valuable complementary information about the proteins. This technique probes both the size and shape of the species under study with a spatial resolution on the order of Angstroms, and can be used to calculate thermodynamic quantities most notably, the excluded-volume interactions.

Still, in this part of book we showed the power of the technique of small angle neutron scattering to characterize the structure of the unfolded proteins. The important possibilities of the method lie not only in the determination, from now on traditional, of the radius of gyration but rather in information which can give the measurements carried out to relatively great values of q. Indeed, the determination of contour length and of statistical length allows highlighting the appearance of local structures along the chain. These elements of structure will cause to shorten the apparent length of this one, to modify its rigidity and to increase its transverse radius of gyration. While using jointly other techniques, such as circular dichroism, Infrared spectroscopy, fluorescence and RMN, it is not impossible that they can be characterized rather precisely.

REFERENCES

[1] D. Shortle, *FASEB J* 10 (1996) 27.
[2] B. Jacrot, *Rep. Prog. Phys.* 39 (1976) 911.

[3] J. Trewhella, *Curr. Struct. Biol.* 7 (1997) 702.

[4] Y.O. Kamatari, et al., *J. Mol. Biol.* 259 (1996) 512.

[5] L. Chen, et al., *J. Mol. Biol.* 261 (1996) 658.

[6] M. Kataoka, Y. Goto, *Fold. Des.* 1 (1996) 107.

[7] D. Russo, et al., *Biochemistry* 40 (2001) 3958.

[8] D. Russo, et al., *Physica* B 276 (2000) 520.

[9] D. Lairez, et al., *Biophy* J 84 (2003) 3904.

[10] A. Paliwal, et al., *Biophys* J 87, 3479 (2004).

[12] P. Calmettes, et al., *J. Mol. Biol.* 231, 840 (1993).

[13] A.J. Petrescu, et al., *Biophys J* 72, 335 (1997).

[14] B. Jacrot, *Rep. Prog. Phys.* 39, 63 (1976).

[15] J. P. Cotton, in : *Neutrons, X-ray and light scattering.* Lindner, P. et Zemb, T., éditeurs, Elsevier Science Publishers B. V. 3-131 (1991).

[16] J. P. Cotton, *Diffusion des neutrons aux petits angles.* J. Physique Paris, Vol. 9 (1999).

[17] Bée M. *Quasielastic neutron Scattering : Principles and Applications in Solid State Chemistry, Biology and Materials Science,* Adam Hilger , Bristol (1988).

[18] B. Mandelbrot, *Les Objets Fractals.* Flamarion, France (1984).

[19] D. Lairez, *Activity Repport LLB CEA Saclay,* France (1995).

[20] E. Leclerc et P. Calmettes, Phys. Rev. Lett., 78, 150, (1997).

[21] A.J. Petrescu, V. Receveur, P. Calmettes, D. Durand, M. Desmadril,B. Roux, and J. C. Smith, *Biophysical Journal* 72 335-342 (1997).

[22] M. Bonetti, , G. Romet-Lemonne, P. Calmettes, and M.-C. Bellissent-Funel, *J. Chem.Phys.* 112 268 (2000).

[23] J. Yvon, « *Les corrélations et l'entropie en mécanique statistique classique* », Dunod, Paris (1966).

[24] P. A., Egelstaff *An introduction to the liquid state,* 2éme édition (Clarendon Press, Oxford)(1992).

[25] A.Guinier, and G.Fournet, *Small-angle scattering of X-rays.* Wiley Interscience, New York (1955).

[26] C. Loupiac, M. Bonetti, S. Pin, and P. Calmettes, *Biochimica et Biophysica Acta* 1764 211 − 216 (2006).

[27] E. Leclerc, and P. Calmettes *Physica* B 234 236 (1997) 207-209.

[28] J. des Cloizeaux, et G.Jannink, *Polymers in solution- their modeling and structure.* Oxford University Press, Oxford (1987).

[29] M. Adam, et D. Lairez, *Fractals* 1, 149 (1993).

[30] P.G. de Gennes, *Scaling Concepts in Polymer Physics,* Cornell University, Ithaca, (1979).

[31] J. Debye, *J Appl Phys*;15:338 (1944).

[32] Rawiso M, Duplessix R and Picot C, *Macromolecules* 20:630 (1987).

[33] Calmettes P, Durand D, Desmadril M, Minard P, Receveur V and Smith JC, *Biophys Chem* 53:105 (1994).

[34] A. Aschi, A. Gharbi, M. Daoud, R. Douillard and P. Calmettes, *Polymer International,* (56) 5, 606-612 (2007).

[35] Flory, P. J. *Principles of polymers chemistry.* Cornell Univ. Press, Ithaca (1979).

[36] J.C. Le Guillou and J. Zinn-Justin, *Phys Rev Lett* 39:95 (1977).

[37] H.Benoit, and P.Doty, *J. Phys. Chem.* 57, 958 (1953).

[38] J. des Cloiseaux, *Macromolecules* 6, 403 (1973).

[39] J. S. Pedersen, and P. Schurtenberg, *Macromolecules* 29, 7602 (1996).

[40] Burchard, W. and Kajiwara, *K. Proc. R. Soc.* London A316, 185 (1970).

[41] Sharp, P. and Bloomfield, V. A. *Biopolymers* 6 1201 (1968).

[42] Utiyama, H., Tsunashima, Y. and Kurate, M. *J. Chem. Phys.* 55,3133 (1971).

[43] G. Porod, Koll. Z. 124, 83, (1951).

[44] O. Kratky and G. Porod, *Recl. Trav. Chim. Holland*, 80, 251 (1949).

[45] M. Kataoka, J.M. Flanagan, F. Tokunaga and D. M. Engelman, *Use of X-ray solution scattering for a protein folding study. In synchrotron Radiation in the Bioscience*, Oxford Science Publication, Clarendon Press, p 187, (1994).

[46] H. Benoit and M. Benmouna, *Polymer* 25, 1059 (1984).

[47] B. H. Zimm, *J. Chem. Phys.* 16, 1093 (1948).

[48] A. Aschi, A. Gharbi, M. Daoud, R. Douillard and P. Calmettes, Physica B 387 179–183 (2007).

[49] H.E. Swaisgood, In: P.F. Fox (Ed.), *Development in Dairy Chemistry*, vol.1, Elsevier, London. 1. (1982).

[50] Douglas JF and Freed KF, *Macromolecules* 18, 201 (1985).

[51] Yamakawa, H. *Modern theory of polymer solutions*. Harper & Row, New York (1971).

[52] Yamakawa, H. *Macromolecules* 25, 1912 (1992).

[53] Yamakawa, H. Abe, F. and Einaga, Y. *Macromolecules* 26, 1998 (1993).

[54] A. Aschi, A. Gharbi, M. Daoud, R. Douillard and P. Calmettes, *Polymer* 50 6024–6031 (2009).

[55] Halperin A. *Macromolecules* 1987;20:2943.

[56] Halperin A. *Macromolecules* 1989;22:3806.

[57] Halperin A. *Macromolecules* 1991;24:1418.

In: Neutron Scattering Methods and Studies
Editor: Michael J. Lyons

ISBN: 978-1-61122-521-1
© 2011 Nova Science Publishers, Inc.

Chapter 4

NEUTRON SCATTERING IN SKIN RESEARCH

***Jarmila Zbytovská**[a]* and Kateřina Vávrová*[b]

[a]Department of Pharmaceutical Technology, Charles University in Prague,
Faculty of Pharmacy, Hradec Králové, Czech Republic
[b]Centre for New Antivirals and Antineoplastics, Department of Inorganic
and Organic Chemistry, Charles University in Prague, Faculty of Pharmacy,
Hradec Králové, Czech Republic

ABSTRACT

The mammalian skin plays a major role in the protection of the body against
pathogen threats from the environment. The unique skin barrier properties are ensured by
the outermost skin layer, the *stratum corneum* (SC). The SC consists of corneocytes
which are embedded in a lipid matrix consisting of ceramides, free fatty acids,
cholesterol, and its derivatives. The SC lipids are organized in lipid membranes arranged
into a lamellar structure. Due to the special physicochemical properties of ceramides,
these membranes are extremely rigid and, therefore, very poorly permeable. A number of
skin diseases, namely atopic dermatitis, psoriasis or recessive X-linked ichthyosis, are
related to an impaired structure of the SC lipid membrane. On the other hand, a reversible
decrease in the skin barrier function connected with the changes in the SC lipid
membrane organisation is often required in the transdermal drug delivery. These facts
encourage looking for new approaches towards investigating the internal membrane
arrangement of the SC lipid matrix.

Recently, several neutron scattering techniques have been applied in this context.
Above all, the neutron scattering on unilamellar vesicles and the neutron diffraction on
oriented multilamellar lipid films allow describing the organisation of SC lipid membrane
models on the molecular level. These promising results open new opportunities in the
skin research.

The present chapter aims to review the recent experience with the neutron scattering
methods concerning the investigation of the SC lipid membrane structure.

---

* Corresponding author: Faculty of Pharmacy, Charles University in Prague, Department of Pharmaceutical
Technology, Heyrovského 1203, 50005 Hradec Králové, Czech Republic, Tel. No.: +420 495067383, E-mail
address: Jarmila.Zbytovska@faf.cuni.cz.

1. INTRODUCTION

The mammalian skin plays a crucial role in the protection of the body against chemical, pathogen and UV radiation threats as well as in thermoregulation and water balance of the body. It must also provide a mechanically strong structure that resists physical stress [1,2,3,4]. The outermost layer of the human skin, the stratum corneum (SC) is responsible for the excellent skin barrier properties which ensure the above mentioned skin functions [5].

SC is formed by dead cells, the corneocytes, which are embedded in a lipid matrix with a unique composition based on ceramides, free fatty acids, cholesterol, and its derivatives. The SC lipids are organized in lipid membranes arranged into a lamellar structure. Due to the special physicochemical properties of ceramides, these membranes are extremely rigid and, therefore, very poorly permeable [6]. Unfortunately, the necessary impermeability of the human skin represents a very strong limitation for the systemic transdermal drug delivery. In order to increase absorption of a drug through the skin, a reversible decrease in its barrier function is needed. Recently, the application of transdermal permeation enhancers has been most commonly used to overcome the SC lipid barrier [7]. These mainly amphiphilic substances incorporate into the SC lipid membranes in order to change the membrane structure that the membrane becomes more fluid and permeable for a drug. This effect was described for a number of substances, for a review, see Ref. [8]. However, only little is known about the molecular background of the permeation enhancers' mode of action so far. Furthermore, decreased concentrations of the skin barrier lipids lead to abnormal function and increased permeability of the skin. A number of skin diseases with elevated or decreased levels of a lipid fraction in the SC, like for e.g. the recessive X-linked ichthyosis, psoriasis or atopic dermatitis, have been reported [9,10,11,12]. Until now, however, there is only little information about the changes in the internal membrane structure of the SC lipid lamellae caused by such an abnormal lipid composition. Nevertheless, topical application of the skin lipids and/or their analogues can recover the skin barrier function [13,14,15].

Currently, there is a number of techniques employed in the characterization of the SC lipid membranes, namely X-ray diffraction [16,17,18], IR- and Raman spectroscopy [19,20,21], nuclear magnetic resonance [22,23,24], calorimetric [23,24], electron microscopic methods etc. Although these methods are very useful in the skin research, their opportunities to describe the internal membrane arrangement of the SC lipid matrix on the molecular level are limited.

In this context, neutron scattering techniques open new way to characterize the structure of the SC lipid membranes. These methods have been firstly developed to characterize the internal membrane nanostructure and hydration processes of phospholipid bilayers [25,26,27,28]. In the last several years, however, neutron scattering from unilamellar vesicles (ULVs) and neutron diffraction from oriented multilamellar lipid films were successfully used to describe membrane properties of SC lipid model systems based on (semi)synthetic lipids as well as on the native SC [29,30,31,32].

The present chapter reviews the recent experience with the neutron scattering techniques in the investigation of the SC lipid membranes to emphasize the possibilities and advantages of these methods.

2. SKIN BARRIER STRUCTURE

The skin barrier resides in the uppermost layer of the epidermis, SC, and is composed of corneocytes filled with keratin embedded in intercellular lipid matrix. This structure is often called "brick and mortar" (Figure 1) [33], forming a tortuous pathway for penetrating compounds. The exceptional barrier properties of the skin are based on the specific content and composition of the SC lipids [34,35] and, in particular, their exceptional arrangement into a multilamellar structure [36]. Moreover, these intercellular lipids are considered, as the only continuous domain through the SC, to be the most important pathway for the diffusion of substances into the body. This lipidic matrix is composed of an approximately equimolar mixture of ceramides, cholesterol and free fatty acids (or ~50:25:10% by weight) [37]. There are also small proportions of cholesterol esters, cholesterol sulphate, and glucosylceramides in the SC lipids, but phospholipids are absent.

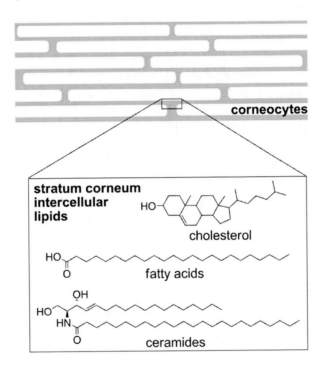

Figure 1: The "brick and mortar" model of the stratum corneum. The "bricks" represent the cells, corneocytes, which are embedded in the lipidic "mortar".

2.1   Ceramides

The most important SC barrier lipids are ceramides. Ceramides, i.e. *N*-acylsphingosines, belong to the class of sphingolipids. In human SC, eleven major ceramide subclasses have been identified (Table 1) [38]. These compounds consist of a sphingoid base – sphingosine (designated as S), 6-hydroxysphingosine (H), phytosphingosine (P) or dihydrosphingosine (DS), which is N-acylated by a long-chain saturated fatty acid (N, C_{16-30}, mostly lignoceric acid), α-hydroxy acid (A, C_{16-30}, mostly 2-hydroxylignoceric acid) or ω-hydroxy acid (O, C_{28-}

₃₂). The ω-hydroxyl is further esterified with linoleic acid (E) or attached to carboxyl side chains on the outer surface of the corneocyte to form a monolayer of covalently bound lipids. The letters in parentheses refer to the Motta's nomenclature [10]; for example, non-hydroxy acyl sphingosine, the most abundant SC ceramide formerly known as ceramide 2, is designated ceramide NS, the phytosphingosine acylated with ω-linoleyloxy acid would be ceramide EOP, etc. For a review on ceramide chemistry and structure-activity relationships, see Ref. [39].

The linoleate-containing ceramides (EO-type) are the most unusual of the SC ceramides. Ceramide EOS was proved to influence dominantly the formation of the long periodicity phase [40], which is essential for skin barrier function. It has been postulated that the ω-hydroxyacyl portion of the ceramide EOS completely spans a bilayer while the linoleate tail inserts into an adjacent bilayer, thus riveting the two together at a molecular level [41]. While it seems likely that ceramide EOP and EOH have similar roles, there is presently little data on this point.

When regarding the molecular shape of ceramides, two conformations were described: a hairpin conformation, i.e. with the two chains pointing in the same direction and a splayed-chain (or extended) conformation, i.e. with the chains pointing in opposite direction (Figure 2) [42]. During the preliminary stages of skin barrier morphogenesis the ceramide precursors, glucosylceramides, should be in the hairpin conformation [43]. During the formation of the SC lipid matrix, however, a transition of the ceramides from hairpin to splayed chain conformation is possible. The possible reason for this change is the length mismatch between the sphingoid base (~18C) and the fatty acid (~24C) of ceramides and their arrangement in separate domains [42]. The existence of the fully extended conformation of ceramide AP in a model SC lipid system was shown by Kiselev [29]. This conformation seems to be advantageous since it allows higher cohesion of the lamellae interconnected by the ceramide chains and the absence of water-swelling hydrophilic interfaces [44].

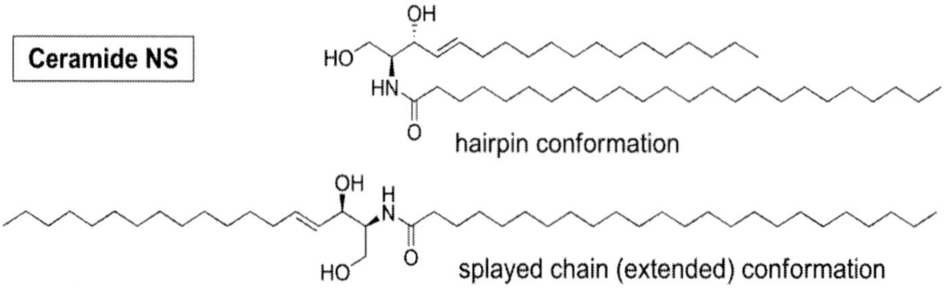

Figure 2: The hairpin and splayed chain (extended) conformations of ceramide NS.

2.2   Organization of SC Lipids

The SC intercellular lipids form multilamellar structures aligned parallel to the skin surface [45,46]. This lamellar organization is determined mostly by ceramides having a polar head and two long hydrophobic chains. Furthermore, they have several hydrogen bonding groups providing cohesivity of the polar heads by a hydrogen-bonding network [47]. The

Table 1. Structures of ceramides in human SC

Acyl part	Sphingoid base: Sphingosine (S)	Phytosphingosine (P)	6-Hydroxysphingosine (H)	Dihydrosphingosine (DS)
HOOC–$C_{23}H_{47}$ Non-hydroxy fatty acid (N)	Ceramide NS (ceramide 2)	Ceramide NP (ceramide 3)	Ceramide NH (ceramide 8)	Ceramide NDS
HOOC–$C_{22}H_{45}$–OH (R)-α-Hydroxy fatty acid (A)	Ceramide AS (ceramide 5)	Ceramide AP (ceramide 6)	Ceramide AH (ceramide 7)	Ceramide ADS
ω-(Linoleyloxy) fatty acid (EO)	Ceramide EOS (ceramide 1)	Ceramide EOP (ceramide 9)	Ceramide EOH (ceramide 4)	

Sphingoid base side chains: Sphingosine (S) $C_{13}H_{27}$; Phytosphingosine (P) $C_{14}H_{29}$; 6-Hydroxysphingosine (H) $C_{12}H_{25}$; Dihydrosphingosine (DS) $C_{15}H_{31}$.

exceptionally low permeability of the SC compared to the other biological membranes is likely a result of a) the unusual length of the free fatty acids and acyl chain in ceramides, b) relatively small polar head of ceramides compared to e.g. phospholipids that enable tighter packing, c) high cohesion by strong hydrophobic interactions and/or hydrogen bonding, and d) presence of more complicated multilayered lamellae instead of a simple bilayer.

In the lamellar regions, both the number of lamellae and the pattern of lamellar organization vary. Near the ends of the corneocytes a three band arrangement with a broad-narrow-broad pattern of lucent bands separated with narrow dark bands is frequently seen. The broad lucent bands are approximately 5 nm wide, and the narrow lucent band is about 3 nm wide. The overall width of the entire broad-narrow-broad unit is 13 nm [48]. The lucent bands correspond to the hydrocarbon chains of the SC lipids and the dark electron dense bands represent the polar head group regions. Between the broad flat surfaces of corneocytes the most common patterns are the 6 to 12-band ones consisting of similar three-band units [49]. The proportions of the lamellar units seen in the electron microscopy were confirmed and specified by small angle X-ray diffraction to be 13.4 nm for the long periodicity phase (broad-narrow-broad pattern), which is unique in biological membranes, and 6.4 nm for the broad lamellae [50,51,52]. This lateral organization was further confirmed by NMR studies [53].

The SC lipids are present both in the crystalline phase with orthorhombic chain packing and in gel phase with looser hexagonal chain packing. The coexistence of these two phases in the SC under physiological conditions was shown in the wide angle X-ray diffraction patterns and electron diffraction [54,55] and by infrared spectroscopy [56].

2.3   Models of SC Lipid Arrangement

Since the discovery of the SC lipid lamellae, several models of their molecular arrangement were proposed. These models attempt to explain the coexistence of the crystalline and gel phase, electron microscopic patterns, X-ray and electron diffraction proportions, thermotropic behavior of the lipids, SC cohesion and elasticity together with the exceptionally low permeability of the skin barrier.

The first model, sometimes referred to as a stacked monolayer model, was proposed in 1989 [49]. It is based on the electron microscopic data and attempts to elucidate the lipid arrangement in the broad-narrow-broad pattern. The ceramides are in the splayed chain conformation with interdigitated chains shared between lamellae. In particular, the EO-type ceramides are thought to act as molecular rivets between the lamellae. Although lipids are described as non-randomly distributed, this model does not address the coexistence of different lipid phases.

In 1994, a domain mosaic model was described by Forslind [57]. According to this model, the bulk of the lipids as segregated into crystalline/gel domains bordered by "grain borders" where lipids are in the liquid crystalline state. However, this arrangement permits the existence of many phase boundaries. Also the continuous liquid crystalline phase, which is more permeable, may be preferred by permeating substances.

A different arrangement of the two lipid phases was proposed by Pascher [58]. In this "laminglass" model, ceramides are in the splayed chain conformation and form a separate crystalline layer with orthorhombic chain packing. These layers are separated by cholesterol-rich liquid crystalline ones. Such alternating crystalline and liquid crystalline layers would combine low permeability with elasticity and resistance towards mechanical stress.

In 2000, Bouwstra et al. described a "sandwich model" of the arrangement of the long periodicity phase, i.e. the ~13 nm broad-narrow-broad pattern [59]. They suggested that in the outer broad lamellae, mainly long chain ceramides are localized having the chains only partially interdigitated. In the narrow central layer, the linoleate tails from ceramide 1, cholesterol and ceramides with shorter and fully interdigitated chains are present. The central layer with its looser hexagonal packing could explain the skin elasticity and it continuously merges into crystalline phase of the outer layers without any phase boundaries. This model uses only the hairpin conformation of ceramides.

In 2003, Hill and Wertz specified the arrangement of the long periodicity phase [60]. Using electron density profiles, they found that this trilamellar repeat unit was actually composed of three units of a similar thickness (4.3 nm). This molecular arrangement is similar to the sandwich model with linoleate tail of EO-type ceramides in the central lamella.

Probably the most controversial is the single gel phase model proposed by Norlén in 2001 [42]. According to this view, the SC lipids exist as a single and coherent gel phase without phase boundaries. The coexistence of the orthorhombic and hexagonal chain packing is proposed to occur within one molecule simultaneously: an orthorhombic packing close to the polar head group and the hexagonal packing near the end of the hydrocarbon chain. Such structure could possess low permeability together with mechanical resistance and very low tendency for phase transitions.

The last improvement explaining the structural alterations of the lamellae under hydration referred to as armature reinforcement model [29] based on the neutron scattering measurements and will be discussed later.

For a detailed overview on these models, see [42,61,62,63].

3.   NEUTRON SCATTERING TECHNIQUES USED IN SKIN RESEARCH

In general, during a scattering experiment a collimated neutron beam of wavelength λ is sent through or on a sample and the variation of the intensity is measured. The scattering intensity can be then plotted as a function of the scattering angle θ or more frequently of the scattering vector q [64], which is given by the difference between the wave propagation vectors of the scattered and incident beam and is related to the scattering angle by:

$$q = |\vec{q}| = |\vec{k}_s - \vec{k}_i| = \frac{4\pi n}{\lambda} \sin \theta \qquad\qquad \text{Eq. 1}$$

where n is the refractive index; for X-rays and neutrons very near to unity.

In the case when the matter scattering the radiation beams does not show a geometrical organization (e.g. particles dispersed in a homogenous medium), the waves scattered travel different distances and so they differ in their relative phases. Such scattering data can give information about the shape, size, and interactions of the individual particles. In this context, diffraction may be regarded as a special type of scattering by which the incident beams streaming on an organized structure (e.g. crystal) are diffracted under a defined angle 2θ

according to the Bragg's law and interference between waves scattered from the parallel planes occurs [65,66].

As it is well known, light and X-rays are scattered by the electron cloud of an atom, while the neutrons by the atomic nucleus [67]. Consequently, the atomic scattering factors are considerably different. The power by which the individual atoms scatter the neutron or X-ray beams is called the scattering length (b_{coh} for neutrons; $f_{X\text{-}ray}$ for X-ray). As it has been described for neutrons, there is a large difference in the coherent scattering length between deuterium and hydrogen that the latter is actually negative. This arises from a change of phase of the scattered wave and results in a marked difference in scattering power between molecules including deuterium or hydrogen [68]. This is the main advantage of the neutron scattering in comparison to the X-rays and most neutron scattering experiments are based on this phenomenon. The deuterium labelling techniques make a molecule or a part of molecule 'visible' for the neutrons. Using either the deuterated molecule or the environment, the signal-to-noise ratio of the measurements can increase by orders of magnitude.

The most essential parameters that can be obtained from the neutron scattering studies of lipid biomembranes are the thickness of the bilayer and the average area occupied by a lipid along the surface of the bilayer (i.e. 'the membrane density'). In principal, there are two approaches to obtain these structural parameters. The first approach is based on diffraction of the beams from multilayer samples (either multilamellar vesicles, MLVs, or oriented multilamellar films). The electron or neutron length density profiles can be determined from the intensities of the diffraction peaks using Fourier transformation. The other approach to obtain the bilayer structure is based on ULV measurements. In comparison to diffraction from the multilamellar structures, the scattering from ULVs is continuous in the scattering vector q [69]. Recently, both of these methods have been applied in the research of the SC lipid membranes.

3.1   Small Angle Neutron Scattering (SANS) on ULVs

A prerequisite for a successful SANS measurements is a well prepared sample consisting solely of unilamellar vesicles (Figure 3). Such a sample should be stable for at least several days. A presence of oligo- or multilamellar structures is undesired because of the strong diffraction signal leading to overlapping the scattering from ULVs. While this requirement is easy to fulfil for phospholipid vesicles, for SC lipid samples based predominantly on ceramides, it is a critical point. The physical-chemical properties of the systems do not enable to create stable ULVs under physiological conditions. As it has been described for a model system consisting of ceramide, cholesterol, free fatty acid, and cholesterol sulphate, the ULVs were stable for several weeks at pH 9, but less than one day at pH 6 [70]. A pH decrease from 9 to 6 activated vesicle fusion and lysis [71]. Unlike at pH values close to human skin, at pH 9, the charged components (free fatty acids and cholesterol sulphate in the present system) are mainly ionized [72]. As a result, vesicles can be prepared without a considerable loss of lipids [30]. The SC lipid ULVs are mainly prepared using conventional 'thin layer method' and extrusion through polycarbonate filter in the fluid state [30,73,74]. Of course, the pH difference between the prepared model membranes and the native SC must be taken into consideration while interpreting the results. Another possibility to obtain stable ULVs is to incorporate an individual SC component (e.g. ceramide) into a phospholipid bilayer and to study its behavior in the membrane [31,75].

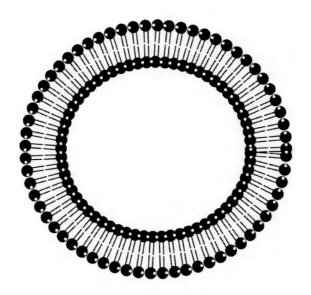

Figure 3: A schematic depiction of a unilamellar vesicle.

Several approaches can be used to evaluate the SANS signal from ULVs (Figure 4 a). Either the curve can be fitted according to certain mathematical model [69,76] or the classical Guinier approximation, which gives the radius of gyration R_g from the Kratky-Porod plot [77,78,79], can be applied for the data analysis. Recently, both ways have been used in the evaluation of the SANS data from SC lipid models [77,78,79]. The shape and internal structure analysis has been determined by the 'model of separated form factors' which was developed for phospholipid vesicles [76] combined with the hydrophobic-hydrophilic approximation of the neutron scattering length density [75,80]. According to this model, the macroscopic scattering cross section of the monodispersed population of ULVs is given by:

$$\frac{\partial \Sigma(q)}{\partial \Omega_{mon}} = NF_s(q,R)F_b(q,d)S(q) \qquad \text{Eq. 2}$$

where N is the number of vesicles per unit volume, $S(q)$ the structure factor in the Debye form, $F_s(q, R)$ is the form factor of the infinitely thin sphere with the radius R

$$F_s(q,R) = \left(4\pi \frac{R^2}{qR}\sin(qR)\right)^2 \qquad \text{Eq. 3}$$

$F_b(q, d)$ is the form factor of the symmetrical lipid bilayer with the thickness d, which can be expressed by

$$F_b(q,d) = \left(\frac{2\Delta\rho}{q} \sin\left(\frac{qd}{2} \right) \right)^2$$

Eq. 4

for the case of a bilayer with a constant scattering length density across the membrane $\rho(x)$=const. $\Delta\rho$ is the neutron contrast.

The average vesicle radius R can be calculated from the scattering curve based on Eq. 3 as $R=\pi/q_{Rmin}$, where q_{Rmin} is the first minimum in the form factor of the infinitely thin sphere after averaging of the population of polydisperse vesicles (Figure 4a) [81].

The membrane thickness parameter d can be directly calculated from the position q_0 of the first minimum of the sine function in the Eq. 4 as $d=2\pi/q_{dmin}$. For a membrane thickness of about 30 Å, the position of q_{dmin} is about 0.2 Å^{-1}.

The Guinier approximation offers another possibility to determine the membrane thickness parameter, d_g, which is a measure of the membrane thickness, from a scattering curve. Also this method is commonly used in the research of phopholipid vesicles [77,78,82]. In the q-range valid for a homogeneous membrane approximation ($\pi/R<q<1/R_g$), the scattering intensity of ULVs dispersed in heavy water can be given by

$$I(q) = 2\pi I(0)q^{-2} \exp\left(-q^2 R_g^2\right)$$

Eq. 5

where $I(0)$ is the scattering intensity to 'zero angle' and R_g is the membrane gyration radius. In this approach, the R_g parameter is the absolute value of the slope of the Kratky-Porod plot $(ln[I(q)q^2]$ vs $q^2)$ (Figure 4b) and the membrane thickness parameter can be calculated as

$$d_g^2 = 12R_g^2$$

Eq. 6

$I(0)$ cannot be measured experimentally but it can be determined by extrapolation of the Kratky-Porod plot to zero value. The value of $I(0)$ is given by the total particle scattering length, namely by the sum of the scattering lengths of all atoms inside the particle. Therefore, the chemical composition being known, the evaluation of $I(0)$ allows the molecular mass per unit of vesicle surface to be determined [78,83]. In the limit of q→0, the mass of the membrane per unit of surface, M_s, can be determined by dividing the scattered intensity $I(0)$ by the total lipid concentration c and the scattering length density per unit mass, $\Delta\rho_m$, according to:

$$I(0) = M_s c \Delta\rho_m^2$$

Eq. 7

The membrane area per molecule, A, in centrosymmetric bilayers can be calculated by:

$$A = \frac{2}{\left(M_s N_A / M_W\right)}$$

Eq. 8

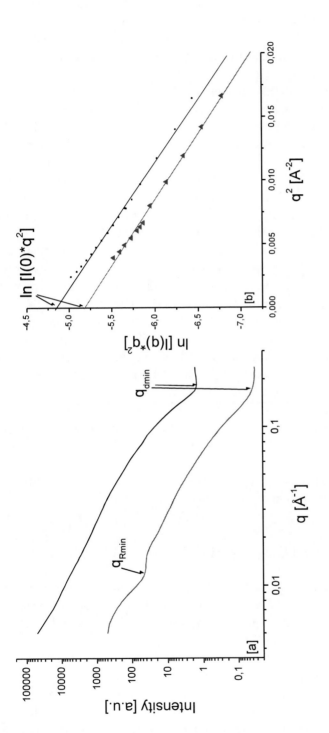

Figure 4: [a] A typical SANS curve from DMPC ULVs (red) and from a quaternary SC lipid model system consisting of ceramide AP/cholesterol/palmitic acid/cholesterol sulphate (55/25/15/5 weight %). The arrows show the minima related to the average vesicle radius and the membrane thickness, respectively. [b] The linear area of the correspondent Kratky-Porod plots.

where M_W is the average molecular weight of the lipids, M_S the determined membrane mass per unit of surface and N_A the Avogadro number. Consequently, the average volume per molecule in the membrane can be calculated by multiplying the average area per molecule by half of the membrane thickness parameter [84]. The average membrane density is determined by dividing the average weight of one molecule by the average volume per molecule in the membrane [30].

Concerning ceramides, two types of ULV samples have been studied using SANS:

A binary system has been studied by Zbytovska et al. [31] who described an influence of phytopshingosine-type ceramides NP and AP on a dimyristoylphosphatidylcholine (DMPC) membrane. In this study both evaluation methods, the model of separated form factors and the Guinier approximation have been used. It has been shown that both the ceramides increase not only the lamellar repeat distance but also the thickness of the DMPC membrane. Also the vesicle radius increases by adding ceramides into the membrane. A fit of the SANS curve of DMPC/ceramide NP ULVs according to the model of separated form factors in combination with the hydrophobic-hydrophilic approximation is presented elsewhere [75].

A quaternary system consisting of ceramide AP, cholesterol, palmitic acid, and cholesterol sulphate imitating the composition of the real SC lipid matrix has been characterized by SANS on ULVs in another study [30]. Despite the handicap of the non-physiological pH, interesting results concerning the influence of cholesterol on the SC lipid membrane are presented. Again both, the model of separated form factors and Guinier approximation have been applied in the evaluation of the SANS curves. As a result, the membrane thickness parameter has been calculated by two ways. Furthermore, it was possible to determine the average area of membrane surface per molecule and the average membrane density. In summary, cholesterol decreases both the membrane thickness and the area of the membrane surface per molecule, which is connected with the increase of the membrane fluidity. SANS curves of the present system have been fitted using the model of separated form factors in combination with the hydrophobic-hydrophilic approximation [75,80]. Compared to phospholipid vesicles, the fitting results, however, are in disagreement with the experimental data especially for small scattering vectors which suggests a strong short-range interaction among vesicles and a formation of cluster structures.

3.2   Neutron Diffraction from Oriented Multilamellar Lipid Films

As mentioned above, the scattering length density profiles of the membranes can be determined using the Fourier analysis in the X-ray and neutron diffraction studies. The profile is a function characterizing the density distribution of the scattering centres in real space and can be interpreted in terms of molecular structure. In the case of X-rays, the scattering centers are the electrons surrounding the atomic nuclei. In the case of the neutron scattering, it is the strength of the neutron-nucleus interaction.

Multilamellar films on the nearly ideally smooth quartz substrate (Figure 5) show a very high degree of ordering and can be described as 'one-dimensional liquid crystals'. Thus, the diffraction signal from such samples is stronger in comparison to the signal from e.g. multilamellar vesicles, and consequently, more diffraction orders can be detected.

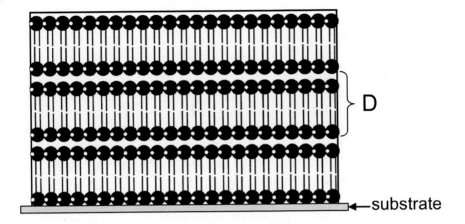

Figure 5: A schematic depiction of a multilamellar lipid film.

In principle, this method is based on the diffraction of neutron beams from such an organized structure according to the Bragg's law:

$$2d_{hkl} \sin\theta = h\lambda \ \text{h=1,2,3...}$$

Eq. 9

where θ is the diffraction angle, d_{hkl} is the distance between the two parallel planes characterized by the Miller indexes hkl, λ is the wavelength of the incident beam, and h is the diffraction order.

In the case of multilamellar lipid samples (multilamellar vesicles or oriented multilamellar lipid films), the membranes are arranged in a lamellar phase and d_{hkl} is usually called the lamellar repeat distance (or lamellar spacing), D, which includes the bilayer thickness and one water layer between the membranes (Figure 5). D can be calculated combining equations 1 and 9:

$$D = h2\pi/q_h$$

Eq. 10

When at least three diffraction orders from the bilayers are successfully measured, Fourier transformation can be applied to calculate the electron or neutron scattering length density profile of the membrane:

$$\rho(x) - \rho_W = \frac{1}{D}F_o + \frac{2}{D}\sum_{h=1}^{h_{max}}\alpha_h F_h \cos\left(\frac{2\pi hx}{D}\right)$$

Eq. 11

where ρ_w is the electron or neutron scattering length density of the environment (mostly water or D_2O), F_h is the bilayer structure factor of the order h and D is the lamellar repeat distance calculated from the position of the diffraction peak.

The structure factor accounts for the statistical distribution of electron/neutron length in the bilayer. For a homogenous centrosymmetric structure, the absolute value of the structure factor can be obtained from the square root of the diffraction intensity I_h according to:

$$F_h = \sqrt{C_h \cdot I_h}$$ Eq. 12

under the hth diffraction peak. C_h is the Lorentz polarization corrector factor. For oriented samples, C_h is nearly proportional to h. The structure factor F_h involves an unknown scale factor so only the absolute values can be measured [84].

This problem is caused by the fact that the measurements of the intensities in diffraction patterns can only give the amplitude, and not the phase, of the structure factor. For this reason, it is impossible to obtain directly the Fourier transform of the structure factor F to determine the electron or neutron length density distributions. This phenomenon is generally known as *the phase problem* of X-ray and neutron scattering [66].

There are several methods to overcome the phase problem directly or indirectly. While in X-ray diffraction it is a difficult question that can be solved only by the Shannon sampling theorem [85,63], the neutron diffraction provides a convenient solution to determine the phases and consequently the membrane structure using the large scattering difference between hydrogen and deuterium. The scattering contrast between different components can be adjusted by simply replacing or mixing H_2O with D_2O. A gradual H_2O/D_2O exchange permits direct observation of phase changes of particular reflections [25,26, 27].

According to this 'isomorphous replacement method', when H_2O is replaced by D_2O, the even-order structure factors will increase but the odd-order structure factors will decrease algebraically. Thus, the linear plots of structure factor versus mole per cent D_2O should have positive slopes for even orders and negative slopes for odd orders [86]. The phases (the signs of the structure factor F_h) can be determined according to this rule when the absolute values of $|F_h|$ are arranged as a linear function versus the concentration of D_2O in the sample environment. Consequently, the neutron length density profiles can be calculated from eq. 11.

In fact, one sample is then measured at least three times at different D_2O/H_2O ratios. In aqueous dispersion with a concentration of 8% D_2O in H_2O, the neutron scattering length density of the dispersion medium is zero and the neutron beam detects only the dispersed substance [29]. Further, the samples are measured mainly at 20 and 50% D_2O. As a result, neutron scattering length density profiles can be reconstructed for various D_2O concentrations. Consequently, the water distribution function across the bilayer can be calculated as a difference between neutron scattering length density profiles obtained at e.g. 50 and 8% D_2O content. This function describes water penetration into the membrane.

A big advantage of neutron relative to X-ray diffraction is the possibility of deuterium labeling. Specific deuteration of a part of a lipid molecule provides information about the exact localization of the label in the membrane. This localization is reflected by the positive difference in the neutron scattering length density profiles between samples containing either deuterated or protonated lipid [87].

As it has been described in detail earlier, the achieved space resolution of the Fourier synthesis depends on the number of the measured diffraction orders [29]. Therefore, highly organized samples with a low degree of mosaicity (about 0.1 degree) are required for the neutron diffraction measurements. The diffractograms from such samples give about 4-5 diffraction orders (Figure 6a). The multilamellar SC lipid membrane layers are mainly prepared by spreading a lipid solution in organic solvent (mainly chloroform/methanol) on the quartz surface according to Seul and Sammon [88]. Afterwards, a subsequent heating and cooling cycle is applied to further decrease the mosaicity of the samples. This annealing procedure improves the signal-to-noise ratio in the experiment [89]. \The proper sample preparation technique is important; however, it alone does not ensure the successful neutron diffraction measurement leading to the Fourier profile of the membrane (Figure 6b). As it has been shown, not every composition of a SC lipid model tends to create highly organised multilamellar samples convenient for the diffraction measurement [32].

The first neutron diffraction studies in the skin research used native SC separated from porcine or human skin to study the hydration process of the SC lipid membranes [90, 91]. The studies have shown that the hydration process of native SC is extremely slow, the lamellar repeat distance increases during hydration which finally results in disruption of the lipid lamellae connected with disappearing of the diffraction peak in the neutron diffractogram.

First, Kiselev et al. [29] used a defined SC lipid model system based on synthetic lipids to determine the neutron scattering length density profile of the membrane. This membrane of the same composition as in [30] has shown low mosaicity which allowed to collect five orders of neutron diffraction. After determination of the phase of the structure factors via the H_2O/D_2O isotopic substitution, the neutron scattering length density profile has been reconstructed. The membrane thickness, the thickness of the membrane hydration layer, the thickness and position of the polar head groups as well as the hydrocarbon chain regions have been determined after fitting the Fourier profile by the sum of four Gauss curves. It has been postulated that cholesterol decreases the lamellar repeat distance of the bilayer, which is in agreement with [30]. Moreover, the bilayer is characterized by a low level of hydration of the intermembrane space and considerably slower water diffusion related to phospholipid membranes. After one hour of the hydration process, drastic changes in the first order structure factor arise, which is connected with structural changes in the polar head groups region of the membrane.

Several studies from Neubert's group follow the investigations of Kiselev et al. and widen the knowledge about the present SC lipid model system. Kessner et al. [92] used for the first time the benefits of the selective deuteration in a SC lipid model membrane. Due to partially deuterated cholesterol, its exact position in the SC lipid membrane has been localized. In another study [89], the present SC lipid model system based on ceramide AP was enriched in long-chain fatty acids to better simulate the real SC lipid matrix composition. Interestingly, the long-chain free fatty acids decreased the membrane repeat distance which is interpreted in terms of partial interdigitation of the longer chains. This assumption has been further confirmed using deuterium labeled long-chain fatty acids [87]. Incorporation of a long-chain ceramide EOS into the system described in [29] did not affect the lamellar repeat distance markedly [32]. Comparison of both neutron scattering length density profiles, however, implies that the ceramide EOS molecules penetrate from one membrane layer into an adjacent layer to span the layers together.

Further on, the effect of variation in the fatty acid chain length in a system consisting of ceramides EOS/AP/cholesterol/free fatty acid was studied [93]. The use of specifically deuterated behenic acid allowed to determine its exact position inside the SC lipid model membrane. The system was suggested to be a reasonable model of the short-periodicity phase of the SC, however, despite the presence of ceramide EOS, it did not succeed to obtain the long-periodicity phase detected in the real SC.

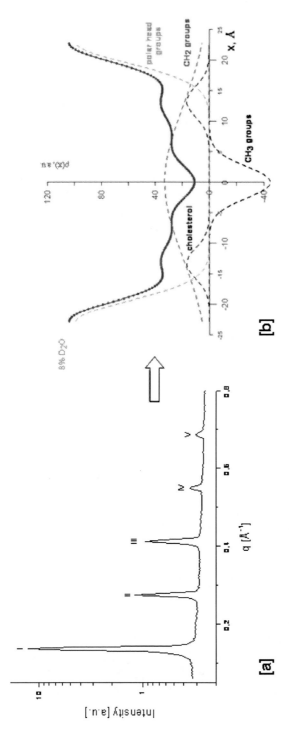

Figure 6: [a] An example of a neutron diffractogram from an SC lipid model system consisting of ceramide AP, cholesterol, palmitic acid, and cholesterol sulphate. Five diffraction orders are detected. D=46 Å. [b] A neutron scattering length density profile of the same system.

In a current study, palmitic acid has been substituted by a multicomponent free fatty acid mixture which led to phase separation [94]. Simultaneously, the effect of cholesterol sulphate concentration on the internal structure of the SC lipid model membrane was studied. It has been postulated that cholesterol sulphate increases the membrane swelling and inhibits the phase separation.

Binary systems based on ceramide AP or cholesterol incorporated in a phospholipid membrane have been characterized by neutron diffraction with a special focus on membrane hydration [95]. In accordance with [30] ceramide increased the membrane thickness and reduced the thickness of the water layer between the membranes.

4. SUMMARY AND PERSPECTIVES

Neutron scattering techniques offer new opportunities in characterizing the SC lipid lamellae. Compared to the X-ray scattering, the advantage of contrast variation in the neutron scattering measurements due to the difference in the coherent neutron scattering lengths for hydrogen and deuterium highly exceeds several disadvantages such as the relative weakness of the neutron radiation or limited availability of the neutron sources. The methods primarily used in the area of skin research, namely the SANS on ULVs and the neutron diffraction on multilamellar lipid films provide unique and complementary information. The neutron scattering length density profile reconstructed via neutron diffraction from the multilamellar lipid films offers a detailed insight into the internal membrane nanostructure with a special regard to membrane hydration. SANS on ULVs, on the other hand, gives information about membrane properties in terms of the membrane density and fluidity.

Concerning the structure of the SC lipid matrix, several important results obtained from both the neutron scattering techniques can be summarized as follows:

1) The thickness of the intermembrane spaces of the studied SC-lipid model systems is extremely thin compared to phospholipid membranes (For comparison: ~1 Å for SC lipids; ~12 Å for DMPC) [29,93].

2) The water diffusion through the SC lipid model membranes is considerably slow relative to phospholipid membranes and the calculated structure factor changes dramatically during this process [29,94].

3) There are structural alterations in the native SC during extremely long hydration, connected with disappearing of the first-order reflection in the diffractogram [90].

4) Ceramide AP creates a super-stable structure of the present SC lipid model membrane which is hardly affected by the membrane composition [32,89].

5) ULVs composed of a SC lipid model system based on ceramide AP aggregate into a cluster structure [80].

Based on these results, a new model of the internal arrangement of the SC lipid matrix has been postulated explaining the structural alterations of the lamellae under hydration [29,96]. According to this 'armature reinforcement model', the lipids are present in both hairpin and splayed chain conformations under low hydration, while under excess of water, chain flip transition from splayed into hairpin conformation occurs allowing water penetration between lamellae.

This finding contributes to the general view on the arrangement of the SC lamellae. As discussed above, all the models describing the SC membrane organization have their pros and cons and are not necessary contradictory. The sandwich model [59] together with the recent findings about the same lamellar spacing within the long periodicity phase [60] and the existence of ceramides in a splayed chain conformation reinforcing the multilamellar structure [29, 96] may be a reasonable explanation of the experimental findings. And perhaps the future experiments will support the single gel phase model [42], which may nevertheless be based on a similar lipid arrangement.

The neutron scattering techniques can play an important role in further elucidation of the SC structure. Concerning the present neutron scattering results, a model of the short periodicity phase of the SC has been relatively well characterized, nevertheless, data about the long periodicity phase are still missing. Thereby, the main goal of the incoming neutron scattering studies should be detailed characterization of the long periodicity phase of the SC. Further on, characterization of alterations in the internal structure of the SC lipid membranes evoked by pathological changes in the SC lipid composition [30] or by incorporated permeation modulators [97] is a promising application of neutron scattering.

ACKNOWLEDGMENTS

The work was supported by the Czech Science Foundation (Project GACR 202/07/P391) and by the Ministry of Education of the Czech Republic (Grant MSM 0021620822).

REFERENCES

[1] Chuong, C.M., Nickloff, B.J., Elias, P.M., Goldsmith, L.A., Macher, E, Maderson, P.F.A., Sunberg, J.P., Tagami, H., Plonka, P.M., Thestrup-Pedersen, K., Bernard, B.A., Schroder, J.M., Dotto, P., Chang, M.H.C., Williams, M.L., Feingold, K.R., King, L.E., Klingman, A.M., Rees, J.L., Christophers, E., *Exp Dermatol* 2002, 11, 159-187.
[2] Menon, G.K., *Adv Drug Del Rev* 2002, 54, Suppl. 1, S3-S17.
[3] Elias, P. M., Feingold, K.R., Fluhr, J. Skin as organ of protection. In: *Fitzpatrick's Dermatology in general medicine*. Edited by Freedberg, I. et al. McGraw – Hill Comp., 2003.
[4] Elias, P. M., *J Invest Dermatol* 2005, 125, 183-200.
[5] Windsor, T., Burch, G.E. *Arch Intern Med* 1944, 74, 428-436.
[6] Wertz, P. W., van den Bergh, B., *Chem Phys Lipids* 1998, 91, 85-96.
[7] Asbill, C.S., Michniak, B.B., *Pharm Sci Technol To* 2000, 3, 36-40.
[8] Vávrová, K., Zbytovská, J., Hrabálek, A. *Curr Med Chem* 2005, 12, 2273–2291.
[9] Williams, M. L., Elias, P. M., *J Clin Invest* 1981, 68, 1404-1410.
[10] Motta, S., Monti, M., Sesana, S., Caputo, R., Carelli, S., Ghidoni, R. *Biochim Biophys Acta* 1993, 1182, 147-51.

[11] Hara, J., Higuchi, K., Okamoto, R., Kawashima, M., Imokawa, G., *J Invest Dermatol* 2000, 115, 406-413.

[12] Choi, M.J., Maibach, H.I., *Am J Clin Dermatol* 2005, 6, 215-223.

[13] Man, M.-Q., Feingold, K.R., Elias, P.M., *Arch Dermatol* 1993, 129, 728–738.

[14] Vávrová, K., Zbytovská, J., Palát, K., Holas, T., Klimentová, J., Hrabálek, A., Doležal, P., *Eur J Pharm Sci* 2004, 21, 581-7.

[15] Vávrová, K., Hrabálek, A., Mac-Mary, S., Humbert, P., Muret, P., *Br J Dermatol* 2007, 157, 704-712.

[16] Bouwstra, J.A., Gooris, G.S., Dubbelaar, F.E., Ponec, M., *J Invest Dermatol* 2002, 118, 606-617.

[17] de Jager, M.W., Gooris, G.S., Dolbnya, I.P., Ponec, M., Bouwstra, J.A., *Biochim Biophys Acta* 2004, 1664, 132-140.

[18] McIntosh, T., *Biophys J* 2003, 85, 1675-168.1.

[19] Caussin, J., Gooris, G.S., Janssens, M., Bouwstra, J.A., *Biochim Biophys Acta* 2008, 1778, 1472-1482.

[20] Moore, D.J., Snyder, R.G., Rerek, M.E., Mendelsohn, R., *J Phys Chem B* 2006, 110, 2378-2386.

[21] Wartewig, S., Neubert, R., Rettig, W., Hesse, K., *Chem Phys Lipids* 1998, 91, 145–152.

[22] Chen, X., Kwak, S.J., Lafleur, M., Bloom, M., Kitson, N., Thewalt, J., *Langmuir* 2007, 23, 5548-5556.

[23] Rowat, A.C., Kitson, N., Thewalt, J.L., *Int J Pharm* 2006, 307, 225-231.

[24] Brief, E., Kwak, S., Cheng, J.T.J., Kitson, N., Thewalt, J., Lafleur, M., *Langmuir* 2009, 25, 7523-7532.

[25] Zaccai, G., Blasie, J.K., Schoenborn, B.P., *Proc Nat Acad Sci USA* 1975, 72, 376-380.

[26] Schoenborn, B.P., *Biochim Biophys Acta* 1976, 457, 41-55.

[27] Worcester, D.L., *Brookhaven Symposia in Biology* 1976, 27, III37-III57.

[28] Worcester D.L., Franks N.P., *J Mol Biol* 1976, 100, 359–378.

[29] Kiselev, M.A., Ryabova, N.Yu., Balagurov, A.M., Dante, S., Hauss, T., Zbytovska, J., Wartewig, S., Neubert, R.H.H. *Eur Biophys J* 2005, 34, 1030-1040.

[30] Zbytovská, J., Kiselev, M.A., Funari, S.S., Garamus, V.M., Wartewig, S., Palát, K., Neubert, R., *Coll Surf A* 2008, 328, 90–99.

[31] Zbytovská, J., Kiselev, M.A., Funari, S.S., Garamus, V., Wartewig, S., Neubert, R., *Chem Phys Lipids* 2005, 138, 69–80.

[32] Kessner, D., Kiselev, M.A., Hauß, T., Dante, S., Lersch, P., Wartewig, S, Neubert, R.H.H. *Eur Biophys J* 2008, 37, 989-999.

[33] Williams, M.L., Elias, P.M., *Arch Dermatol* 1993, 129, 626–628.

[34] Elias, P.M., *J Invest Dermatol* 1983, 80, 44s-49s.

[35] Grubauer, G., Feingold, K.R., Harris, R.M., Elias, P.M., *J Lipid Res* 1989, 30, 89–96.

[36] Potts, R.O., Francoeur, M.L., *J Invest Dermatol* 1991, 96, 495–499.

[37] Long, S. A., Wertz P. W., Strauss, J. S., Downing, D. T., *Arch Dermatol Res* 1985, 277, 284-287.

[38] Masukawa, Y., Narita, H., Shimizu, E., Kondo, N., Sugai, Y., Oba, T., Homma, R., Ishikawa, J., Takagi, Y., Kitahara, T., Takema, Y., Kita, K. *J Lipid Res* 2008, 49, 1466-1476.

[39] Novotný, J., Hrabálek, A., Vávrová, K., *Curr Med Chem* 2010, 17, 2301-2324.

[40] Bouwstra, J. A., Gooris, G. S., Dubbelaar, F. E., Weerheim, A. M., Ijzerman, A. P., Ponec, M., *J Lipid Res* 1998, 39, 186-196.

[41] Wertz, P.W., Cho, E.S., Downing, D.T., *Biochim Biophys Acta* 1983, 753, 350-355.

[42] Norlén, L., *J Invest Dermatol* 2001, 117, 830-836.

[43] Corkery, R.W., Hyde, S.T., *Langmuir* 1996, 12, 5528-5529.

[44] Corkery, R.W., *Colloid Surf B*, 2002, 26, 3-20.

[45] Garson, J.-C., Doucet, J., Leveque, J.-L., Tsoucaris, G., J *Invest Dermatol* 1991, 96, 43–49.

[46] Bouwstra, J.A., Gooris, G.S., Salomons-de Vries, M.A., van der Spek, J.A., Bras, W., *Int J Pharm* 1992, 84, 205–216.

[47] Williams, M.L., Elias, P.M., *CRC Crit Rev Ther Drug Carrier Syst* 1987, 3, 95–122.

[48] Madison, K.C., Swartzendruber, D.T., Wertz, P.W., Downing, D.T., *J Invest Dermatol* 1987, 88, 714–718.

[49] Swartzendruber, D.T., Wertz, P.W., Kitko, D.J., Madison, K.C., Downing, D.T., *J Invest Dermatol* 1989, 92, 251-257.

[50] Bouwstra, J.A., Gooris, G.S., van der Spek, J.A., Lavrijsen, S., Bras, W., *Biochim Biophys Acta* 1994, 1212, 183–192.

[51] Bouwstra, J.A., Gooris, G.S., Bras, W., Downing, D.T., *J Lipid Res* 1995, 36, 685–695.

[52] White, S.H., Mirejovsky, D., King, G.I., *Biochemistry* 1988, 27, 3725–3732.

[53] Fenske, D.B., Thewalt, J.L., Bloom, M., Kitson, N., *Biophys J* 1994, 67, 1562–1573.

[54] Cornwell, P.A., Barry, B.W., Stoddart, C.P., Bouwstra, J.A., *J Pharm Pharmacol* 1994, 46, 938–950.

[55] Bouwstra, J., Pilgram, G., Gooris, G., Koerten, H., Ponec, M., *Skin Pharmacol Appl Skin Physiol* 2001, 14, suppl. 1, 52-62,

[56] Gay, C. L., Guy, R. H., Golden, G. M., Mak, V. H. W., Francoeur, M. L., J *Invest Dermatol* 1994, 103, 233–239.

[57] Forslind, B., *Acta Dermato Venereol.* 1994, 74, 1-6.

[58] Pascher, I., pers. commun. 1996, In: Norlén, L., *Skin Pharmacol Physiol* 2003, 16, 203-211.

[59] Bouwstra, J.A., Dubbelaar, F.E., Gooris, G.S., Ponec, M., *Acta Derm Venereol Suppl* 2000, 208, 23-30.

[60] Hill, J.R., Wertz, P.W., *Biochim Biophys Acta* 2003, 1616, 121-126.

[61] Bouwstra, J.A., Pilgram, G, Ponec, M., *J Invest Derm* 2002, 118, 897–898.

[62] Norlén L. *Skin Pharmacol Appl Skin Physiol* 2003, 16, 203-211.

[63] Kessner, D., Ruettinger, A., Kiselev, M.A., Wartewig, S., Neubert, R.H. *Skin Pharmacol Physiol* 2008, 21, 58-74.

[64] Grossmann, J.G., In: *Scattering and inverse scattering in pure and applied science.* Pike, R., Sabatier, P. Academic Press, San Diego, San Francisco, NY, Boston, London, Sydney, Tokio, 2002. pp. 1123-1139.

[65] Jenkins, R., Snyder, R.L. *Introduction to X-ray powder diffractometry*, Willey, NY, 1996, pp. 1-94.

[66] Michette, A.G., In: *Scattering – Scattering and inverse scattering in pure and applied science.* Pike, R., Sabatier, P. Academic Press, London, San Diego, 2002. pp.: 1099-1107.

[67] King, S.M. Small angle neutron scattering. In: *Modern Techniques for Polymer Characterisation, Pethrick, R.A., Dawkins, J.V., John Wiley, 1999*

[68] Wignall, G. D. and Melnichenko, Y.B., *Rep Prog Phys* 2005, 68, 1761-1810.

[69] Kučerka, N., Nagle, J.F., Feller, S. E., Balgavý, P., *Phys Rev E* 2004, 69, 051903, 1-9.

[70] Hatfield, R.M., Fung, L.W.M., *Biophys J* 1995, 68, 196–207

[71] Hatfield, R.M., Fung, L.W.M., *Biochemistry* 1999, 38, 784–791.

[72] Lieckfeldt, R., Vallalaín, J., Gómez-Fernández, J.-C., Lee, G., *Pharm Res* 12, 1995, 1614–1617.

[73] New, R.R.C., *Liposomes—A Practical Approach*, IRL Press at Oxford University Press, Oxford, NY, Tokyo, 1990.

[74] MacDonald, R.C., MacDonald, R.I., Menco, B.Ph.M., Takeshita, K., Subbarao, N.K., Hu, L, *Biochim Biophys Acta* 1991, 1061, 297-303.

[75] Zemlyanaya, E. V., Kiselev, M. A., Zbytovska, J., Almasy, L., Aswal, V. K, Strunz, P., Wartewig, S., Neubert, R., *Crystallogr Rep* 2006, 51, Supp. 1, S22-S26.

[76] Kiselev, M.A., Lesieur, P., Kisselev, A.M., Lombardo, D., Aksenov, V.L., *Appl Phys A* 2002, 74, S1654-S1656.

[77] Knoll, W., Haas, J., Stuhrmann, H.B., Fuldner, H.-H., Vogel, H., Sackmann, E., *J Appl Cryst* 1981, 14, 191-202.

[78] Feigin, L.A., Svergun, D.I., 1987. *Structure analysis by small-angle X-ray and neutron scattering.* Plenum Publishing Corporation, New York.

[79] Balgavý, P., Dubničková, M., Kučerka, N., Kiselev, M. A., Yaradaikin, S. P., Uhríková, D., *Biochim Biophys Acta* 2001, 1512, 40-52.

[80] Zemlyanaya, E.V., Kiselev, M.A., Neubert, R., Kohlbrecher, J., Aksenov, V.L., *J Surf Inf* 2008, 2, 884-889.

[81] Kiselev, M. A., Zbytovská, J., Matveev, D., Wartewig, S., Gapienko, I. V., Perez, J., Lesieur., P., Hoell, A., Neubert, R., *Coll Surf A* 2005, 256, 1-7.

[82] Kiselev, M.A., Zemlyanaya, E.V., Aswal, V.K., Neubert, R., Eur *Biophys J* 2006, 35, 477-493.

[83] Garamus, V.M., Pedersen, J.S., Kawasaki, H., Maeda, H., *Langmuir* 2000, 16, 6431-6437.

[84] Nagle, J.F., Tristam-Nagle, S., *Biochim Biophys Acta* 2000, 1469, 159–195.

[85] Shannon, C.E., *Proc Inst Radio Eng* 1949, 37, 10-21.

[86] Franks, N. P., Lieb, W. R., *J Mol Biol* 1979, 133, 469-500.

[87] Schroetter, A., Kiselev, M.A., Hauss, T., Dante, S., Neubert, R., *Biochim Biophys Acta* 2009, 1788, 2194-2203.

[88] Seul, M., Sammon, J. *Thin Solid Films* 1990, 185, 287-305.

[89] Ruettinger, A., Kiselev, M.A., Hauss, T., Dante, S., Balagurov, A.M., Neubert, R.H.H. *Eur Biophys J* 2008, 37, 759-771.

[90] Charalambopoulou, G.Ch., Steriotis, Th.A., Hauss, Th., Stefanopoulos, K.L., Stubos, A.K., *Appl Phys A* 2002, 74(Suppl.), S1245–S1247.

[91] Charalambopoulou, G.Ch., Steriotis, Th.A., Hauss, Th., Stubos, A.K., Kanellopoulos, N.K., *Physica B* 2004, 350 e603–e606.

[92] Kessner, D., Kiselev, M.A., Hauß, T., Dante, S., Wartewig, S., Neubert, R.H.H., *Eur Biophys J* 2008, 37, 1051-1057.

[93] Schroeter, A., Kessner, D., Kiselev, M.A., Hauss, T., Dante, S., Neubert, R.H.H., *Biophys J* 2009, 97, 1104-1114.

[94] Ryabova, N.Yu., Kiselev, M.A., Dante, S., Hauss, T., Balagurov, A.M., *Eur Biophys J* 2010, 39, 1167-1176.

[95]   Ryabova, N.Yu., Kiselev, M.A., Balagurov, A.M., *Crystallogr Rep* 2010, 55, 516-525.

[96]   Kiselev, M.A., *Crystallography* 2007, 52, 572-576.

[97]   Zbytovská, J., Vávrová, K., Kiselev, M.A., Lessieur, P., Wartewig, S., Neubert, R.H.H., *Coll Surf A* 2009, 351, 30-37.

In: Neutron Scattering Methods and Studies
Editor: Michael J. Lyons

ISBN: 978-1-61122-521-1
© 2011 Nova Science Publishers, Inc.

Chapter 5

PRECIPITATE ANALYSIS OF MICROALLOYED STEELS USING SMALL ANGLE NEUTRON SCATTERING

J. Barry Wiskel

Dept. of Chemical and Materials Engineering, University of Alberta,
Edmonton, Alberta, Canada

ABSTRACT

Precipitates (2nd phase particles) are a major strengthening mechanism in many types of steels. In particular, it is the size (and distribution) and volume fraction of very fine precipitates ($\approx$<10 nm) that contribute significantly to the strengthening of these materials. This important precipitate size range coincides well with the precipitate analysis capabilities of Small Angle Neutron Scattering (SANS). The major challenges in characterizing precipitates in microalloyed steels are the relatively low volume fraction of nano size precipitates (<0.20wt%) and both the wide range of composition and sizes of precipitates that can form during the TMCP process. Given these challenges, characterizing precipitates in these types of steels requires a diversity of techniques including TEM, SEM, EDX, XRD and SANS. This paper will detail the analysis of nano sized precipitates in microalloyed steels using SANS. The precipitate data (e.g. size and distribution of the nano sized precipitates) measured with SANS was compared with other measurements techniques (TEM and XRD) for validation. In addition, the importance of the SANS precipitate data generated and how it can be used to correlate nano precipitates size and volume fraction with steel processing parameters will also be presented. This relationship can be used to identify steel processing conditions that would enhance fine precipitate evolution and ultimately the strength of the steel.

INTRODUCTION

Microalloyed steels are a type of steel characterized by a medium level of yield and tensile strength, excellent toughness (e.g. low ductile to brittle transition temperature) and good weldability (ability to weld the steel without the formation of a brittle phase). Because of these properties, microalloyed steels are used in a wide range of important engineering applications including bridge structures, transport trailers and for hdyrocarbon pipelines. The advantageous mechanical properties discussed above are achieved through a combination of thermo mechanical controlled processing (TMCP) [1] and composition. The former consists of rough and finish rolling at relatively high temperature followed by a rapid reduction in temperature in a laminar water cooling system (i.e. runout table) [2] where the steel undergoes a transformation from the high temperature austenite phase to the low temperature ferritic phase . The composition of the microalloyed steel is tightly controlled with a relatively low carbon content (typically < 0.1wt%) [3] and includes the addition of niobium, and titanium (and less frequently vanadium) at levels also less than 0.1% wt - hence the name microalloyed steel. In addition to the ubiquitous additions of manganese and silicon (common to most steels), small amounts (typically < 0.4%) of Ni, Cu, Mo and Cr can be present in the steel. It is the presence of all these alloying elements and in particular, carbon, niobium, vanadium, copper, molybdenum and chromium which can result in the formation of 2nd phase particles (precipitates) in the steel. The presence of precipitates in a TMCP microalloyed steel can indirectly increase strength (and toughness) via grain refinement/austenite conditioning during the rolling process [3] and also directly increase strength from the presence of nano sized precipitates. It is these latter precipitates which will be the focus of this paper and in particular the application of SANS (small angle neutron scattering) in quantifying them.

Strengthening mechanisms for high strength microalloyed steels used in pipelines includes: grain refinement, dislocation strengthening and precipitation strengthening. Precipitates that are produced in microalloyed steels can include titanium carbo-nitrides (Ti(C,N)), Nb carbo-nitrides (Nb(C,N) and complex niobium (Nb,Mo,Ti,V) carbides [4-11] depending on the exact composition of the steel and processing route. It is also possible for copper precipitates to form during the final TMCP processing [8]. In addition to composition, the TMCP processing parameters (e.g. homogenization temperature, rough rolling temperature, coiling interrupt temperature etc) can effect the size and number of precipitates that form. However, to achieve any significant increase in strength via precipitation strengthening, precipitate diameters should be on the order of <10 nm [4] in diameter. In addition to size, the volume fraction of these nano size precipitates has a direct effect on the magnitude of strengthening. An equation [12,13] correlating the increase in yield strength and the size and volume fraction of precipitates exhibits the following general form:

$$\Delta\sigma_{ppt}(MPa) = K \cdot \left(\frac{f^{1/2}}{D} \right) \tag{1}$$

where f is the volume fraction and D is the diameter of the precipitate. Therefore to increase yield strength either (or both) the volume fraction of the precipitates must be increased or the diameter of the precipitate decreased. It should be noted that Equation (1) does not consider a change in strengthening mechanism effect if D is decreased significantly below a critical size. Regardless, an understanding of the relationship between the number density (i.e. distribution) of fine nano-size precipitates that form and their relation to processing is essential for the production of higher strength microalloyed steels.

Precipitation events during thermomechanical processing can be grouped according to which stage in the TMCP process the precipitates form [9]. In a simplistic categorization, large (>0.5 μm) primarily TiN precipitates form during the solidification process and subsequent cooling to the reheat furnace temperature, medium sized Nb,Ti(C,N) precipitates in the size range from 50 to 500 nm form during the hot rolling schedule and very fine precipitates <10 nm form following laminar cooling [5-6]. The number and size distribution (which effects strengthening) of the fine precipitates is a complicated function of many factors including the steel composition, TMCP processing conditions and non-equilibrium amount of alloying elements (e.g. Nb) in solution prior to laminar cooling [14]. The nano sized precipitates will typically form in the low temperature ferrite phase [5, 15] following the transformation to a ferritic phase from the high temperature austenite phase. Therefore specific processing conditions which may influence the size distribution of these fine precipitates can include finish rolling temperature (FRT) and cooling interrupt temperature (CT). In particular, cooling interrupt temperature (i.e. temperature at the end of water cooling) can have a profound effect on the number, size and type of fine precipitates as the nucleation and growth of these precipitates are a strong function of temperature [5]. Theoretical nucleation and growth calculations of the nano size precipitation event have been undertaken [16] but the direct measurement of the size and distribution of these precipitates in commercially processed steels has been limited due to the difficulty in properly characterizing these precipitates.

A number of analytical techniques exist by which fine precipitates in low alloyed steels can be analyzed; these include scanning electron microscopy (SEM), transmission electron microscopy (TEM) [6-7,17-18], small angle x-ray scattering (SAXS) [18-19] and some small angle neutron scattering (SANS) studies [17, 20-22]. Traditional SEM is generally limited to large size precipitates (>50 nm), although field emission SEM has the capabilities of examining finer precipitates (<10 nm) [23]. TEM analysis has been widely used to provide direct information on the composition, morphology and size of individual precipitates, the spatial distribution of these precipitates and, to a lesser extent the precipitate size distribution. These precipitates characteristics have been obtained from TEM samples that have been either electro-polished samples [4,9,17-19,21] (i.e., precipitates remain in matrix) or extracted via a carbon replica technique [7,9]. However, TEM analysis does not lend itself easily to volume fraction determination due to the projection of a three dimensional structure into two dimensions. SAXS has been used to study fine precipitation in Fe. However, SAXS analysis of precipitates in steel has generally been limited to experimental alloys and/or simple commercial alloys in which the precipitate composition is well known [18-19].

SANS is a versatile technology for quantifying the size-distribution of precipitates, whose size is on the order of 1-100 nm, and its application to microalloyed steels would complement the techniques discussed above particularly in quantifying the size, volume fraction and distribution of the fine precipitates. SANS has been used to characterize copper precipitates in a Fe-Cu binary system [20], NbC precipitates in experimental Fe-Nb-C alloys [21], Nb(C,N) precipitates at high temperature (900-1200°C) [22], M_2C carbides in high alloy Co-Ni steels [24], Precipitates in maraging steels [25], determination of the solubility limit for high chromium steels [26] and precipitation of in low carbon steels [27].

Challenges associated with applying SANS to commercially produced microalloyed steels include: 1) the low volume fraction of the very fine precipitates and 2) the wide range of precipitate sizes (see earlier discussion) resulting from precipitation events associated with discrete stages in the TMCP process. Given the challenges, The aim of the work was twofold: Firstly, to establish a methodology for applying SANS to microalloyed steels and validating the results via a comparison to precipitate information obtained from alternate techniques and secondly, to correlate the SANS generated precipitation information for 4 types of microalloyed steel including X-70 and X-80 (A.P.I. specifications) and to plate steel Grades 80 and Grade 100 with processing parameters to identify processing condition that will modify the precipitate characteristics in these type of steels.

BACKGROUND

SANS is a versatile technology for quantifying the size-distribution of 1-100 nm precipitates, therefore its application to microalloyed steels would complement the techniques discussed in the introduction particularly in quantifying the size, volume fraction and distribution of the fine precipitates in these types of steels. To place the SANS analysis of a microalloyed steel into context it is necessary to review the fundamentals of small angle neutron scattering in the presence of small second phase particles and secondly to correlate the fundamental scattering equations to precipitate characteristics in these type of steels. This review will provide an understanding of the some of the main features of SANS application to steel precipitate analysis and potential limitations when applied to TMCP microalloyed steel.

Macroscopic differential scattering cross section

The general differential scattering cross section of a dispersion of precipitate particles for a dilute system can be calculated as follows [25]:

$$d\Sigma/d\Omega = (\Delta\eta)^2 \cdot \int_0^\infty V(R)^2 \cdot n(R) \cdot |F(Q,R)|^2 \cdot |S(Q,R)| \cdot dR \qquad (2)$$

where R is the radius of any precipitate, $V(R)$ is the particle volume, $\Delta\eta$ is the difference in scattering length densities of the particle and the matrix, $Q = 4 \cdot \pi \cdot \sin(\theta/2)/\lambda$ where θ is the scattering angle and λ is the neutron wavelength, $n(R)$ dR is the number density of precipitates between R and $R + dR$, $F(Q,R)$ is the form factor and $S(Q,R)$ is the interparticle structure factor. A number of assumptions have been made in this work to make the application of Equation (2) tractable for analysis of precipitates in a commercial microalloyed steel.

TEM characterization of the nano precipitates in commercial microalloyed steels [7,9] has shown that this size of precipitate is primarily spherical in morphology. Thus, the particle volume in Equation (2) can be calculated as the volume of a sphere ($4/3\pi R^3$). In addition, the form factor for a spherical precipitate can be used as follows:

$$|F(Q,R)| = 3\frac{\sin(Q \cdot R) - Q \cdot R \cdot \cos(Q \cdot R)}{(Q \cdot R)^3} \qquad (3)$$

The value for $n(R)$ was calculated using the density function of a log normal precipitate size distribution of the following form [18,21,25]:

$$n(R) = N_o \left(\frac{1}{\alpha \cdot R \cdot \sqrt{2 \cdot \pi}} \cdot \exp\left[-\frac{1}{2} \cdot \left(\frac{\ln(R/R_o)}{\alpha} \right)^2 \right] \right) \tag{4}$$

where N_o is the total number of scattering bodies in the distribution per volume (μm^3), R_o is the precipitate radius at the maximum of the distribution and α is a fitting parameter related to the width of the distribution via the relationship:

$$\alpha = \frac{\beta}{R_o} \tag{5}$$

where β is the width of the distribution. The rationale for applying a log-normal distribution has not been developed on a fundamental level in the literature but its application is based on experimental justification [18].

The value of the difference in nuclear scattering length density $(\Delta\eta_{nuc})$ between the precipitate and the matrix (for our system predominantly iron atoms with 1.8% Mn in solid solution) can be determined if the exact composition of the precipitates were known. As discussed in the introduction, the precipitates generated during TMCP processing can exhibit a variety of compositions and thus would require $(\Delta\eta_{nuc})$ to be determined for each possible composition. However, as the number of each precipitate type is limited (i.e. low volume fraction) the magnitude of the differential scattering of each precipitate type would be relatively low. To circumvent this problem associated with a low volume fraction, a magnetic field is applied to the steel sample such that a magnetic differential scattering cross section $(\Delta\eta_m)$ is used in Equation (1). Therefore the value of $(\Delta\eta_{nuc})$ can be calculated as follows:

$$(\Delta\eta)^2 = (\Delta\eta_{nuc})^2 + (\Delta\eta_m)^2 \cdot \sin^2\alpha \tag{6}$$

where α is the azimuthal angle on the detector plane and $\alpha = \pi/2$ correspond the maximum magnetic scattering.

This modification assumes that all the precipitates are non-magnetic [28] and that the magnetic size is the same as the chemical size. Though individual types of precipitates cannot be separated out from each other, the grouping of all precipitates as nonmagnetic holes increase the number density of particles and hence the magnitude of the differential scattering cross section. Using the methodology of Staron et al [25], $(\Delta\eta_m)$ was calculated to have a value of $5.0 \times 10^{10}/cm^2$ for the microalloyed steels studied.

Combining Equations 2 to 6 results in the following overall equation:

$$d\Sigma/d\Omega = (\Delta\eta_m)^2 \int_0^\infty \left(\frac{4}{3} \cdot \pi \cdot R^3 \right)^2 \cdot N_o \left(\frac{1}{\alpha \cdot R \cdot \sqrt{2 \cdot \pi}} \cdot \exp\left[-\frac{1}{2} \cdot \left(\frac{\ln(R/R_o)}{\alpha} \right)^2 \right] \right) \cdot |F(Q,R)|^2 \cdot |S(Q,R)| \cdot dR \tag{7}$$

where the independent variables are R_o, N_o and α (i.e. specific precipitate characteristics). It is these values which will be determined when applying Equation 7 to the experimental SANS data.

EXPERIMENTAL

The compositions of the TMCP steels studied in this work are summarized in Table I. All the steel samples have relatively low carbon contents (< 0.06 wt%), a Nb content less than 0.10 wt%, and a relatively low Ti content. Other alloying elements present in these steels may include Cr, Ni and Mo to varying weight percentages but typically less than 0.2 wt% each. The composition of Cu has been included due to the possibility of Cu precipitate formation at compositions greater than 0.15 wt% [29]. Also included in Table I are the process variables deemed important in the nucleation and growth of the fine ($\approx$ 5 nm) precipitates - the finish rolling temperature (FRT) and the cooling interrupt temperature (CT). Both the FRT and CT are expressed as relative values (i.e. the actual value divided by a constant) to protect the intellectual property of the steel manufacturer who sponsored the research. In general, except for the differences in FRT, CT and composition, each group of steels (e.g. X80 or X70) was processed in a similar manner.

Table I. Composition of Steels Studied

	%C	%Nb	%Ti	%Cu	FRT	CT
X80-462	0.03	0.09	0.013	0.27	0.94	1.04
X80-A4B	0.04	0.09	0.017	0.34	1.05	0.93
X80-B4F	0.05	0.08	0.010	0.15	1.00	1.00
X80-A4F	0.05	0.04	0.010	0.15	1.00	0.90
X70-AOL	0.043	0.08	0.025	0.24	1.01	0.64
X70-AOJ	0.045	0.07	0.019	0.22	1.00	0.63
G80-B3A	0.056	0.089	0.032	0.41	1.07	1.04
Grade 100	0.06	0.09	0.060	0.40	1.07	1.09

SANS samples ranged in thickness between 4.5 mm and 6.6 mm, with a surface area of 10 mm x 10 mm and were extracted from the original steel skelp (strip) material both above (T) and below (B) the centerline of for each sample shown in Table I. Each samples was prepared (i.e. cut and polished) to avoid including material from both the centreline of the skelp (where possible segregation may alter the nominal composition and hence the volume fraction of 2nd phase particles) and the surface of the skelp. The samples were mechanically polished to 1.0 μm diamond and were oriented in the SANS neutron beam with the rolling plane (i.e., rolling surface) perpendicular to the beam orientation.

SANS Testing

SANS experiments were conducted on the NG-3 Neutron Guide at the National Institute of Standards Center for Neutron Research [30]. Neutrons were monochromated by a velocity selector to a mean wavelength of 0.54 nm with the wavelength spread being $\Delta\lambda/\lambda = 10\%$. The samples were magnetized to saturation in a 2 Tesla magnetic field. Three detector distances (1.9 m, 5.55 m and 13.65 m) were used, giving access to scattering vector (Q) values ranging from 0.023 nm to 2.1 nm^{-1}. The scattered neutrons were recorded with a 640 mm x 640 mm ^{3}He position sensitive proportional counter with a 5 mm x 5 mm resolution.

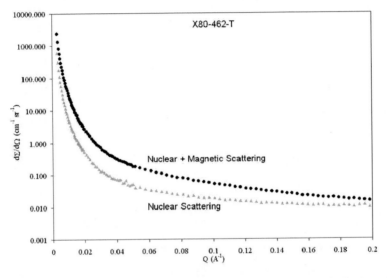

Figure 1. Differential scattering cross section as a function of scattering vector for both nuclear and nuclear + magnetic scattering for sample X80-462-T.

The measured neutron beam intensity at each Q value was converted to a differential scattering cross section ($d\Sigma/d\Omega$) value by scaling the measured neutron beam intensity with sample thickness and transmission, response of the detector and background scattering [24]. The differential scattering cross section obtained for Sample x80-462-T for both the nuclear and magnetic scattering (obtained at $\alpha = \pi/2$) and the solely nuclear scattering ($\alpha = 0$) is show in Figure 1. In this work the respective scattering magnitudes at each Q were calculated over sectors for $\alpha \pm 10°$. The difference in magnitude between the two curves is the contribution of the magnetic scattering component.

The magnetic differential scattering cross section for each sample tested was obtained by subtracting the value of the horizontal (i.e., nuclear) differential scattering cross section measured from the value of the vertical (i.e., nuclear + magnetic) differential scattering cross section [25]. Figure 2 is a plot of the calculated magnetic differential scattering cross section for sample X80-462-T as a function of Q. The magnetic differential scattering cross section calculated for each sample was used for all subsequent size distribution calculations and analysis. The magnetic scattering data shown in Figure 2 includes the scattering effects of not only the fine precipitates ($\approx$5 nm) of interest, but also of larger precipitates that occur in these microalloyed steels. To circumvent the added complexity of including these large precipitates in the SANS analysis the assumption was made that the precipitates (in a commercial microalloyed steel) do not form a continuous distribution of particle sizes but discrete precipitate distributions associated with the individual stages of the TMCP processing. As discussed earlier, TEM [7,9] analysis of both X70 and X80 microalloyed steels has identified groupings of precipitates associated with the different TMCP stages including: 1) precipitates in the size range from 500 nm to 100 nm (formed during the rough rolling operation), 2) precipitates in the 30 – 50 nm range (during finish rolling) and 3) fine precipitates (less than 20 nm with an average size of 5 nm for X70 and 15 nm for X80) have been observed that are generated during the transformation and cooling operations.

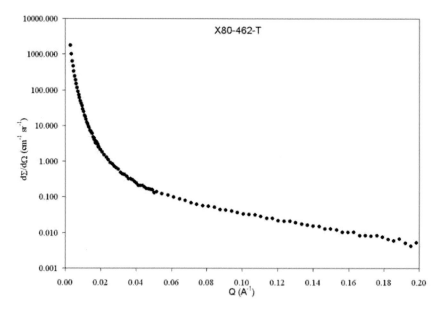

Figure 2. The magnetic component of the differential scattering cross section as a function of scattering vector for sample X80-462-T.

Based on the concept of discrete size distributions, the effect on the differential magnetic scattering cross section of the larger precipitates (i.e., > 30 nm) was removed from the scattering associated with the very fine precipitates (around 5 nm). This was achieved by assuming that the contribution to the magnetic differential scattering cross section of these larger particles followed Porod behaviour [31-32] (i.e., $d\Sigma/d\Omega \propto 1/Q^4$) at large Q values, whereas the finer nano precipitates are following Guinier behaviour. The effect of the Porod scattering behaviour for the large precipitates was removed from the measured magnetic differential scattering cross section by subtracting an extrapolation (numerically fitted to the data) of the Porod behaviour at scattering angles $Q < 0.3$ nm. An example of this scattering adjustment is illustrated in Figure 3 where the solid line represents the estimated Porod behaviour of the relatively large precipitates. The resultant magnetic differential scattering cross section (following subtraction of the Porod scattering of the large particles) is shown in Figure 4 for sample X80-462-T. This data manipulation was applied to the data collected for all the samples.

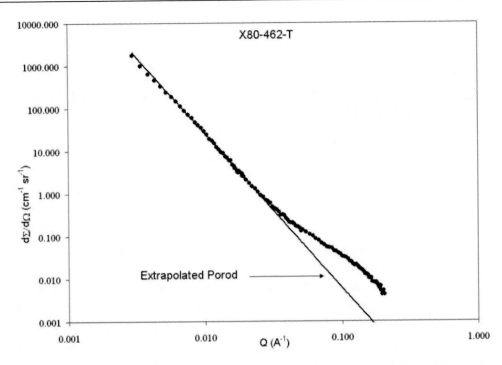

Figure 3. Comparison of extrapolated Porod behaviour and measured magnetic differential scattering cross section for sample X80-462-T.

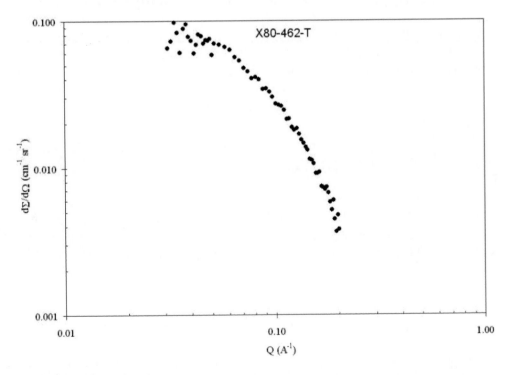

Figure 4. Modified (following subtraction of extrapolated Porod behaviour) differential scattering cross section of magnetic scattering for sample X80-462-T.

RESULTS

This section will present the application of Equation 7 to the differential scattering results obtained for each sample tested and in particular the calculation of the mean size, number and distribution width of the nano size precipitates for each sample. This values for these unknown precipitate characteristics were achieved by sequentially:

a) estimating initial values for R_o, N_o and α in Equation 7

b) numerically integrating Equation 7 for all Q values

c) comparing the predicted magnetic differential scattering cross section with the measured values

d) applying a nonlinear Newton-Raphson technique to calculate new values for N_o, R_o and α that would improve the fit.

This approach was repeated until a suitable convergence was obtained (i.e. minimization of the difference between the predicted and measured data.) The value of N_o, R_o and α obtained at a suitable convergence represent the actual distributions of the fine precipitates for a particular sample.

Calculation of N_o, R_o and α for a Single Log Normal Distribution

Using the method discussed above, the measured and predicted scattering data for X80-462-T is shown as a Kratky plot in Figure 5. Included in Figure 5 are the calculated values for N_o, R_o and α and the calculated volume fraction (based on the total number of particles and their volume) for this distribution. The predicted log normal average radius R_o (1.07 nm) is on the same order of magnitude as the Guinier radius (R_g) value (1.53 nm) predicted using the following equation:

$$R_g = \frac{A}{Q_{max}}$$

(6)

where $A = 1.7$ [18] and Q_{max} is the value of Q at the peak (≈ 1.11 nm^{-1}) in Figure 5. As will be discussed later sample X80-462-T (and B) are unique in that a single log normal distribution could be used to successfully predict precipitate behaviour. This uniqueness will be discussed later.

Calculation of N_o, R_o and α for a Bimodal Log Normal Distribution

The predicted differential scattering cross section - calculated using a single log normal distribution of precipitates - is compared to the measured value for X80-B4F-T in Figure 6. Unlike Figure 5, where the correspondence between the predicted (unimodal distribution) and measured differential scattering cross section was reasonably good, Figure 6 exhibits a marked deviation of the predicted behaviour from the measured. This deviation may indicate the existence of a bimodal precipitate population (i.e., the occurrence of two distinct precipitation events).

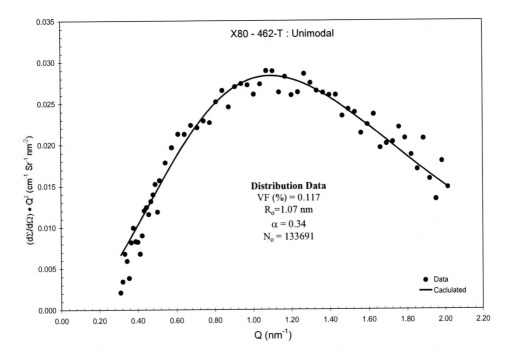

Figure 5. Measured and predicted (unimodal log-normal distribution) differential scattering cross sections for X80-462-T.

The calculations for X80-B4F-B were repeated but were undertaken using a bimodal precipitate distribution. The number of variables used for the calculation included two sets of distribution data (R_1, N_1 and α_1) and (R_2, N_2 and α_2) where the subscript 1 indicates the distribution of one set of precipitates and the subscript 2 the behaviour of a second precipitation distribution. A comparison between the measured and predicted differential scattering cross sections calculated using a bimodal distribution of fine precipitates, is shown in Figure 7. Unlike Figure 6, the bimodal distribution calculations show a reasonably good fit with the measured data.

Comparison of SANS results with TEM Analysis for Grade 100

The precipitates in the Grade 100 material were examined using TEM [33] and quantitative XRD [34]. The former included manual counting of the size and number of the fine precipitates in a carbon replica film taken from this steel. The normalized (the volume of carbon replica was not known) size distribution values observed are compared to the predicted size distribution values obtained with SANS in Figure 8. Also included Figure 8 is the mean precipitate size determined using quantitative XRD on extracted precipitate residues. The measured results of the both other techniques lie within the distributions predicted using SANS. This reasonable agreement validates the veracity of the SANS precipitate characterization methodology described above.

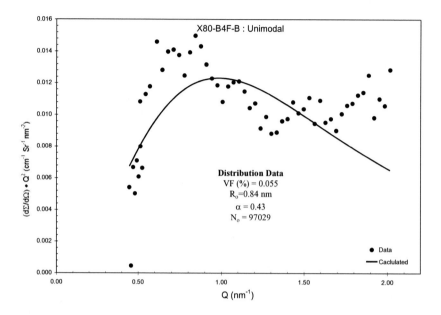

Figure 6. Measured and predicted (unimodal log normal distribution) differential scattering cross sections for X80-4BF-B.

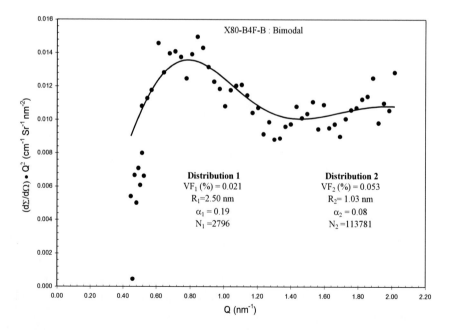

Figure 7. Measured and predicted (bimodal log normal distribution) differential scattering cross sections for X80-4BF-B.

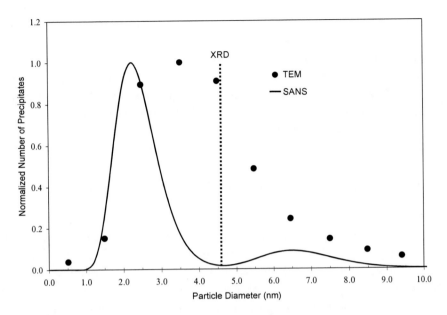

Figure 8. Comparison of the normalized number distribution of precipitates measured by TEM, Quantitative XRD and SANS for Grade 100.

Calculation of SANS Precipitate Data for all Samples

The SANS data analysis described in the previous section was repeated for all the samples listed in Table I. The results of the analysis are tabulated in Table II and include the log normal bimodal precipitate distribution parameters, the number of particles per volume, the volume fraction of each size distribution and the total volume fraction of the fine precipitates. To maintain consistency, the SANS scattering calculations for both X80-462-T and X80-462-B were repeated using a bimodal distribution. The justification (in terms of fit between the measured and the calculated differential scattering cross section) of the unimodal vs. bimodal for these two samples is not as strong as in the other samples, however, this assumption was undertaken to ensure consistency in the analysis that follows. The data points for both X80-462-T and X80-462-B are explicitly noted in all the figures. For most samples (except the Grade 100), the volume fraction of the finer (i.e. 2nd) precipitate distribution is larger than that observed for the larger precipitate distribution. In addition, both the values of R_1 and R_2 are observed to vary from sample to sample. These differences between samples are attributed to changes in both processing and composition. This concept is be elaborated on in the discussion.

DISCUSSION

The size distribution data presented in Table II was analyzed in conjunction with the processing data for each sample (Table I), including the effects of finish rolling temperature (FRT) and cooling interrupt temperature (CT). In general, many factors will affect the precipitation of nano-sized particles including temperature, applied strain, composition, etc.; [16] therefore, the lack of a definite correlation of the precipitation distribution data with a single specific processing variable (e.g., FRT) should not be construed as this variable having no effect. The objective of the following discussion is to illustrate how the precipitate information generated using SANS can be used to establish what variable(s) have both a direct and significant effect on the precipitation phenomena and use this information to establish guidelines in the processing of these steels as a means of enhancing precipitation strengthening in TMCP microalloyed steels.

Effect of Finish Rolling Temperature (FRT)

Though our procedure for SANS analysis is confirmed by other methodologies, the importance of this work is to relate the precipitate distribution characteristics (i.e. size and volume fraction) with TMCP processing parameters and to establish a correlation with these processing parameters (in particular FRT and CT which intuitively would have a significant effect on the fine precipitates) as a means to alter the actual process to enhance fine precipitate production and hence strength. For the work presented in this paper, interactions of processing variables with the compositional differences is not considered.

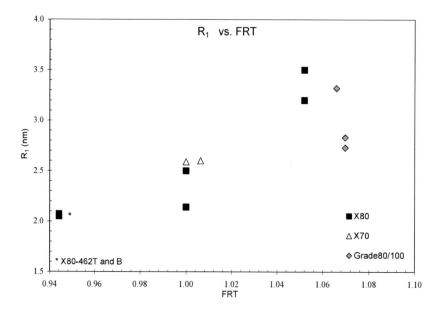

Figure 9. Lognormal radius (R_1) for X70, X80 and Grades 80 and 100 as a function of normalized finish rolling temperature (FRT).

The values of R_1 and R_2 are plotted as a function of the normalized finish rolling temperature (FRT) in Figures 9 and 10, respectively, for all the steels tested. It is observed in Figure 9 that R_1 increases in value as finish rolling temperature (FRT) increases. The trend

Table II. Calculated Precipitate Distribution Characteristics

Sample	R_1 (nm)	α_1	$\#_1$	VF_1(%)	R_2 (nm)	α_2	$\#_2$	VF_2(%)	VF_{total} (%)
X80-462T	2.05	0.23	3979	0.018	1.19	0.25	100350	0.096	0.114
X80-462B	2.07	0.3	4401	0.024	1.22	0.25	89350	0.091	0.115
X80-A4BT	3.20	0.18	1003	0.016	1.06	0.06	120128	0.051	0.067
X80-A4BB	3.50	0.17	558	0.011	0.90	0.06	115692	0.036	0.047
X80-B4FT	2.50	0.18	2760	0.02	1.01	0.08	113728	0.05	0.070
X80-B4FB	2.50	0.19	2796	0.021	1.02	0.08	113781	0.053	0.074
X80-A4FT	2.14	0.25	1861	0.051	1.02	0.07	106619	0.049	0.100
X80-A4FB	2.50	0.33	1096	0.012	0.98	0.10	80727	0.036	0.048
X70-A0JT	2.59	0.21	1472	0.013	0.93	0.06	90139	0.031	0.044
X70-A0LT	2.60	0.22	666	0.006	0.8	0.06	101466	0.022	0.028
G80-B3AT	2.73	0.21	3926	0.041	1.27	0.07	53935	0.047	0.088
G80-B3AB	2.83	0.22	3405	0.04	1.22	0.07	57303	0.044	0.084
G100	3.32	0.15	4939	0.083	1.18	0.24	31668	0.029	0.112

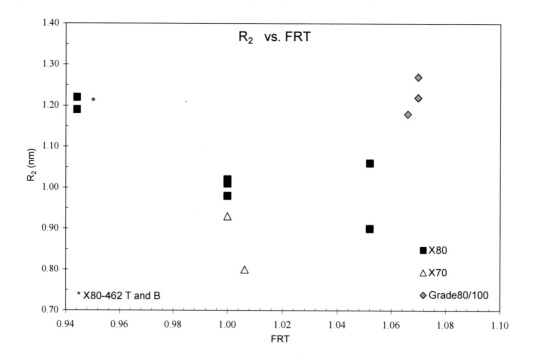

Figure 10. Lognormal radius (R_2) for X70, X80 and Grades 80 and 100 as a function of normalized finish rolling temperature (FRT).

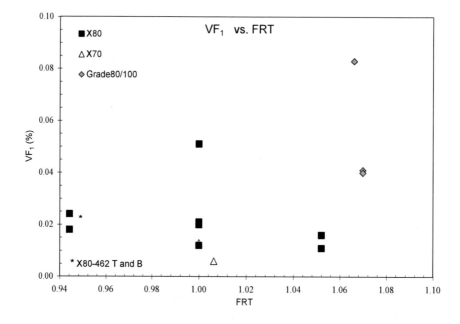

Figure 11. Volume fraction (VF_1) for X70, X80 and Grades 80 and 100 as a function of the normalized finish rolling temperature (FRT).

observed suggests that the first distribution precipitation event (i.e., the larger of the fine precipitates) is associated with the finish rolling practice. The larger mean size observed with increasing FRT can be qualitatively linked with increased coarsening of these precipitates at the higher temperatures at which they form. Conversely, in Figure 10 (R_2 vs. FRT), there is not a definite trend between the size of the precipitate for the second precipitate distribution and the finish rolling temperature. This data suggests that FRT may have only secondary effect on the 2nd distribution of precipitates that form. As will be shown later, a combination of FRT and CT have a strong influence on the size of the second precipitation distribution.

Figures 11 and 12 compare the volume fraction of each precipitation distribution (VF_1 and VF_2) with FRT. A direct correlation of volume fraction (of either size distribution) with FRT is not apparent. As FRT is an indicator of overall finish rolling temperature and, hence, solubility of the alloying elements, the lack of a correlation is surprising. However, it is more likely that other variables are influencing volume fraction amounts in conjunction with FRT.

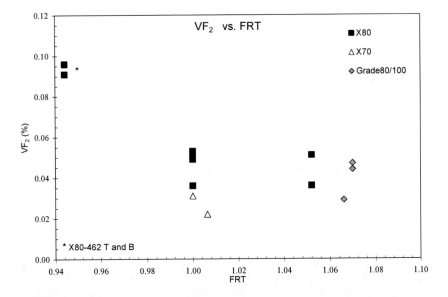

Figure 12. Volume fraction (VF_2) for X70, X80 and Grades 80 and 100 as a function of the normalized finish rolling temperature (FRT).

Effect of Cooling Interrupt Temperature (CT)

The value of R_1 is plotted versus the normalized cooling interrupt temperature (CT) in Figure 13. The lack of a distinctive correlation between R_1 and CT suggest that the (first) distribution of precipitates is not associated with the cooling interrupt process. However, as observed in Figure 9, R_1 is directly affected by the finish rolling temperature and, hence, precipitation of the first (larger distribution) is occurring during the final stages of rolling and not during cooling.

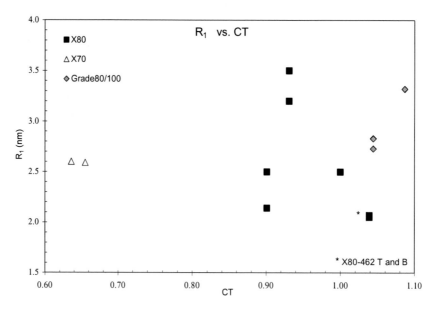

Figure 13. Lognormal radius (R_1) for X70, X80 and Grades 80 and 100 as a function of the normalized cooling interrupt temperature (CT).

A plot of R_2 as a function of CT is shown in Figure 14. Unlike, the R_1 vs. CT figure, a definite decrease in the log normal mean radius (R_2) with decreasing CT is observed. This correlation may indicate that this second precipitation event is strongly influenced by the cooling interrupt temperature. From an understanding of nucleation and growth of precipitates in ferrite [20], the lower value of CT, the smaller the overall size of precipitate that forms (i.e., less coarsening of these fine precipitates would occur due to the lower temperature). However, CT can also affect the amount of the fine precipitates that form.

Figure 15 plots the volume fraction (VF_2) of the second size distribution as a function of CT. It is observed in this figure that the volume fraction of fine precipitates decreases with decreasing CT. This suggests a kinetic limitation on the formation of these fine precipitates as the precipitation temperature (i.e., CT) is lowered. Thus, processing conditions that are manipulated to produce very fine precipitates (i.e., smaller sized precipitates) may be counter productive towards the objective of increasing strength, since the volume fraction of the precipitates that forms also decreases.

Combined Effect of FRT and CT on Precipitation

As observed in Figure 13, and to a lesser extent Figure 10, more than one process variable can affect the precipitation events. To illustrate this effect, the values of R_2 and VF_2 are plotted as a function of the ratio of CT/FRT. Though the value of CT/FRT does not have a specifically recognized phenomenological effect on precipitation, it does include in a single variable the effects of both FRT and CT on precipitation. Figure 16 shows the effect of this ratio on R_2. It is observed that R_2 decreases as CT/FRT decreases - a combination of a low CT value and a high FRT value results in a lower R_2 value. A plot of VF_2 vs. CT/FRT (Figure 17) exhibits a similar trend indicating the dependence of fine precipitation on both CT and FRT.

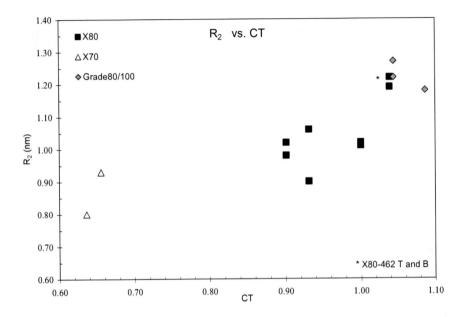

Figure 14. Lognormal radius (R_2) for X70, X80 and Grades 80 and 100 as a function of the normalized cooling interrupt temperature (CT).

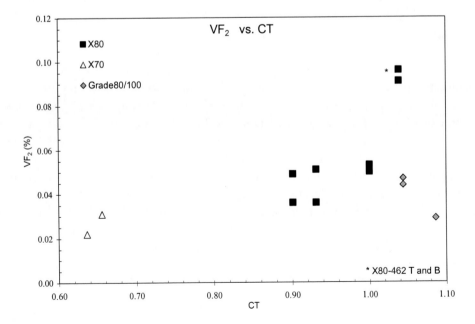

Figure 15. Volume fraction (VF_2) for X70, X80 and Grades 80 and 100 as a function of the normalized cooling interrupt temperature (CT).

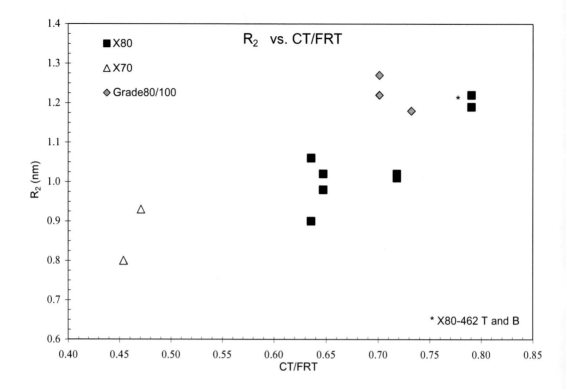

Figure 16. Lognormal radius (R_2) for X70, X80 and Grades 80 and 100 as a function of the ratio of CT/FRT.

Unimodal vs. Bimodal

The anomaly in the SANS analysis has been the X80-462 steel in which a clear separation into two distinct precipitation events is not apparent. Interesting, this steel also exhibits the highest ratio of CT/FRT (i.e., high CT and relatively low FRT). As discussed earlier a high CT value will result in a larger R_2 value, while a lower FRT results in a finer R_1 value. As the log normal mean values (R_2 and R_1) associated with each precipitation event approach each other the distributions will overlap to a greater and greater extent, such that it becomes difficult to distinguish between the two distinct distributions. X80- 462 is unique in that it lies at the upper end of our data for CT and at the lower end of the data for FRT. Hence, the apparent unimodal precipitation observed with these samples maybe due to the unique combination of processing conditions, such that the low fraction of precipitation at the finish rolling stages and the enhanced precipitation at the cooling interrupt temperature mask the bimodal nature of the precipitation events.

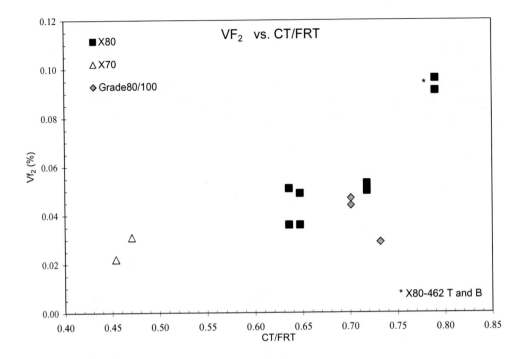

Figure 17. Volume fraction (VF_2) for X70, X80 and Grades 80 and 100 as a function of the ratio of CT/FRT.

ACKNOWLEDGMENTS

The author would like to thank both EVRAZ INC NA and NSERC for financial support. Also the authors would like to thank L. Collins of EVRAZ INC NA for feedback and support and J. Barker of NIST in Gaithersburg, Md. for his help with the SANS testing. The SANS facilities at the National Institute of Standards and Technology are supported by the National Science Foundation under Agreement No. DMR-9986442.

CONCLUSIONS

1] SANS has been successfully used in quantifying precipitation characteristics of fine (nano scale) precipitates in commercially produced microalloyed steels.

2] The precipitate sizes measured by SANS for a Grade 100 microalloyed steel compare favourably with precipitate sizes measured using TEM and quantitative X-ray Diffraction.

3] The precipitation characteristics obtained from the SANS analysis can be correlated with TMCP processing conditions. In particular, the size and volume fraction of the very fine nano sized precipitates observed in these steels are effected by both the cooling interrupt and the finish rolling temperatures applied in the industrial TMCP process.

REFERENCES

[1]   T. Tanaka, Controlled rolling of steel plate and strip, *Int. Met. Rev.*, Vol 26, 1981, p 185-212

[2]   A. Mukhopadhyay and S. Sikdar: *J. Mat. Proc. Tech.*, 169, 2005, pp. 164-172.

[3]   A. Nowotnik and T. Siwecki, *J. Micro.*, Vol. 237 (3), 2010, pp. 258-262.

[4]   T. Gladman: *The Physical Metallurgy of Microalloyed Steels*, Institute of Metals, London, UK, 1997, pp. 53-55.

[5]   A.J. Deardo: *International Metals Review*, 2003, vol. 48 (6), pp. 371-402.

[6]   S. Akhlaghi and D.G. Ivey: *Canadian Metallurgical Quarterly*, Vol. 41 (1), 2002, pp. 111-119.

[7]   J. Lu, *PhD Thesis*, University of Alberta, 2009.

[8]   M.S. Gagliano and M.E. Fine: *Met. Trans. A*, Vol. 35A (8), 2004, pp. 2323-2329.

[9]   U. Sharma, M.A. Sc. Thesis, University of Alberta, 2001

[10] A. Fatehi, J. Calvo, A.M. Elwazri, S. Yue: *Mat. Sci and Eng. A*, 527, 2010, pp. 4233–4240.

[11] W. Lee, S. Hong, C.Park, and S. Park: *Met. Trans. A*, Vol. 33A, 2002, pp. 1689-1698.

[12] R. Schnitzer S. Zinnerb and H. Leitnera: *Scrip. Mat.*, 62, 2010, pp.286-289. _YieldStrength_Ppt

[13] M. Charleux, W.J. Poole, M. Militzer and A. Deschamps: *Met. Trans. A*, Vol. 32A (7), 2001, pp. 1635-1647.

[14] C. Klinkenberg, K. Hulka and W. Beck: *Steel Research Int.*, Vol. 75 (11), 2004, pp. 744-752.

[15] H.J. Kestenbach, S.S. Campos and E.V. Morales, *Mat. Sci Tech.*, Vol. 26, no. 6, 2006, pp. 615-626.

[16] Dutta, B, Palmiere, E.J. and Sellars, C.M., *Acta Mater.*, 49, 2001, pp. 785-794.

[17] S.Q. Yuan, S.W. Yang, C.J. Shang and X.L. He: *Materials Science Forum*, Vols. 426-432, 2003, pp. 1307-1312.

[18] A. Deschamps, M. Militzer and W.J. Poole: *ISIJ Int.*, Vol 41 (2), 2001, pp. 196-205.

[19] A. Deschamps, M. Militzer and W.J. Poole: *ISIJ Int.*, Vol 43 (11), 2003, pp. 1826-1832.

[20] K. Osamura, H. Okuda, S. Ochiai, M. Takashima, K. Asanao, M. Furusaka, K. Kishida and F. Kurosawa: *ISIJ Int.*, Vol 34 (4) , 1994, pp. 359-365.

[21] F. Perrard, A.Deschamps, F. Bley, P. Donnadieu and P. Maugis: *J. Appl. Cryst.*, 39, 2006, pp. 473-482.

[22] N.H. Van Dijk, S.E. Offerman, W.G. Bouwman, M. Rekvedlt, J. Sietsma, S. Van Der Zwagg, A. Bodin and R.K. Heenan: *Met. Trans. A*, Vol. 33A (7), 2002, pp. 1883-1891

[23] A. M. Elwazri, R. Varano, F. Siciliano, D. Bai and S. Yue: *Mat. Sci. Tech.*, Vol. 22 (5), 2006, pp. 537-541.

[24] A.J. Allen, D. Gavillet and J.R. Weertman, *Acta Metal.*, Vol. 41 (6), 1993, pp. 1869-1884

[25] P. Staron, B. Jamnig, H. Leitner, R.Ebner and H. Clemens: *J. Appl. Cryst.*, 36, 2003, pp. 415-419.

[26] F. Bergner, A. Ulbricht and C. Heintze: *Scrip. Mat.*, 61, 2009, pp.1060-1063.

[27] B. S. Seong, Y. R. Cho, E. J. Shin, S. I. Kim, S.-H. Choi, H. R. Kima and Y. J.Kima:, *J. App. Cry.*, 41, 2008, pp. 906-912.

[28] M. Bischof, P.Staron, D. Caliskanoglu, H. Leitner, C. Scheu and H.Clemens: *Mat. Sci and Eng. A*, 472, 2008, pp. 148–156.

[29] M. Perez, F. Perrard, V. Massardier, X. Kleber, A. Deschamps, H. DeMonestrol, P. Pareige and G. Covarel: *Phil. Mag.*, Vol. 585 (20), 2005, pp. 2197-2210.

[30] C.J. Glinka, J.G. Barker, B. Hammouda, S. Krueger, J.J. Moyer, and W.J. Orts , *J. Appl. Cryst.*, 31, 1998, pp. 430-445.

[31] A. Guinier and G. Fournet, *Small Angle Scattering of X-rays*, John Wiley and Sons, New York, NY, 1955, pp. 127-128.

[32] Riello, S. Polizzi, G. Fagherazzi, T. Finotto and S. Ceresara: *Phys. Chem.*, 3 2001, pp. 3213-3216.

[33] K. Poorhaydari-Anaraki, PhD Thesis, University of Alberta, 2005

[34] J.B. Wiskel, X. Li, J. Lu, D.G. Ivey and H. Henein: *Int. Pipeline Conf.* (IPC2010), 2010, in publication.

In: Neutron Scattering Methods and Studies
Editor: Michael J. Lyons

ISBN: 978-1-61122-521-1
© 2011 Nova Science Publishers, Inc.

Chapter 6

SMALL ANGLE NEUTRON SCATTERING FROM PROTEINS IN SOLUTION

Maria Grazia Ortore*
Dipartimento SAIFET and CNISM,
Università Politecnica delle Marche, Via Brecce Bianche,
Ancona, Italy

Abstract

The study of biomolecules in solution is quite important in order to reproduce environmental conditions quite resembling those existing in-vivo. Small Angle Neutron Scattering (SANS) from proteins in solution provides information concerning size and shape, over a large length scale. Performance of SANS experiments on samples at different deuteration grades enables a deeper knowledge both of protein structure and of protein solvation shell. The chance to perform numerical simulation before carrying on a SANS experiment can lead the user to a wiser choice of the experiment preparation and greatly improve the structural resolution. A brief discussion about the scattering theory, isotopic substitution and thermodynamic equilibria between water and cosolvent molecules in the protein solvation shell, will be presented. This chapter will discuss theory and practical aspects with protein conditions in order to guide potential users to successfully apply SANS experiments to answer peculiar biological questions involving proteins in solution.

Introduction

The complex reactions which make up living processes are mainly dependent on proteins role: they are storage and carrier molecules, structural molecules, catalysts and molecular motors. Hence, every function requires a three-dimensional structure correctly folded from the amino-acid chain, in order to reproduce a native biologically active structure. That is why the efforts in the analysis of mechanisms driving the protein folding/unfolding processes origin from several decades ago[1]. In fact, conformational changes and stability of

---

*Tel.: (+39)071-2204664, Fax: (+39)071-2204605

proteins are crucial issues for what concerns molecular biology, medicine and biotechnology.

Small Angle Neutron Scattering (SANS) is a technique easily applied to biological molecules as it probes the size, shape and conformation both of macromolecules and of macromolecular complexes in aqueous solution on a length scale from ten to several thousand Angstrom [2]. Because to study macromolecules in solution means both to investigate the issue at biologically relevant conditions and to perform experiments on samples difficul or impossible to crystallize, SANS has increasingly become an important technique for the structural characterization of biological macromolecules. The advantage of using SANS in respect to SAXS (Small Angle X-ray Scattering) is due to the fact that the scattering of neutrons by hydrogen and deuterium is different in magnitude and phase, and thus neutrons open up the possibility to perform contrast variation experiments as an additional tool for structural studies to highlight the components of a multimacromolecular complex. Contrast variation experiments are routinely performed both for small angle neutron scattering and for neutron protein crystallography. The success of the contrast variation in a SANS experiment can be appreciated considering that several biological issues were solved using this approach, such as the structure of the elongating ribosome [3]. Together with the contrast variation method, another advantage provided by the use of neutrons is due to the recent production of cold neutrons, whose wavelengths are 10 - 20 times larger than thermal neutrons, which allow SANS to examine macromolecular complexes, such as living cells.

This chapter will discuss theory and pratical aspects with protein solution conditions to stress the capability of SANS technique.

In particular, a recent method to quantitatively determine the composition of the solvation shell of a protein dissolved in a solvent mixture[4–6] by a set of SANS experiments will be presented. The method is based on the global analysis of a series of in solution SANS experiments on a set of different protein experimental conditions. Hence to the achievement of the method, it is necessary to wisely choose protein concentration, cosolvent amount and deuteration grade. The global analysis of a large number of SANS curves allows to estimate a thermodynamic parameter (K) that, according to the solvent exchange model [7], describes the 1:1 exchange between water and cosolvent molecules in the bulk solution and in a region close to the protein surface.

A Simple Approach to a Small Angle Neutron Scattering Experiment on Protein Solutions

The Experiment

SANS measurements can be carried out at nuclear research reactors or at spallation neutron sources. The neutrons are transported from the source to the instrument by neutron guide tubes which can have translatable guide sections and apertures that may be moved in and out of the neutron beam in order to defin the incident beam collimation. Then, there is an accessible section at the sample position to accommodate temperature-controlled or pressure-controlled sample changers, fl w cells, etc. A detector is positioned in the post-sample flight-tube like the incident neutron guides, this is normally evacuated to reduce air scatter, which would otherwise be strong.

The neutron wavelenght λ and the sample-detector distance have to be chosen in order to cover the convenient scattering vector (Q) range ($Q = 4\pi \sin\theta/\lambda$, being 2θ the scattering angle) related to protein dimensions. In particular, quite often it is necessary to adopt two different sample-detector distances to obtain a wider Q range able to provide information on the different levels of the protein structure[8]. Protein acqueous solutions are usually contained in 1 mm thick quartz cells and the small angle scattering signal is revailed by the detector. It results to be useful to repeat several short runs for each sample, with the aim to check of any possible degradation of the sample during the neutron scattering measurements. Scattered intensities have to be radially averaged in the case of a bidimensional detector, and corrected for background, buffer contribution, detector efficien y and sample transmission[9].

According to this data reduction procedure, the differential scattering cross sections $d\Sigma/d\Omega(Q)$ can be reported in units of (cm^{-1}) by comparison with the intensity scattered by 1 mm of light water. It is essential to emphasize the importance of obtaining intensity data on an absolute scale: while the use of absolute units can be not essential for the measurement of the spatial dimensions, such as the radius of gyration of a protein, it is necessary to avoid artifacts to which this technique can be vulnerable during data analysis[10].

The Basic Theory

In a SANS experiment, the crucial function to be analysed is the macroscopical differential cross section $\frac{d\Sigma}{d\Omega}(Q)$. In the case of monodisperse and randomly oriented particles in solution, it can be expressed as

$$\frac{d\Sigma}{d\Omega}(Q) = n_p S(Q) P(Q) + B \qquad (1)$$

where n_p is the particle number density (the protein number density, if SANS experiment concerns proteins in solution), $S(Q)$ the structure factor, $P(Q)$ the particle averaged squared form factor and B is a fla background. The structure factor is mainly responsible for particle-particle interactions, while the form factor describes the particle structural features.

In the case of a SANS experiment on proteins in solution, in a dilute concentration range chosen in order to avoid protein aggregation, hence to have monodisperse particles, the structure factor can be successfully modeled according to a two-body interaction and under the Random Phase Approximation [6, 11, 12]. The interaction potential $U(r)$ includes a hard sphere potential ($U_{HS}(r)$), a coulombic screened potential ($U_C(r)$) and an attractive Yukawian term ($U_a(r)$) [4, 13]:

$$U(r) = U_{HS}(r) + U_C(r) + U_a(r). \qquad (2)$$

U_{HS} is given by

$$U_{HS} = \begin{cases} +\infty & r < \sigma \\ 0 & r > \sigma \end{cases} \qquad (3)$$

with σ being the diameter of the particle. U_C is the screeened Coulomb potential:

$$U_C(r) = \frac{Z^2 e^2}{\epsilon(1 + \frac{\kappa_D \sigma}{2})^2} \frac{exp[-\kappa_D(r - \sigma)]}{r}, \qquad (4)$$

where Ze is the protein charge, ϵ is the dielectric constant of the acqueous solution and κ_D is the reciprocal Debye-Huckel screening length that determines the range of the interaction. The attractive potential is given by:

$$U_a(r) = -J\sigma \frac{exp[-(r-\sigma)/d]}{r}, \qquad (5)$$

where J is the depth of attractive potential at contact ($r = \sigma$) and d, its range. The resulting analytical function, obtained via the Fourier transform of $U_a(r) \rightarrow U_{aY}(Q)$ and of $U_C(r) \rightarrow U_{CY}(Q)$, is:

$$S(Q) = S_0(Q)[1 + \beta n_p S_0(Q)\{U_{aY}(Q) + U_{CY}(Q)\}]^{-1}, \qquad (6)$$

where $\beta = 1/k_B T$ and $S_0(Q)$ is the solution of the Ornstein-Zernike integral equations of the liquid state theory under the mean spherical approximation of the hard sphere potential [14] :

$$S_0(Q) = \left\{1 - \frac{12\eta[\eta(3-\eta^2)-2]}{(1-\eta)^4}\frac{j_1(\sigma Q)}{\sigma Q}\right\}^{-1}, \qquad (7)$$

being $\eta = \frac{4\pi}{3}(\sigma/2)^3 n_p$ the volume fraction of the hard spheres, and $j_1(x)$ the 1st-order spherical Bessel function.

The form factor $P(Q)$ of monodisperse, randomly oriented particles in solution whose crystallographic structure is known, can be evaluated taking into account effects due to the hydration layer, which is quite important in the case of proteins in solution, because water molecules at protein surface result to be more dense in respect to bulk molecules [15]. Hence three different scattering domains have to be considered: protein (p), local domain (d) and bulk (b):

$$P(Q) = [(\rho_p - \rho_b)^2 V_p^2 P_{pp}(Q) + (\rho_d - \rho_b)^2 V_d^2 P_{dd}(Q) + \qquad (8)$$
$$2(\rho_p - \rho_b)(\rho_d - \rho_b)V_d V_p P_{pd}(Q)]$$

where ρ_p, ρ_d and ρ_b refer to each domain scattering length density, V_p is the protein scattering volume and V_d is the local domain volume, define as a protein shell of thickness R_d (see Fig. 1). The difference between the scattering length density of the solute and the scattering length density of the solvent is usually define as contrast. In this case, two different contrasts determine the form factor expression: the contrast between the protein and the bulk solution and the one between the local domain and the bulk. Since hydrogen and deuterium cross sections are very distinct and different in sign, it is possible to modify the contrast in neutron scattering experiments using samples at different deuterium grade. It follows that two protein solutions prepared with the same amount of cosolvent and the same protein content, but with different deuteration grades, will provide different contrasts between protein and bulk and between bulk and the local domain. Hence the possibility to study the same system at different deuterium content enables to focus step by step on different features of the system. Further, because ordinary hydrogen has a large incoherent neutron cross section, which is nil for D, the substitution of hydrogen atoms for deuterium atoms reduces the incoherent scattering background.

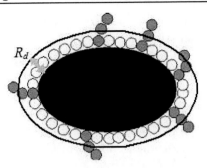

Figure 1. Pattern of a protein in solution: the black ellipsoid represents a protein, white circles water molecules and dark grey circles cosolvent molecules. Dashed ellipsoid represents the local domain.

Because the study of proteins dissolved in binary mixtures is still an interesting issue under debate, we will consider here after proteins in aqueous solutions in presence of a cosolvent.

The scattering length density of a protein can be calculated for each deuterium percentage in solution, considering the number of exchangeable hydrogen atoms present in the molecule, according to Jacrot's method[16]. On the other side, the local domain scattering length density can be estimated as:

$$\rho_d = \frac{n_c^d v_c \rho_c + n_w^d v_w^d \rho_w}{V_d},$$

(9)

where n_w^d and n_c^d are the number of water and cosolvent molecules owing to the local domain, respectively, v_c is cosolvent molecular volume, v_w^d is the volume occupied by a water molecule owing to the local domain and ρ_c and ρ_w are cosolvent and water scattering length densities. Bulk scattering length density has to be calculated considering that the number of water and cosolvent molecules in the bulk solvent are simple functions of n_w^d, n_c^d, protein concentration and the known number of water and cosolvent moles in solution.

The partial form factor $P_{pd}(Q)$ is the isotropic Fourier transform of the probability $p_{pd}(r)$ (normalised to the unit integral) that a vector of length r has one end inside the protein p and the other end in the local domain d[17]. $P_{pp}(Q)$ is the averaged squared form factor of the protein in vacuo evaluated from the atomic coordinates contained in the Brookhaven Data Bank, $P_{dd}(Q)$ is the form factor of the local domain. Both V_p and V_d can be estimated by a Montecarlo approach[17]; V_p is define by the envelope of the van der Waals spheres centered on each atom and V_d is calculated according to the fi ed thickness R_d (see Fig. 1).

This approach can be applied to the study of proteins dissolved in binary mixtures as well to the study of proteins in water, considering that in the local domain denser water molecules[15] are present.

The thermodynamic model and the global fit. In order to trace an unique interpretation of the different experiments, comprehending several protein and cosolvent concentration conditions in solution, it is necessary to look for an unique model able to describe the mixed solvent exchange on protein surface.

According to the mixed solvent exchange model discussed by Schellman [7], the protein hydration process can be described as a thermodynamic equilibrium, in which solvent molecules in contact with the protein surface slide and exchange with solvent molecules in the bulk (one cosolvent molecule replaces just one water molecule, and *viceversa*). This thermodynamic model does not consider any effects due to counter-ions specificall bound to the protein surface. Therefore, as the exchange mechanism is 1:1, the protein surface hosts m sites, which can be occupied either by water or by co-solvent molecules and none of them can be unoccupied. In the framework of this model, the local domain corresponds to the region where all the molecules of water and co-solvent in contact with the protein are located, and includes the solvent molecules fillin the gaps between the protein bound co-solvent molecules (see Fig. 4 in [4]). The water molar fraction in this region is indicated as $x_{w,l}$ (see Eq. 11 in ref. [4]).

Accordingly, the process of cosolvent and water exchange over the m sites of the protein surface can be described as: $c_s + w_b \, c_b + w_s$ where c_s, c_b and w_s, w_b represent cosolvent, c, and water, w, molecules in the bulk phase, b, and in contact with the protein surface s, respectively. Note that while the local domain d is define as the protein shell of thickness R_d (see Fig. 1), the domain s contains all the molecules (water and cosolvent) in contact with the protein surface, hence it is not an uniform shell of the protein.

The thermodynamic equilibrium can be translated into the equation $K = \frac{x_{w,s}}{1-x_{w,s}} \frac{1-x_{w,b}}{x_{w,b}}$ where $x_{w,s}$ and $x_{w,b}$ are the molar fraction of water in direct contact with protein surface and in the bulk, respectively.

Therefore, the thermodynamic equilibrium constant K is the only common observable for all the SANS investigated experimental conditions: the constant is responsible for the process of solvent exchange between the bulk and the local domain. It follows that the thermodynamic equilibrium constant K determines the composition of the bulk and the local domain. This allows the calculation of the scattering length density of each domain and hence the SANS macroscopical differential cross section. Thus, all the SANS curves can be simultaneously analysed by fittin the value of the constant K.

To calculate the scattering length density of the local domain (l) and of the bulk phase (b) at each experimental condition, it is necessary to estimate the water and cosolvent partial molecular volumes $\nu_{w,i}$ and $\nu_{c,i}$ ($i = l, b$), respectively. The cosolvent and water volume as a function of solvent composition, can be obtained from available density data of water-cosolvent mixtures [18] and assuming that the cosolvent volume is the same both in the bulk and in the local domain ($\nu_{c,b} = \nu_{c,l}$). The water partial molecular volume in contact with protein surface ($\nu_{w,s}$) is instead enabled to be different from the water volume in the bulk phase ($\nu_{w,b}$), as already suggested in literature [4, 15]. This feature, which results as a variation of mass and scattering length density of water in the local domain, can also account for the presence of counter-ions eventually bound to the protein surface.

Sample Preparation

To perform a SANS experiment, sample preparation is quite simple. In general, protein solutions at weight concentrations c between $\simeq 5$ and $\simeq 150 \, \text{g L}^{-1}$ are prepared by dissolving the requested amount of protein powder in a certain buffer. Several values of cosolvent concentrations in buffer can be chosen, starting from 0 M, with the intent to study the protein before the cosolvent can induce structural modificatio to the protein itself. By using proper amounts of light and heavy water, as well as light and deuterated cosolvent, the sample deuteration grade can be fi ed at a certain value of x_D, according to the number of exchangeable hydrogens in the cosolvent. Hence, different deuteration grades and equal sample compositions for what concerns protein and cosolvent amounts, can be chosen. This strategy provides the possibility to perform experiments on samples with the same physical chemical features and different contrast between the domains (see Fig. 1).

Results

According to the theoretical approach here described, it was possible to perform SANS experiments on proteins in solution in presence of both stabilizing and denaturing agents. A set of SANS experimental curves regarding lysozyme with one of the most common denaturants, urea, and their theoretical fit are shown in Fig. 2. It can be appreciated that the signal to noise ratio is satisfactory and the theoretical fit reproduce the experimental data.

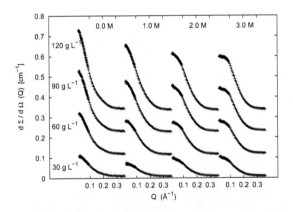

Figure 2. Experimental SANS data and fittin curves referring to samples with different urea contents and protein concentrations, as indicated on the top and on the left, respectively. For the sake of clarity, one experimental point every two has been shown, and each curve has been scaled by a factor multiple of $0.1 \ cm^{-1}$.

In consideration of the wide and recent debate about the possible presence of protein clusters in solution that may affect Small Angle Scattering spectra at low Q values [19–22], it has to be stressed that even at the higher protein concentration we investigated, there is no evidence of protein clusters, as it is discussed in Ref. [6].

Since the aim of this approach is to determine both the local domain composition and to understand protein-protein interactions in different experimental conditions, the global fi strategy [4, 7] has been adopted in the two cases. The main advantage of this kind of procedure mainly consists of reducing the number of fittin parameters for a set of exper-

imental curves. Also the global fi needs to be based on a model, hence it can validate
or not its reliability. According to the model, the global fi procedure considers the local
domain composition strictly dependent on experimental conditions (protein concentration
and cosolvent molar fraction in solution) via the unique thermodynamic exchange constant
K. Hence, a set of parameters common to all the experimental curves and only a few parameters referred to a single experimental curve have been used. In particular, in both the
experiments the thermodynamic constant K, the local domain thickness R_d, the protein
volume V_p and the water partial molecular volume at protein surface $\nu_{w,s}$, were considered
common to all the experimental curves. On the other side, the parameters related to protein-
protein interactions were considered common to all the investigated sample conditions or to
linearly vary with water molar fraction in the local domain according to the different cosolvent. In particular, the presence of cosolvent is expected to modify the attractive potential
[23], hence we considered the depth J to linearly vary with water molar fraction in the local
domain and to be independent on protein concentration.

**Table 1. Common fittin parameters obtained by SANS data analysis of lysozyme
dissolved in glycerol-water mixtures [4] and in urea-water mixtures [6]. Symbols as
in the text.**

	V_p ($\mathring{A}^3$)	R_d ($\mathring{A}$)	$\nu_{w,s}$ ($\mathring{A}^3$)	K
Glycerol	17060 ± 100	5.83 ± 0.04	28.81 ± 0.04	1.87 ± 0.03
Urea	16950 ± 100	3.8 ± 0.4	27.5 ± 0.5	0.52 ± 0.08

Looking at the fittin curves superimposed to experimental data in Fig. 2, it is evident
that the quality of the performed analysis is quite good. Common parameters resulting
from the global fi are reported in Table 1. The thermodynamic constant result to be $K =
0.52 \pm 0.08$ in the case of urea, which indicates that the local domain is enriched in urea
with respect to bulk, and $K = 1.87 \pm 0.03$ in the case of glycerol, which indicates that
the local domain is enriched in water with respect to the composition of the bulk. These
results provide the estimation of water molar fraction in the local domain at each water
molar fraction in solution, as reported in Fig. 3.

These results are very interesting, as they fully agree with previous results obtained at
infinit protein dilution (see the comparison in Ref.[4, 6]). Hence, this approach is able to
extend the literature results[24, 25] about protein solvation to finit protein concentrations,
approaching a more *in vivo* like situation.

A recent study demostrated that it is possible to quantitatively determine the number of
additional cosolvent molecules that are preferentially recruited during the a protein unfolding transition by a wise combination of SAXS and SANS experiments, with a modelfree
approach [26]. However, this approach can be applied just to specifi sample conditions and
cannot be easily extended to both stabilyzing and denaturing cosolvents as the one above
described, based on a thermodynamic exchange constant.

Also, the thermodynamic constant K is the main result of this new approach, but the
other fittin parameters provide interesting information. In particular, the fitte protein

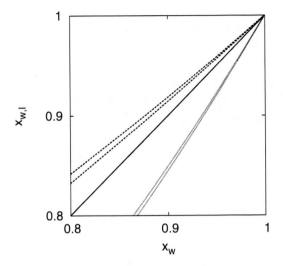

Figure 3. Water molar fraction in the local domain as a function of water nominal content: continuous line represent the values corresponding to the ideal case $K = 1$, dashed lines correspond to lysozyme dissolved in water-glycerol mixtures and dotted lines to lysozyme dissolved in water-urea mixtures. Double dashed and dotted lines refer to the lower and higher protein concentrations.

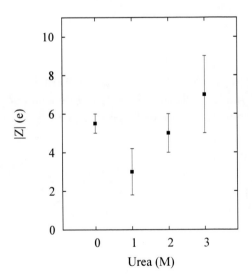

Figure 4. Modulus of the lysozyme effective charge determined from SANS data analysis. Values here reported are the average of the effective charges at different protein concentrations (see Ref.[6])

volume V_p results always comparable to the nominal value of 17400 Å^3[15] and the local domain thickness δ_i agrees with previous MD results [27]. The molecular volume of water in contact with the protein surface $\nu_{w,s}$ results comparable to that found in previous works [15], i.e. smaller than the water volume in the bulk. The terms in the attractive potential agree with literature data[23]. It is possible to note that the modulus of the protein charge, obtained by global fi analysis, is shown to vary as function of urea concentration in Fig. 4. Even if a practically constant value around 5 cannot be excluded, data seem to suggest a decrease towards a minimum, followed by a small increase, indicating that $|Z|$ could be modifie by the presence of urea. It could open new perspectives in the study of urea unfolding processes.

Conclusion

In a binary aqueous solvent, the description of protein solvation shell, both in terms of thickness and composition, can be obtained by SANS experiments considering a convenient combination of experimental conditions. This chapter reviews two successful examples of the application of this approach to a set of SANS experiments [4, 6]. Since a thermo-dynamic law is necessary to connect the investigated samples of each set of experiments, the application of Schellman solvent exchange model to the global fi analysis of SANS experimental curves, has been described.

The results of thess studies can deserve two types of comments: we validated several hypothesis already discussed in literature in a wider range of protein concentrations and we suggest new questions and new approaches for what concerns the proteins dissolved in binary mixtures. In particular, the thermodynamic exchange constants K derived from our experiments clearly state if a preferential binding of cosolvent to lysozyme in the whole set of protein and cosolvent concentration investigated, happens or not. Also, we note that together with cosolvent preferential binding or exclusion, protein protein interactions are affected by the presence of cosolvent. However, as the resolution is low, SANS cannot in-dicate in detail specifi binding sites and evaluate different amounts of cosolvent molecules in the surroundings of some specifi sites on protein surface.

In order to determine the mechanisms of an unfolding process, it appears necessary to study proteins at finit concentration in solution, in a in-vivo-like situation, looking for the reasons of the phenomenon both at protein-solvent interphase and at the modificatio of protein-protein interactions. Particular attention should be devoted to protein-protein interaction phenomena, that should be further investigated both via different experimental approaches and via new emerging theoretical support[28].

References

[1] Levinthal C 1968 *J. Chem. Phys.* **65** 44–45

[2] Fitter J, Gutberlet T and Katsaras J 2006 *Neutron Scattering in Biology* (Springer-Verlag Berlin, Germany: Springer) ISBN 3-540-29108-3

[3] Nierhaus K H, Wadzack J, Burkhardt N, Jnemann R, Meerwinck W, Willumeit R and Stuhrmann H B 1998 *Proc. Natl. Acad. Sci. USA* **95** 945–950

[4] Sinibaldi R, Ortore M G, Spinozzi F, Carsughi F, Frielinghaus H, Cinelli S, Onori G and Mariani P 2007 *J. Chem. Phys.* **126** 235101–9

[5] Sinibaldi R, Ortore M, Spinozzi F, Funari S S, Teixeira J and Mariani P 2008 *Eur. Biophys. J.* **37** 673–681

[6] Ortore M G, Sinibaldi R, Spinozzi F, Carsughi F, Clemens D, Bonincontro A and Mariani P 2008 *J. Phys. Chem. B* **112** 1288112887

[7] Schellmann J A 2003 *Biophys. J.* **85** 108–125

[8] Feigin L A and Svergun D I 1987 *Structure analysis by small-angle X-ray, neutron scattering.* (New York, Plenum Press)

[9] Carsughi F, D'Angelo D and Rustichelli F 1993 *Journal de Physique* **3** 515–518

[10] Hayter J B and Penfold J 1983 *J. Colloid Polym. Sci.* **261** 1022–1030

[11] Narayanan J and Liu X Y 2003 *Biophys. J.* **84** 523–532

[12] Barbosa L R S, Ortore M G, Spinozzi F, Mariani P, Bernstorff S and Itri R 2010 *Biophys. J.* **98** 147–157

[13] Javid N, Vogtt K, Krywka C, Tolan M and Winter R 2007 *Chem. Phys. Chem.* **8** 679–689

[14] Liu Y, Chen W R and Chen S H 2005 *J. Chem. Phys.* **122** 044507–13

[15] Svergun D, Richard S, Koch M H J, Sayers Z, Kuprin S and Zaccai G 1998 *Proc. Natl. Acad. Sci. USA* **95** 2267–2272

[16] Jacrot B 1976 *Reg. Prog. Phys.* **39** 911

[17] Spinozzi F, Carsughi F, Mariani P, Teixeira C V and Amaral L Q 2000 *J. Appl. Cryst.* **33** 556–559

[18] Lide D R 1996 *Handbook of Chemistry and Physics, 77th ed.* (Cleveland, Ohio: CRC Press)

[19] Stradner A, Sedgwick H, Cardinaux F, Poon W C K, Egelhaaf S U and Schurtenberger P 2004 *Nature* **432** 492–495

[20] Stradner A, Cardinaux F and Schurtenberger P 2006 *Phys. Rev. Lett.* **96** 219801

[21] Liu Y, Fratini E, Baglioni P, Chen W R, Porcar L and Chen S H 2006 *Phys. Rev. Lett.* **96** 219802

[22] Shukla A, Cola E M E D, Finet S, Timmins P, Narayanan T and Svergun D I 2008 *Proc. Natl. Acad. Sci.* **105** 5075–5080

[23] Niebuhr M and Koch M H J 2005 *Biophys. J.* **89** 1978–1983

[24] Gekko K and Timasheff S N 1981 *Biochem.* **20** 4667–4676

[25] Timasheff S N and Xie G 2003 *Biophys. Chem.* **105** 421–448

[26] Gabel F, Jensen M R, Zaccai G and Blackledge M 2009 *J. Am. Chem. Soc.* **131** 8769–8771

[27] Baynes B M and Trout B L 2003 *J. Phys. Chem. B* **107** 14058–14067

[28] Enright A J, Iliopoulos I, Kyrpides N C and Ouzounis C A 1999 *Nature* **402** 86–90

Reviewed by Raffaele Sinibaldi
Institute for Advanced Biomedical Technologies
University "G. D'Annunzio" Chieti Pescara
Italy

In: Neutron Scattering Methods and Studies
Editor: Michael J. Lyons

ISBN: 978-1-61122-521-1
© 2011 Nova Science Publishers, Inc.

Chapter 7

Neutron and X-ray Diffraction Studies of Nanoparticles Confined within Porous Media

I. V. Golosovsky[*]
Petersburg Nuclear Physics Institute,
188300, Gatchina, St. Petersburg, Russia

PACS 61.46.+w, 1.12.Ld, 61.10.Nz.

Keywords: Nanoscale materials, neutron and X-ray (synchrotron) diffraction.

1. Introduction

The practical needs and technology achievements make available large diversity of nanostructured systems. They can be arranged into several more or less definite classes. First, there are isolated nanoparticles, usually weakly interacting, sometimes immersed into any suspension (liquid), or/not covered with some coating. They are of interest from fundamental point of view as initial blocs of any nanostructured magnetic systems. From practical point of view they are of importance for medical applications. Another class is multicomponent systems. One can divide them onto two sub-classes: composite systems, where nanoparticles are embedded (synthesized) within different type of porous media, for example, within a porous glass. Another class is heterogeneous systems, layered [1, 2] or "core-shell" structures [3, 4] with properties which are engineered by controlling the microstructure and chemical profile of the layers.

All nanostructured materials can be treated as ordinary compounds, which are synthesized within artificial boundaries, in the so-called conditions of "restricted geometry". All these objects have common peculiarities. First, their dimensions are comparable to the correlation lengths of atomic, magnetic and other characteristic interactions. Such restrictions lead to a change in the nature of the phase transition, sometimes to its complete suppression. Secondly, the quantity of atoms on the surface, which exist in the conditions of violation of local symmetry, is comparable with the total number of atoms in a system that significantly

---

[*]E-mail address: golosov@mail.pnpi.spb.ru

modifies physical properties of nanostructured system. Therefore, in such systems the usual consideration when surface atoms are simply ignored, is not valid.

The subject of the presented chapter is diffraction studies of nanocomposites, materials which obtained by the chemical synthesis (or other methods) of oxides, metals and other compounds inside porous media. The study of the physical properties of confined nanoparticles has become a very active field of research during the last decade. The reason to investigate such materials is fundamental since the confined geometry and the influence of the surface yields unusual properties as compared with the bulk. Confined nanoparticles also result in new applications, for example, in the field of catalysis, high-density magnetic memory, etc.

The most common application of nanocomposite systems is the chemical catalysis. In this case, the embedded compounds, "attached" to the inner walls of porous medium, "work" as catalysts or have some specific activity. Since the yield is directly proportional to the catalyst surface, the host-matrix should have as much as possible the inner surface. A good example is the matrix MCM-41 (Mobil Company Material), first synthesized by Exxon Mobil in 1992, known as well as mesoporous molecular sieve [5, 6]. This matrix has a record value of the internal surfaces up to 1500 m^2/g, which is several times greater than the well-known microporous zeolites.

Over the past years the interest in magnetic nanoparticles greatly increased because of the potential applications in the super-dense data recording devices. It is obvious that for reliable operation of such gadgets the magnetic order should be stable over time. However, with the particle size decreasing the magnetic anisotropy energy responsible for retention of the magnetic moment becomes comparable to thermal energy. When this happens, the thermal fluctuations cause random "roll-over" of the magnetic moment. Magnetic system loses stability and becomes superparamagnetic. This is a fundamental physical constraint known as the "superparamagnetic limit" in magnetic recording media [7, 8]. Overcome this limit, which is the main obstacle to practical application, is the subject of many studies. It was reported that in the magnet, deposited on nanoporous media, it is possible to overcome the "superparamagnetic limit" [9]. Therefore the magnetic nanocomposites are also considered as potential candidates for using in the high-density recording devices. Note, that the use of magnetic nanoparticles in a device for recording information is not the only application. Nanocomposites are widely used in medicine as contrast agents in the tomographic diagnosis for targeted delivery of active drugs in the treatment of many diseases and in other cases.

Most published studies on the physical properties of compounds within the porous media is related to embedded liquids (see, e.g., [10, 11, 12]) or condensed gases [13, 14], which well wet the inner surface of porous matrices. Structural studies of the embedded nanoparticles by diffraction methods are usually limited to standard phase characterization and estimation of the size from the peak broadening. Structural researches by diffraction over a wide range of momentum transfer (so-called wide-angle diffraction) performed little. This state is caused by the difficulty of synthesizing nanoparticles in an amount sufficient for neutron diffraction. Since usually the quantity of the introduced material is negligible and the diffraction reflections are strongly broadened due to the size effect, the diffraction signal is very weak. Such studies are possible only on high-flux reactors.

In spite of large diversity of experimental methods, employed for studying of nanoparticles, diffraction methods and especially neutron scattering remains the primary tool for magnetic structure and dynamic determination. Only neutron diffraction method can produce direct information on the magnetic order, magnetic moment and magnetic phase transition. It should be noted that integral methods such as neutron small-angle scattering and neutron reflectometry, are now widely used to study multilayered magnetic structures. Nevertheless, as noted in the review of Fitzsimmons et al [15], "The unambiguous determination of the spin structure in thin films and small particles, both at their interfaces and surfaces as well as in their interior, still remains one of the most challenging experimental questions". Therefore, any systematic study of the magnetic phenomena is impossible without neutron diffraction.

Obviously, neutron and x-ray (synchrotron radiation) scattering are complementary. The difference in scattering power of neutrons and photons, the isotopic contrast make the combined use of neutron diffraction and synchrotron radiation very attractive. A very productive technique is using the neutron polarization analysis to pick out the magnetic scattering [16]. However, this method has limited luminosity and nanostructured material studies are possible only by small-angle scattering method.

2. Samples

2.1. Porous host-matrices

Modern technologies use a variety of porous host-matrices, which differ in composition and topology of pores [17]. The most popular matrix for studies of materials in the "restricted geometry" is vycor^{TM}-type glass [18, 19]. The thermal treatment of borosilicate glass results in separation into SiO_2 rich phase and B_2O_3 rich phase. After the glass has been heat treated and annealed, the boron-rich phase can be removed by leach, leaving an almost pure SiO_2 (silica) skeleton. In Figure 1 the microphotograph of porous glass is shown. Changing the parameters of thermal treatment one can fabricate glasses with different pore size [20, 21]. Pore volume is usually 20-40 % of total matrix volume.

Other matrices used are mesoporous matrices MCM-41 and SBA-15 [5, 6]. During the synthesis surface-active molecules in the presence of aluminum-silicate salts, self-assemble in "nanorods". Inorganics condensed around these "nanorods" (or micelles) can be removed during calcination, resulting in an aluminosilicate (or silicate) matrix with a system of parallel, hexagonally packed channels with different diameters 20-100 Å. Channel diameter can be tuned by changing the length of organic molecules and adding special agents to promote "swelling" of organics.

Honeycomb packing of parallel nanochannels gives rise to a system of diffraction reflections from the two-dimensional hexagonal lattice at small momentum transfers. The first reflex 10 is strong enough, and its intensity is directly dependent on the diffraction contrast between the matrix and the embedded material. This fact is widely uses in studies of physical properties and phase transitions in liquids, which fill the channels of mesoporous matrix by capillary effect [22]. In the case of neutron scattering the contrast also depends on the magnetic moment, therefore, a reflexes from the honeycomb structure can be used for studies of ferromagnetic compounds synthesized within the channels.

In our researches the matrices prepared by technology Mobil-Exxon (denoted as MCM-

41), with diameters of channels 24 and 35 Å, and a matrices prepared by the technology developed at the University of Santa Barbara (USA), known as SBA-15 [23], with diameters of channels 47, 68 and 87 Å were used. The matrices were prepared at the Laboratory of Chemical Physics, University of Paris-VI, France, (C. Alba-Simionesco). Our experiments revealed no differences in physical properties of embedded materials associated with the type of matrix MCM-41 or SBA-15. A section of MCM grain, made by electron tomography, is shown in Figure 2.

In a family of mesoporous silica matrices there is an object with surprising properties - the matrix of MCM-48 [5]. In this matrix the channels penetrating the matrix body have the symmetry of the cubic space group $Ia\bar{3}d$ [25, 26, 27]. The inner surface of the channels corresponds exactly to the so-called periodic minimal surface, forming a giroidal three-dimensional network of channels (see review [28]). Specific form of the inner surface is a result of a balance between the boundary energy, which is minimal for a surface with constant curvature and the minimum elastic energy, which is minimal on a flat surface. A simple example of such mesophase surface - a thin layer of small quantity of surfactant in a mixture of oil and water [29]. In Figure 3 a portion of the canal giroidal system is shown.

The amorphous silica matrices are very attractive for diffraction studies because they are free from Bragg reflections. Moreover, silica is an inert material with respect to the embedded compounds and has high melting temperature that, first, makes possible high-temperature synthesis, and secondly, allows us to investigate the physical properties at high temperatures, for example, high-temperature melting-freezing processes.

A synthesis of different compounds within pores is usually performed by the chemical method (so-called "bath deposition method") from a solution. For example, classical anti-ferromagnet MnO is synthesized from a manganese nitrate solution. Fusible metals can be easily crystallized from the melt under the application of external pressure about 10 kbar (Y. A. Kumzerov).

High internal surface of the matrix and good wetting of the walls ensures that the embedded oxides occupy inner voids. Indeed, the combined analysis of neutron and X-ray diffraction, electron spin resonance and magnetization measurements showed that even in the matrices, which exist in the form of powders as MCM-matrices, embedded oxides occupy only the voids. The presence of oxides on the surface of the granules was not detected.

Apart from the above matrices based on amorphous silica other porous media are often used. For example, natural minerals - chrysotile asbestos, which is quasi-single crystal $Mg_3Si_2O_5(OH)_4$ with the monoclinic space group C2/m. This mineral forms a system of parallel nanochannels with an average diameter of about 70 Å. Since this is a quasi-single crystal, it can be oriented in different ways with respect to the incident beam, gives an unique opportunity to investigate the anisotropy of physical properties of the embedded material [30]. However, in contrast to the amorphous silica asbestos gives strong diffraction reflections, which are often superimposed on the weak diffraction peaks of the implanted material. This is not the only problem. Our studies showed that a significant amount of oxide enters to the cavity, cracks and other structural defects, which complicates the interpretation of results.

Among other matrices the synthetic opals with a regular system of identical spheres with diameters $\sim$ 100-300 nm, should be mentioned. The space between them can be filled with different compounds, however, the size of voids usually appeared to be large, about

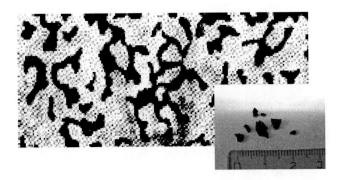

Figure 1. The fragment of a typical micrograph of pore network (in dark) is shown (from [18]). In inset the typical samples used in experiments are displayed.

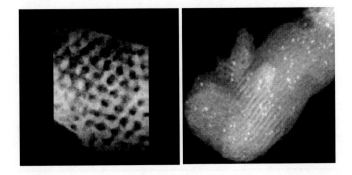

Figure 2. Channels in MCM matrix. Field of view - 100 nm, channel diameter - 3 nm. Micrographs were made using scanning electron microscopy (STEM) by HAADF (high angle annular dark field) (from [24]).

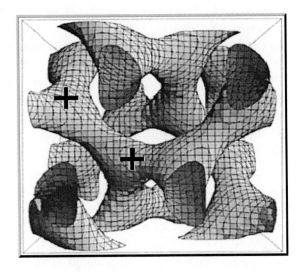

Figure 3. 3D-giroidal channel system of MCM-48 matrix (from [25]).

50-100 nm, which limits the possibility of synthesis of very small objects. Very similar in topology to the opals there are so-called nanodiamonds or carbon nanocomposites [31]. They represent the diamond nanoparticles (or carbides) with a constant size. Each particle is coated with a thin layer of graphite-like material. Between the nanoparticles the voids of complex shape remain, which, in principle, can be filled.

In recent years the matrices based on porous aluminum oxide (alumina) with a regular system of channels, a similar matrix type MCM become popular. These matrices are especially attractive because there is a developed technology that allows to create metal nanowires inside the channels by electroplating method [32] or nanowires from semiconductors, for example, CdTe [33]. However, for diffraction analysis by the classical method the amount of embedded metal is insufficient, because the matrices are thin films.

Many other porous matrices are known: zeolites, porous aerogels, porous silicon [34], "track" membranes [35], porous polymer films and others (see review on porous matrices [17]). However, such matrices are beyond of scope the present study.

2.2. Embedded compounds

2.2.1. Antiferromagnets with NaCl crystal structure

3d-metal oxides have always attracted attention in physics research. The discovery of high temperature superconductivity, "colossal magnetoresistance" and others are connected with them. Among them the oxides MnO, CoO, NiO and FeO, with the crystal structure of NaCl, had always been the classical objects for the study of magnetism. Determination of the magnetic structure of MnO was the first neutron diffraction experiment, showed the power of neutron scattering [36].

Collinear magnetic structure in these oxides consists of alternating ferromagnetic layers, which are antiferromagnetic with respect to each other, i.e., their configuration symmetry is similar. However, the direction of the magnetic moments with respect to the crystallographic axes is different. In MnO the magnetic moments lie in the plane (111) [37, 38, 39], in NiO and FeO - perpendicular to the plane (111) [40, 38], and in CoO, they are inclined to this plane at angle about 10^0 [37, 38, 41].

For real studies only two oxides, MnO and CoO are suitable. Indeed, stoichiometric FeO is unstable [40], while NiO has too high Néel temperature (T_N = 520 K).

Since opening in 1949, the antiferromagnetic ordering by neutron scattering manganese oxide attracts great interest of both experimenters and theorists. He was also very suitable for studies of magnetism in the confined geometry. First, this oxide has a simple antiferromagnetic structure, for which magnetic and nuclear Bragg reflections are separated. Secondly, MnO good wets of amorphous silica and can be easily synthesized in the voids. Thus, one can introduce a large amount of oxide sufficient for neutron scattering research. Thirdly, manganese has a negative nuclear scattering length, while the oxygen has a positive scattering length. This provides a good contrast and allows controlling the stoichiometry of the oxide. Fourth, the ion Mn^{2+} has a large magnetic moment of about 5 μ_B/ion. Finally, the bulk MnO is well explored.

The magnetic moments in the oxides with face centered cubic (fcc) lattice are frustrated in the first coordinate sphere. Therefore, the magnetic order is stabilized with structural distortion that occur at the Néel temperature. In MnO, NiO and FeO the rhombohedral (trigonal) distortion of the crystal lattice, in CoO - the tetragonal distortion, accompanied by a weak monoclinic distortion [42] were observed. Different crystal symmetry leads to different behavior of the order parameter: in MnO, NiO and FeO magnetic order appears discontinuous first-order transition, in CoO a continuous magnetic transition was observed [43, 44].

It should be noted that the nanostructured magnetic oxides of transition metals, especially CoO, have long been objects of intensive investigations in connection with the "superparamagnetic limit" in magnetic memory devices. CoO was used as the antiferromagnetic layer in multilayer heterostructures: Fe_3O_4/CoO [45], CoO/NiO [46], CoO/SiO_2 [47, 48], and in the form of single-layer films of CoO[49].

2.2.2. Ferrimagnetic ferrous oxides

Classic magnetic iron oxides magnetite (Fe_3O_4), maghemit (γ-Fe_2O_3) and hematite (α-Fe_2O_3) are known since ancient times and are widely distributed[1]. They are present in the nature: in bacteria, mollusks, even in birds and fish, which are believed to serve for the purposes of navigation. Iron oxides are responsible for the paleomagnetism of rocks. Today, iron oxides have very wide practical application: in the magnetic tapes, computer disks and other magnetic storage media ultra-thin needle-like particles of maghemite used (see the survey [51]).

The emergence of new technologies gave additional impetus to fundamental research of nanostructured oxides, since the properties of these materials were studied very little. Despite the rich history of research, our understanding of the magnetic behavior of nanoparticles of iron oxides is far from be completed.

2.2.3. Metals with low melting point

For investigations of atomic motion in the confined nanoparticles the low-melting metals, as Ga, In, Bi, Se, Pb, Sn and others are very attractive, because they demonstrate high amplitude of atomic vibrations. Usually the melt wets well the inner surface of silica matrices, therefore such metals can be embedded into the porous media, usually, porous glass, in the melting state under the external pressure [52].

3. Dimensions and shape of nanoparticles within the porous matrices with different topology

Profiles of diffraction reflections from nanostructured objects within porous media are strongly broadened with respect to the instrumental line, indicating the finite size of the diffracting object. The observed broadening of peaks results from two factors: size-effect and internal stresses. However, due to different angular dependence of these factors they can be separated. In the first case the broadening of the peak in momentum space $\Delta Q(Q)$ does not depend on momentum transfer (Q), in the second case the broadening is proportional to

---

[1]Also known oxide ϵ-Fe_2O_3 c orthorhombic crystal structure [50], but it is little explored.

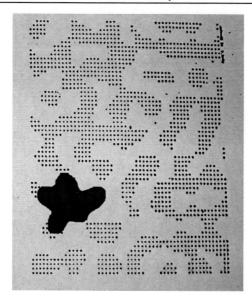

Figure 4. Schematic drawing of nanoparticles in a porous glass.

Q. Therefore, if the measurements were carried out for sufficiently large momentum transfers, the separation of contributions from the size effect and internal stress can be is easily performed.

In most cases, the profile is well described by a function known as Voigtian, which is convolution of the Lorentzian, describing the atomic order in the diffracting object, with a Gaussian instrumental line. Therefore, the analysis of the shape of the observed diffraction peaks is convenient to carried out in the approximation pseudo-Voigtian, the algebraic sum of two lines with Lorentzian and Gaussian forms or in the so-called approximation Thompson-Cox-Hastings with independent variation of the contributions from Gaussian and Lorentzian line shapes [53].

Based on the parameters of the pseudo-Voigtian, in the isotropic approximation, the integral parameter, known as "breadth", proposed by J. Langford [54] can be calculated. This parameter is defined as a width of a triangle with the same amplitude and the area that diffraction reflection. After correction for instrumental resolution the diameter (size) of nanoparticle can be calculated by the Scherrer formula, with only difference that instead of the full width at half maximum (FWHM) a parameter breadth is used. It should be emphasized that in powder diffraction the volume averaged diameter can be defined only.

3.1. Dimensions and shape of nanoparticles within a porous glass

In most cases, the sizes of confined nanoparticles do not vary with temperature. Their sizes for different compounds confined to different porous glasses measured at room temperature are given in Table 1. Nanostructured particles of lead have an anisotropic shape, so there are two dimensions. Listed in the table values greatly exceed the average pore diameter. It is common for embedded nanoparticles and it manifests the simple fact that the crystallization expands at least several adjacent pores. It is known that in the case of compounds that

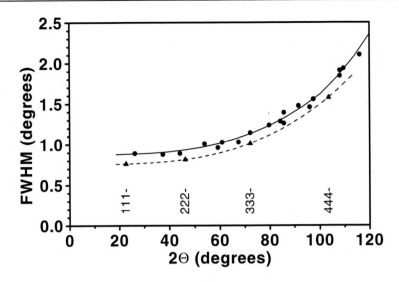

Figure 5. Full width at half maximum (FWHM). The reflections hhh type are shown by the triangles. The errors do not exceed the sizes of the symbols. The lines are guides for the eye (from [63]).

do not wet amorphous silica, such as mercury, the characteristic size of nanoparticles is close to the average pore diameter of [55]. Analysis of the shape and size of nanoparticles embedded into a porous glass shows that the physico-chemical features play crucial role, for example, a wetting of the matrix walls by the liquid (melted) compounds, from those a nanoparticle crystallizes. Therefore, the nanoparticles have a shape of interconnected aggregates rather than spherical particles (see Figure 4). However, in powder diffraction, it cannot be distinguishable.

3.2. Anisotropy of shape in Pb nanoparticles, embedded in a porous glass

From diffraction theory it is well known that the anisotropy of shape of the scattering particle leads to a systematic change in the width of the reflections. This effect was observed in the case of lead nanoparticles inside the porous glass [63].

Profile analysis (Rietveld procedure) [66, 67] shows that the observed broadening of the diffraction peaks is due to the size effect without contributions from internal stresses. In Figure 5 the full width at half maximum (FWHM) as a function of diffraction angle 2Θ is shown. It is seen that reflections 111, 222, 333 and 444 have systematically smaller FWHM, than others. This means that the average dimension of the nanoparticles along the [111] direction is larger than in the perpendicular direction, i.e., nanoparticles have a shape elongated along the [111] axis.

Calculated for all temperatures the average size of nanoparticles along and perpendicular to the axis [111] are shown in Figure 6. It is seen that, on approaching the melting point, particle size increases rapidly, showing a spreading ("percolation") lead within a connected network of pores.

Profile analysis, taking into account the uniaxial anisotropy along the crystallographic

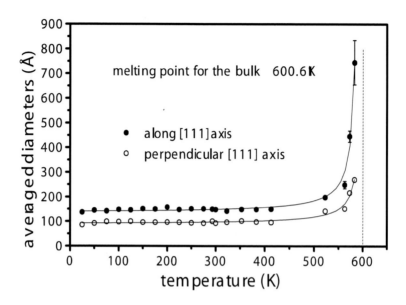

Figure 6. Averaged diameters of nanoparticles along (solid circles) and perpendicular (open circles) to the [111] direction the (pore axis). The errors do not exceed the size of the symbols, if not shown. The solid lines are guides for the eye (from [63]).

axis [111] shows that the shape of the peaks corresponding to the smaller size, is close to Lorentzian, which means the fuzzy boundary of the nanoparticle. At the same time, the peaks, which correspond to the larger size (reflection type *hhh*), are well described by Gaussian, which corresponds to a certain boundary. Indeed, the Fourier transform of the particle with a sharp edge is a product of functions of $\sin(Qx)/Qx$, where x - size, and Q - the momentum transfer [68]. However, such function, known as the interference function, in the region of small momentum transfer is well approximated by Gaussian. This gives a basis to associate the Gaussian profile with a sharp boundary.

Thus, we can conclude that the nanoparticles are elongated along the axis of the pores and a smaller diameter is determined by the pore walls. This conclusion is consistent with the results of scanning electron microscopy, [69], from which it follows that directional crystallization of nanometer-sized lead particles in the equilibrium state occurs along the [111] axis.

3.3.   Features of diffraction from the objects inside the channel matrices.

The shape of MnO nanoparticles inside the channels of mesoporous matrices MCM-41 and SBA-15 appeared to be very different from that of nanoparticles in porous glass.

Already first neutron diffraction experiments with nanoparticles of MnO, synthesized within the nanochannels showed that shape of diffraction line of nuclear reflections in some cases has a specific "saw-tooth" asymmetric profile, characteristic of diffraction from two-dimensional structure. This was clearly demonstrated in x-ray diffraction studies using synchrotron radiation (Figure 7a).

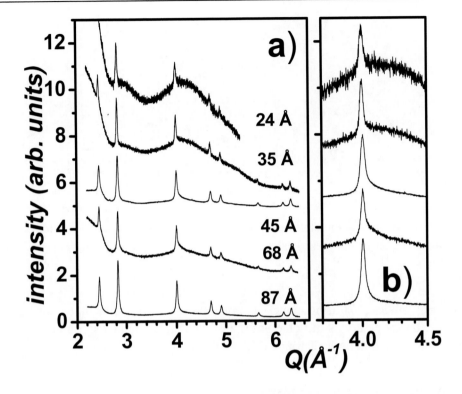

Figure 7. a) X-ray patterns of MnO within the nanochannels (scaled); b) reflection 220 in an enlarged scale.

In matrices with the large channel diameters (47-87 Å) nuclear peaks were asymmetric, whereas in the matrices with the smaller channel diameters (24 and 35 Å) peak shape was symmetrical. However, further studies showed that the channel diameter is not a critical factor. In the samples with small diameter of the channel, but with a greater degree of filling of oxide a line shape appeared to be the same "saw-tooth" profile as in the matrices with large diameters.

The asymmetric shape of the line has a fast rise at the smaller diffraction angles, to the left of the maximum, and a long "tail" on the large angles. This profile is due to two-dimensional periodicity and is well known in the diffraction of the carbon nanotubes [70], layers of graphite or thin layers of various adsorbates on graphite [71, 72, 73]. In Figure 8 the profile calculated for an ideal two-dimensional lattice [74] is shown.

Diffraction from two-dimensional objects inside the nanochannels has many similar features with the diffraction from carbon nanotubes, which can be regarded as the folded two-dimensional objects. Most publications on diffraction from complex objects can be divided into two groups. First, work on electron diffraction. These experiments are characterized by small momentum transfers, so one can use direct methods of calculation [75]. Direct calculations of the x-ray diffraction were used to describe the diffraction from zeolite nanocrystals with a complex unit cell, which includes a large number of atoms. However, in this case to calculate a structure factor a numerical integration of the interference function in reciprocal space was used [76].

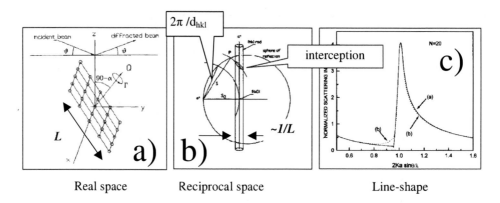

Real space   Reciprocal space   Line-shape

Figure 8. a) Two-dimensional lattice, b) one of the rods in reciprocal space and its intercep-
tion with the scattering sphere, c) calculated profile of the diffraction line (from [74]).

Another large group of works is based on the classical theory of diffraction by low-
dimensional lattices [77] using numerical integration. These works on carbon nanotubes
[70, 78], monatomic layers of noble gases adsorbed on a graphite substrate [71, 72, 73],
nanoparticles with randomly distributed planar defects [79].

3.4. Analysis of diffraction profile

In the pioneering work of Warren [77] it is shown that the diffraction from two-dimensional
grating has features that lead to the specific form of the line. Indeed, the Fourier transform
of the ideal two-dimensional lattice (Figure 8) in reciprocal space produces an ensemble
of endless rods, whose thickness is inversely proportional to the lattice characteristic size.
Averaging over all possible orientations of the scattering vector leads that the measured
intensity is proportional to the area of intersection of the scattering sphere with the rod.
Obviously, the minimum intensity corresponds to the contact of the sphere and of the rod.
With further increase of the scattering vector the area of intersection is growing rapidly,
then slowly falls down.

There are two specific features. First, there is an asymmetric "saw-tooth" profile, which
is in the ideal case is described by the integral of Warren. Second, there is the shift of the
diffraction peak maximum from Bragg position to the large diffraction angles 2Θ. In Figure
9 the profile of the diffraction reflection 111 from MnO within the mesoporous matrix is
shown, as measured by neutrons with long wavelengths, where the shift of the peak can
be clearly seen. Displacement effect, which we shall call as "Warren shift", leads to an
"effective" unit cell parameter, which is less than the parameter of the cell corresponding to
three-dimensional case.

Comparison of experimental profiles with profiles calculated in accordance with the theory of Warren for an ideal two-dimensional lattice, shows that the intensity in the measured
profile decreases with increasing diffraction angle is much faster due to the finite thickness
of the real structure. However, the position of the diffraction peak does not change strongly,
as this position is determined by the characteristic dimensions of two-dimensional lattice.
Therefore, to estimate the dimensions of nanoparticles, one can use the "effective" lattice

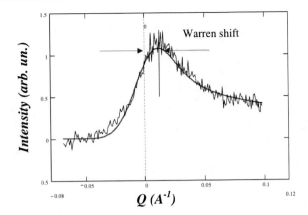

Figure 9. Fragment of neutron diffraction pattern with the 111 reflection, which has a characteristic "saw-tooth" profile.

parameter, calculated from the positions of the diffraction peak.

From the theory of diffraction from an ideal two-dimensional lattice the diffraction peak displacement is given by [77]:

$$\Delta\left(\sin\left(\theta\right)\right) = 0.16 \cdot \lambda/L, \tag{1}$$

where L - the characteristic size of a two-dimensional lattice, λ - the incident wavelength. Since the modulus of the scattering vector $Q = 4\pi\sin\Theta/\lambda$, the equation ((1) can be rewritten as $\Delta Q = 2.01/L$. Change the position of the maxima of Bragg reflection in the Q-space corresponds to a change in the lattice parameter from the true (three-dimensional) to some "effective" (two-dimensional) parameter, which can be calculated by the profile analysis (Rietveld method).

To perform profile analysis with highly asymmetric reflections, a special form of the line, the so-called "split profile "[67] was used. In this option, a peak is divided into left and right sides relative to the maximum, i.e., the position determined by a lattice with an "effective" parameter. Each part of the peak is approximated by the individual pseudo-Voigtian function whose parameters vary independently. Parameter describing the broadening of reflections due to the size effect remains the same for the two parts of the peak.

It turned out that this approximation describes of the observed X-ray diffraction quite satisfactory (Figure 10). As expected, the right side of the profile was almost a Lorentz shape, while in the left side of the profile the contribution from Lorentzian is much less. Variation of the common parameters shows that the observed peak broadening caused only by the size effect and the contribution of internal stress is negligible. No preferred orientation nanoparticles were found. This means that the samples are isotropic in the sense of a powder diffraction.

In Figure 11 the refined lattice parameters as a function of channel diameter are shown. The unit cell parameter for bulk MnO [80] (three-dimensional lattice), shown by a horizontal line was used as a reference value. For matrices with the narrow channels: 24 Å and 35 Å, the refined "effective" unit cell parameter appeared to be close to three-dimensional size, in contrast to matrices with thick channels. Thus, the difference in the asymmetry of

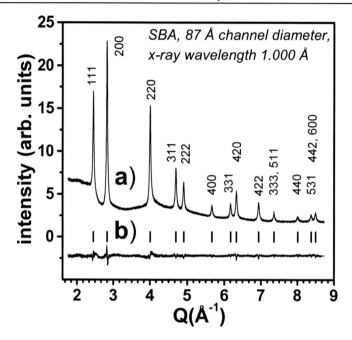

Figure 10. (a) Observed x-ray diffraction pattern of MnO confined to SBA matrix with 87 Å channel diameter; (b) difference (calculated-observed) pattern. Diffraction reflections are shown by vertical bars (from [81]).

the profiles and different "effective" lattice parameters correspond to two different forms of nanoparticles, which will henceforth be called as the "nanoribbons" and "nanowires", implemented in the matrices with larger and smaller channel diameter, respectively.

3.5.   Numerical modeling and an estimation of objects within the matrices.

There are two different approaches in modeling of diffraction from the complex objects: an analytical methods and direct numerical calculations. Direct numerical calculation of the scattering intensity from powder samples $I(q)$ based on the classical Debye formula [82, 74]:

$$I(Q) \sim \frac{1}{N} \sum_{i,j=1}^{N} A_{i,j} \frac{sin(R_{i,j}q)}{R_{i,j}q} \tag{2}$$

where $R_{i,j}$ - the distance between the i-th and j-th atoms, $A_{i,j}$ - the number of identical distances in a cell $R_{i,j}$, N- total number of all possible distances in volume and Q - momentum transfer. Such universal approach is suitable for objects of any shape and assumes that orientation of the diffracting objects is random.

Since one is interested only in the shape of the diffraction line, and not the intensity, the atomic form factor was not taken into account. To simplify the calculations, a monatomic lattice was considered. This assumption is valid in the case of x-ray diffraction, because the

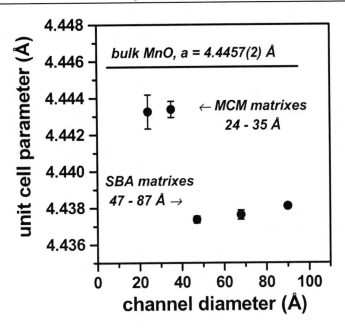

Figure 11. The "effective" unit cell parameters, refined from the profile analysis (from [81]).

large differences in charges of Mn^{2+} and O^{2-} ions, Mn contribution to the x-ray pattern is dominant.

A direct algorithm used for computing all possible distances Ri, j, includes a recursive procedure of quick sorting and calculation of $A_{i,j}$. This is a fast method and the calculation is actually limited only by available computer memory. Debye formula is often uses for numerical simulation of the diffraction profile of nanoparticles. Usually a special algorithm to increase computational speed for large nanoparticle calculation is used [83].

3.6. Two-dimensional "nanoribbons" in the matrices with the large channel diameter and an estimation of their dimensions

By comparing the calculated profiles to those observed, one can conclude that the diffracting objects are most likely consist of planar fragments in the form of strips or ribbons. Using the definition of "nanoribbons" does not assume the objects with distinct shapes; one can only speak about highly anisotropic, quasi-two-dimensional objects. Importantly, the profile analysis shows a lack of internal stresses, which means that the fragments are relatively free inside the channels.

However, the profile "smearing" is possible if one takes into account the distribution of nanoparticle sizes. However, the analysis shows that the "saw-tooth" profile quickly transforms into symmetrical profile with a slight increase in layer thickness. Numerical calculations of the profiles showed that the size dispersion has no significant effect up to $\Delta L/L \sim 30$-40 %. In the work [78] a similar result is obtained for different distributions of diameters of nanochannels.

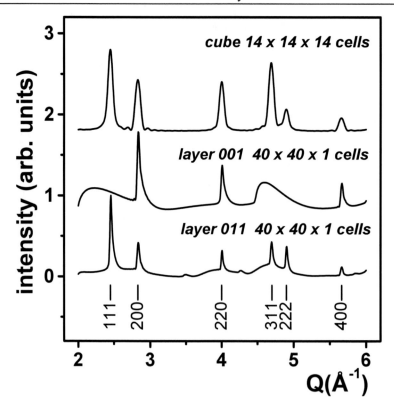

Figure 12. Numerical simulation of diffraction patterns from different diffracting objects: cube and two thin layers with different orientation with respect to the crystallographic axis. The labels "layer 001" and "layer 011" mean that the layers are perpendicular to the [001] and [011] axis, respectively. All unit cells have fcc lattice with a parameter a of 4.44 Å (from [81]).

From diffraction theory it is well known that in the case of anisotropic objects the peak broadening depends on the angle between the scattering vector and the anisotropy axis [84] and is inversely proportional to the so-called "apparent size", namely object thickness along the scattering vector [85]. If an object has a cylindrical or layered shape, the "apparent size" is different for different scattering vectors. It is evident that peak broadening is the same for all reflections for which the scattering vector has the same angle with the anisotropy axis. Such an example had been discussed above in connection with nanostructured anisotropic nanoparticles of lead inside a porous glass [63].

The effect of the "apparent size" in powder diffraction pattern demonstrates Figure 12. At the top of the figure the diffraction pattern, calculated for a cube with size $14 \times 14 \times 14$ in the units of lattice parameter (4.44 Å for MnO), which is close to that for three-dimensional crystal (the bulk), is shown. Two diffraction patterns shown below are calculated for two cases of the same layer with dimensions in the plane 40×40 of unit cell parameters and the thickness of a single unit cell parameter, but with a different orientation to the layer plane to crystallographic axes.

Diffraction patterns from the layers with different orientations significantly differ. In-

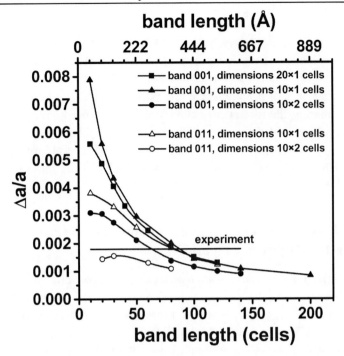

Figure 13. Warren shift: deviation of the "effective" lattice parameter from the lattice parameter of an infinite lattice, calculated for thin layers of different sizes and orientations (from [81]).

deed, for a layer with a thickness of one unit cell parameter a, oriented perpendicular to the [001] direction, the thickness of ("apparent size") along all directions of the type $\{111\}$ is equal to $a\sqrt{3}$, while the "apparent size" along [100] [010] and [001] directions are equal to $40a$, $40a$ and a, respectively. Therefore, since the peak broadening is inversely proportional to the "apparent size", a profile of reflection $\{111\}$ (the sum of all overlapping reflections of the type $\{111\}$) will be wider than the profile of reflection $\{200\}$ (the sum of two narrow 200, 020, and one wide 002 reflections).

We do not know a real contribution to the diffraction reflection from the layers with different crystallographic orientation, because we do not know the direction of nanocrystal growth inside the channels. It is therefore impossible to obtain quantitative data directly from the peak broadening. However, "Warren shift" is not very sensitive to a peak shape, and some estimates of the dimension of the diffracting object can be done by comparing the experimental measured deviation of the "effective" unit cell parameter $\Delta a/a$ from the unit cell parameter for three-dimensional object with numerically calculated deviations for model objects with different sizes and orientations.

In Figure 11, the measured "effective" unit cell parameters for "nanoribbons" inside the channels are shown. Their averaged value shown by a horizontal line in Figure 13. The intersection of this line with the calculated curves gives an estimate of the dimensions of nanoribbons. Numerical calculations were made for layers with fcc lattice, with two possible orientations of the layer with respect to the crystallographic axis (Figure 13).

From the analysis of quantities $\Delta a/a$, calculated for different layers, it follows that for the layers with equal thickness of the layer a change of the transverse dimension in the range of 20-10 unit cell parameters, which corresponds to the diameter of channels, has a small effect, if a length of the layer is more than 80 unit cell parameters. In contrast, the thickness of the layer and its orientation, i.e., the "apparent size" effects strongly. It is evident that, regardless of orientation, a layer thickness can not be more than 2 unit cell parameters. Comparing the numerical calculations with experimental data (straight line), the following rough estimations can be obtained for nanoparticles in matrices with large channels diameters: the "nanoribbons" have a thickness of several unit cell parameters (4-9 Å), width $\sim$ 10-20 unit cell parameters ($\sim$ 44-88 Å) and the length $\sim$ 60-80 unit cell parameters ($\sim$ 270-350 Å).

3.7. Estimations of the dimensions of "nanowires" in the matrices with the small diameter channels

Diffraction reflections from MnO within matrices with narrow channels do not show obvious "saw-tooth" profile (Figure 7). Since the diameters of the channels in MCM matrices are small, 24 and 35 Å, that correspond to 4-8 lattice parameters of MnO, it is natural to assume that the diffracting objects are in the form of narrow cylinders. Diffraction reflections from such objects are symmetrical, so their FWHM is a "good" parameter for the determination of the nanoparticle dimensions.

The standard treatment by profile analysis method shows that the observed peak broadening is due to size effect, without contribution of internal stresses. Averaged over the volume maximal sizes of nanoparticles were found to be 168(2) Å and 204(2) Å for matrices with 24 Å and 35 Å diameter channels, respectively. Because the diameters of the channels are much smaller the values obtained should correspond to the height of the cylinder (length of "nanowires").

These values are identical to those obtained by comparing the experimentally measured FWHM with those obtained by numerical calculation by the Debye formula. This procedure is identical to one described above for the "nanoribbons" if one replaces the FWHM values for the "Warren shift".

Numerical simulation of the diffraction line shape from thin cylinders (nanowires) shows that the "Warren shift" exists also for such objects. As expected, this parameter appeared to be very sensitive to the cylinder diameter and can be used to estimate the size. In figure 14 the value $\Delta a/a$, calculated for cylinders with different diameters, but with a height of 40 unit cell parameters is shown. Experimentally measured "Warren shift", averaged for the two MCM samples, is represented by a horizontal line with a confidence range. Its intersection with the calculated curve gives an estimation of the diameter of "nanowires" $\sim$ 4-6 unit cells (18-27 Å).

Let us consider three reflections 200, 020 and 002, each of which correspond to a different "apparent size" and contribute to the total reflection 200. If one assumes that the cylinder axis coincides with the axis [001], when the width of the reflections 200 and 020 is determined by the diameter while the width of reflection 002 is determined by the height of the cylinder. Since in our case the diameter is much smaller than the height, the resulting line shape is actually determined by the narrow line of 002. Indeed, the shape of the reflection 200, measured for MnO within the matrix MCM with a diameter of 35 Å,

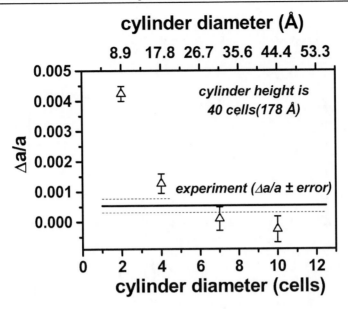

Figure 14. Deviation of the "effective" lattice parameter from the parameter of an infinite 3D-lattice $\Delta a/a$, calculated for thin cylinders (from [81]).

is well described in the analytical theory developed by Langford for powder diffraction on cylindrical objects [84] (see figure 15).

However, for very narrow cylinder Langford's formula is inapplicable. In Figure 16 some profiles calculated by the Debye formula for cylinders of different diameters are shown. Two effects are clearly seen. First, for very narrow cylinders the calculated profile has a "saw-tooth" form, as in the case of quasi-two-dimensional objects. This is an expected result, since the powder averaging in a reciprocal space is similar for a thin layer and to a thin cylinder. Second, a clearly visible broad pedestal, is determined by the diameter of the cylinder, while the narrow peak is mainly determined by the height of the cylinder. Experimental profile measured for a matrix with a diameter of Å, shown in Figure 16, better corresponds to the diameter of the cylinder in 5 unit cell parameters, or 22 Å.

The resulting diameter of the embedded nanoparticles appeared to be smaller than the diameter of the channel 35 Å, i.e., the crystallized nanoparticle does not fill the channel completely. There are two possible reasons for this effect: the roughness of the wall, estimated to be several Å and/or boundary layer of amorphous manganese oxide. Marked difference in the diffuse background in the diffraction patterns (see Figure 7), which does not coincide with the diffuse background of the unfilled matrix, confirms the presence of the amorphous phase. The existence of large amount of amorphous phase with disordered spins was confirmed in ESR experiments [86].

A similar effect was observed in studies of oxygen inside the channel matrices [11] and in a porous glass [12]. It was found an immiscibility of the amorphous layer close to pore walls, and capillary condensation in the center. Perhaps there is a similar effect, since MnO crystallized in the channels from a liquid phase. Obviously, the oxidation processes that depend on the preparation and storage methods, determine the appropriate amounts of amorphous and crystalline phases.

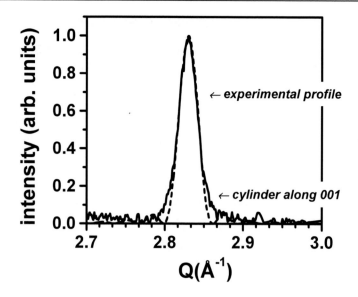

Figure 15. Experimental profile of 200 reflection (solid line) and profile calculated for cylinder with the dimensions of ⌀ 5 height of 40 cells aligned along [001] direction (dash line) from [81].

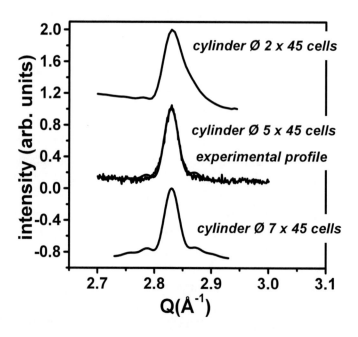

Figure 16. Simulated profiles of 200 reflection for cylinders of different diameters. At the centre, the observed profile of MnO in MCM matrix with 35 Å channel diameter is shown for comparison (from [81]).

Table 1. The average diameter of nanoparticles in porous glass, as measured by the diffraction method

	diameter nanopore (Å)	diameter nanoparticle (Å)	reference
MnO	20	94(3)	[56]
MnO	70	143(3)	[57]
MnO	80	156(5)	[56]
CoO	70	100(5)	[58]
Co_3O_4	70	91(2)	[56]
CuO	70	110(5)	[56]
γ-Fe_2O_3	70	106(2)	[59]
α-Fe_2O_3	70	157(2)	[60]
Ni	70	127(2)	[61]
Fe	70	110(5)	[56]
Se	70	183(6)	[62]
Pb	70	96(3)/142(3)	[63]
Ga	70	120-150	[56]
Bi	70	180-240	[56]
In	70	150(10)	[56]
$NaNO_2$	70	450(10)	[64]
KD_2PO_4	70	180(5)	[65]

4. Atomic order in the nanoparticles synthesized within matrices with different topology

4.1. Crystal structure and stoichiometry in the "restricted geometry"

Structural studies of the embedded compounds show a wide diversity of different cases. For example, selenium has several structural modifications [87], of which only the trigonal phase is stable, other modifications, being metastable, slowly transforming to trigonal phase [88, 89]. Indeed, just such a phase with space group $P3_121$ was found in Se, nanostructured into a porous glass [62]. A similar situation was observed with metal gallium. Crystalline Ga also has several polymorphic modifications [87]. However, from neutron and x-ray diffraction it follows that in dependence on cooling rate, at room temperature Ga is in the liquid state, nanostructured gallium crystallizes into the orthorhombic space group Cmca, well-known for massive Ga, or in the space group I4/mmm. Note, that very often the samples of Ga, as well as Bi, embedded in a porous glass appeared to be textured.

The opposite example is known, when a structure unstable under normal conditions is realized in the "restricted geometry". This is ferroelectric KD_2PO_4, nanostructured in porous glass [65].This compound in the bulk form, undergoes a phase transition from the ferroelectric phase (orthorhombic space group Fdd2) to the paraelectric phase (tetragonal space group $I\bar{4}2d$) at a temperature of 223 K [90]. However, for nanostructured KD_2PO_4 no phase transitions were observed in the temperature range 90-308 K. Moreover, the profile analysis showed that the neutron diffraction patterns are not described in the orthorhombic nor tetragonal space groups. From a comparison of the neutron diffraction patterns calculated for the possible crystal structures, it was found that the observed pattern is best suited to the monoclinic group $P2_1$ with the unit cell parameters, which are close to the parameters of the bulk. It is known that monoclinic form of KD_2PO_4 exists at room temperature [91, 92, 93, 94], however, it turns to tetragonal phase for a few days [95]. In contrast, nanostructured samples show surprising stability.

Stoichiometry of the studied oxides of 3d-metals depends on procedure of synthesis. Indeed, in MnO and CoO, the bulk examples of which are quite stable compounds, the crystal structure, type of structural distortion caused by magnetostriction effect are preserved. Stoichiometry does not change from the that in the bulk with an accuracy better than a percent.

Iron oxides are more active in the synthesis process, so the stoichiometry of the nanoparticles changes from that in the bulk. For example, a profile analysis of hematite (α-Fe_2O_3) within a porous glass shows that 4.9(1) % Fe crystallographic sites are empty, i.e., embedded hematite is nonstoichiometric [60].

A more complex situation was observed in maghemite (γ-Fe_2O_3), nanostructured within a porous glass [59]. First neutron diffraction studies showed that the embedded oxide has a cubic spinel structure. Parameter x, that determines the position of oxygen ions and the degree of distortion of the octahedra, was found to be $x = 0.258(1)$, which corresponds exactly to the inverse spinel structure.

There are two types of iron oxides with a spinel structure: maghemite (γ-Fe_2O_3) and magnetite (Fe_3O_4). It is generally accepted that maghemite with the chemical formula Fe_2O_3 has the structural formula $(Fe^{3+})[Fe^{3+}_{5/6} \square_{1/6}]_2\{O^{2-}\}_4$. In this notation, the paren-

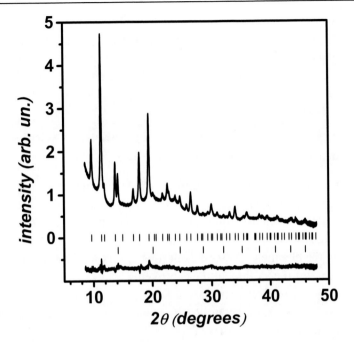

Figure 17. Observed x-ray pattern and the difference of measured and calculated patterns of nanostructured oxide γ-Fe_2O_3. Bars show the position of the reflections in the structure maghemite and in the structure of α-Fe (bottom row). Measurement on a diffractometer ID31 (ESRF) at 200 K (from [59]).

theses refer to the tetrahedral voids, and the square - to octahedral. Symbol $\square$ corresponds to vacancies. Note that the octahedral vacancies in maghemite can arrange, in this case, the crystal structure becomes tetragonal [96, 97]. Magnetite Fe_3O_4 has the structural formula $(Fe^{3+})[Fe^{3+}Fe^{2+}]\{O^{2-}\}_4$, where, unlike maghemite, Fe^{2+} ions present.

Both oxides can be converted into each other. Thus, at slow oxidation magnetite can transform to maghemite, which goes back to magnetite when heated in vacuum [98]. Although the bulk maghemite is not stable even at room temperature, the repeated measurements of nanostructured maghemite at synchrotron source showed its remarkable stability, surprising taking in mind the large total surface of the nanoparticles. On the x-ray pattern (Figure 17) apart from the lines of the cubic spinel structure there are only weak lines of α-Fe. Its volume fraction is less than 4%. Traces of hematite α-Fe_2O_3, which is often obtained in the synthesis of iron oxides were not detected. Measurements showed the absence of any additional superstructure reflections. This means that vacancies remain disordered at all temperatures.

Profile analysis of x-ray patterns, taking into account the electroneutrality shows that nanostructured maghemite has the formula $(Fe^{3+}_{0.81}\square_{0.19})[Fe^{3+}_{0.91}\square_{0.09}]_2\{O^{2-}_{0.97}\square_{0.03}\}_4$. The accuracy of occupation factors in this formula is about one percent. Unlike the bulk maghemite with vacancies only in the octahedral sites, nanostructured oxide has vacancies in tetrahedral positions as well. The lattice parameter, extrapolated to 300 K, is equal to 8.380 (1) Å, which differs markedly from the lattice parameter in the bulk 8.339 (1) Å. This difference attributed to the difference in the stoichiometry of the bulk and nanostructured maghemite.

Because Fe^{2+} ions are absent, Verwey transition at 120 K, associated with the ordering of two-valence ions [99], should absent. Indeed, the measurements of temperature dependence of resistance (Yu. Kumzerov) and structural parameters showed the absence of any phase transition.

4.2. Short-range atomic order in MnO nanoparticles within the matrix of MCM-48 with giroidal system of channels

The above examples show that the nanoparticles embedded in various porous media have a long-range atomic order. Anyway, there are Bragg reflections, which are well described in the theory of diffraction from a regular lattice, i.e., the diffraction correlation length is of order a nanoparticle size. However, it was found that in the case of nanoparticles of MnO, synthesized within the channels of the mesoporous matrix MCM-48 with giroidal channel system, the regularity of the atomic lattice is broken.

In figure 18 the system of reflexes, measured at small angles of diffraction is displayed. Indexing of these reflections corresponds to the space group Ia$\bar{3}$d, which corresponds to that known in literature for MCM-48 matrix [25, 27].

It is known that in the x-ray pattern of the carbon replica of the channel system of MCM-48 matrix 110 reflection at $Q = 0.11$ Å^{-1} was observed, which corresponds to the space group I4$_1$32 without an inversion center [100]. In experiments on matrices with embedded MnO, this reflex has not been observed [101], i.e., the inversion is present.

Profile analysis at small angles gives the unit cell parameter of the giroidal system of channels $a_0 = 79.70(5)$ Å. From the 211 peak broadening a correlation length was estimated of about 310 Å, which indicates a defect system, typical for mesophase systems [29]. While for matrix MCM-41, with the system of the parallel channels of the same diameter, the corresponding correlation length, calculated from the 10 peak broadening, is smaller, namely, 233(3) Å. It means that giroidal system is more perfect than a system with parallel channels. One can say that the higher symmetry "supports" more quality.

In figure 19 the x-ray patterns from the filled matrices of MCM-41 and MCM-48 with parallel and giroidal system of channels, respectively, and the same channel diameter of $\sim$ 33-35Å, are shown. The observed Bragg reflections are due to MnO, crystallized inside the channels, whereas the diffuse background originates from the amorphous silica and some amount of amorphous MnO.

The two patterns have some remarkable differences and similarities. First, the large difference in diffuse background is caused by the different ratio of voids to the total volume of amorphous silica, and various amount of amorphous MnO. Second, asymmetric, "sawtooth" profile of the diffraction lines observed in both samples, reflecting a two-dimensional nature of the lattice MnO, i.e., the nanoparticles have a shape of "nanoribbons".

However, there is a significant difference. In the MCM-41 matrix with a system of parallel channels the asymmetry of different reflections are different, that means the strong shape anisotropy of the nanoparticles. Profile analysis shows that the contribution to the peak broadening of internal stress is negligible. Estimation of the nanoparticle length from the peak broadening gives value of 260(4) Å.

A completely different line shape is observed in the case of the matrix MCM-48 with

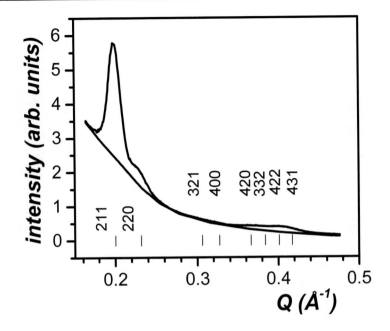

Figure 18. Low-angle x-ray diffraction pattern from the MCM-48 matrix. Bars mark the reflection positions in SG $Ia\bar{3}d$.

giroidal channel system (Figure 20b). Obviously, such a line shape with a sharp maximum can not be described by the usual pseudo-Vojtian. Comparing the "integral breadths" for MCM-41 and MCM-48 (see section 3), we can estimate the maximum length of the nanoparticles embedded within MCM-48 as 53(3) Å. The same estimate is obtained from the profile analysis of the diffraction patterns, starting from reflex 311, where the asymmetry is not so noticeable, assuming Lorentzian line shape.

It turns out that the obtained value of 53(3) Å is close to the distance between two nearest branching points in the matrix of MCM-48 (marked in Figure 3 by crosses), which in our case is $a_0/\sqrt{2} = 56.3$ Å. It means that the crystallization of MnO nanoparticles ends somewhere at these points. Thus, in contrast to other topologies in the giroidal channel system a natural limitation of the nanoparticle length exists. Indeed, since the instrumental line in the experiments with synchrotron radiation is very narrow, the presence of the sharp maximum mean a very little spread in the size of the embedded nanoparticles.

Obviously, the transverse dimension of the nanoparticles can not be larger than the diameter of the channel, namely, 33 Å in the MCM-48. Diffraction profiles for the two nanoparticles with sizes $50 \times 30 \times 20$ Å and $50 \times 30 \times 5$ Å, limiting the possible dimensions within the channels, were calculated using the Debye formula. In figure 20b, these profiles are marked in numbers 2 and 1. It is seen that the experimental profile can not be described in the standard diffraction on a regular lattice.

It can be rigorously shown that in an ideal two-dimensional lattice long-range order is impossible [102]. On the other hand, there are many examples of partially disordered structures giving rise to Bragg-like peaks in the x-ray diffraction patterns. Shining examples are the diffraction from monolayers of rare gases adsorbed on graphite [71, 73, 72]

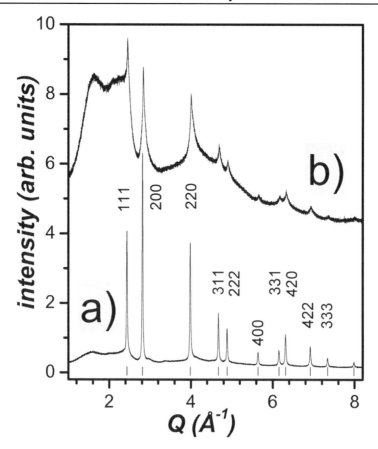

Figure 19. X-ray diffraction patterns of MnO confined within the matrices MCM-41 (a) and MCM-48 (b), (in the last case the intensity was multiplied by a factor 6) measured at room temperature. Bars mark the reflection positions (from [101]).

or the diffraction from a smectic liquid crystal [103], where the diffraction line shape is described in the frame of phonon mechanism, which destroys the atomic periodicity in the low-dimensional lattices [104].

For such a mechanism the pair correlation function can be described as [105] $\langle \mathbf{U}(\mathbf{R}), \mathbf{U}(0) \rangle \sim R^{-\eta}$. Here $\mathbf{U}(\mathbf{R})$ is the deviation of an atomic position from its average lattice position $\mathbf{R}$. Such a dependence, known as the algebraic decay of correlations, leads to the scattering function $S(\mathbf{Q}) \sim |\mathbf{Q} - \mathbf{q}|^{-2+\eta}$, which well describes the peak "tails" in the vicinity of a Bragg reflection with a momentum transfer $\mathbf{q}$, while the "divergence" at $\mathbf{Q} = \mathbf{q}$ is removed by the finite-size effect [72, 104].

The scattering function $S(\mathbf{Q})$, measured in powder diffraction experiment results from the power averaging of all possible orientations of the vector $\mathbf{q}$. This averaging depends on the dimensionality of diffracting objects. However in the case of the algebraic decay of correlations, the power law still plausibly fits the tails of the line shape [72].

Since the phonon mechanism is temperature dependent, the line shape is expected to evolve with temperature. However we did not observe any change of the line shape in the temperature interval 10 - 300 K. A fit of the experimental profile with a power law (after 2D

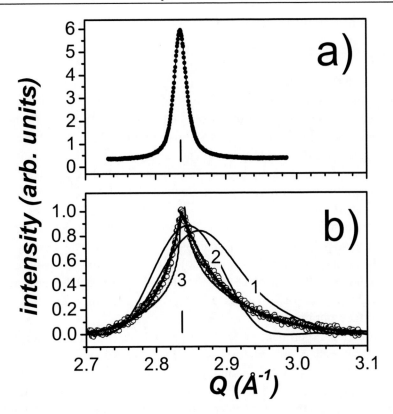

Figure 20. (a) Profile of the 200 reflection from MnO confined within the matrix MCM-41 measured at 130 K. Thin solid line corresponds to a fit with pseudo-Voigtian. (b) Profiles of the 200 reflection from MnO confined within the matrix MCM-48 measured at 130 K: open circles - experiment, solid line through the experimental points is a fit with an exponential law; solid lines: (1) - a numerical calculation for the ribbon of $50 \times 30 \times 5$ Å; (2) - a numerical calculation for the ribbon of $50 \times 30 \times 20$ Å, (3) - a fit with a power law. Bar marks the reflection position (from [101]).

powder averaging) or by the sum of two profiles resulting from the power law line shape and the Lorentzian line shape, which correspond to partially ordered and fully disordered phases, respectively, as was proposed in ref. [73, 72] was not successful.

Surprisingly, an excellent fit of $S(Q)$ around the Bragg position can be achieved with an exponential law, namely, $S(Q) \sim \exp(-|Q - q| \cdot \eta)$ (the thick line through the experimental points in Figure 20b). Different values of η at $Q < q$ and $Q > q$ were used, namely 21.86 Å and 9.5 Å, respectively, to take into consideration the peak asymmetry due to the ribbon-like form of nanoparticles.

Two important conclusions follow: first, the embedded anisotropic nanoparticles are so small that they cannot be considered as objects with a long-range ordered lattice. The results shown suggest that short-range positional order is an intrinsic feature of small confined nanoparticles. Note that the presence of extended structural defects in the nanoparticles of gold with 3 nm size suspended in water was confirmed by x-ray diffraction and computer simulation [106].

4.3. Amorphous component of the embedded compounds

4.3.1. Diffuse background and its constituents

Enormous inner surface of the embedded nanoparticles leads to new features, which absence in the bulk. For example, because one cannot avoid the presence of air into the nanovoids, even when samples are stored in an inert gas, crystallized MnO gradually transforms into amorphous one. This process is almost unnoticeable in the case of porous glass, but in the matrices of the channel type almost half of crystallized MnO after about of 2 years appears to be amorphous. It should be noted that in the "restricted geometry" MnO oxidation occurs directly to an amorphous phase, in contrast to the bulk, when an intermediate phase Mn_3O_4 occurs [107].

Another situation was observed in the case of nanostructured CoO, which after about a year, without special precautions, almost completely oxidized to stoichiometric Co_3O_4, (in notation for the spinel structure $(Co)_{0.92(1)}[Co]_2O_4$), without any formation of amorphous phase. Sometimes, as in the case of Se (which will be discussed below), an amorphous component appears by natural way. In all cases, the process of amorphization in the nanovoids due to large surface amplifies.

In the diffraction experiment crystallized part of the embedded compound gives the characteristic Bragg reflections, whereas as a diffuse background is composed of two components: diffuse scattering on amorphous silica and diffuse scattering on amorphous component of an embedded compound. For example, for CoO (Figure 21) the total diffuse scattering pattern repeats the pattern on amorphous silica, as measured on a unfilled matrix [58]. Consequently, one can conclude that there is no amorphous CoO in the nanopores. A similar pattern is observed in the case of MnO, but only within the porous glass. However, in the case of MnO within the channel-type matrices a fraction of the amorphous component is significant, especially for matrices with small diameter [86].

It should be noted that the synthesis of crystalline compounds into porous media is not always possible. Very often it turns out that only an amorphous fraction obtains. For example, it was reported that for Co_3O_4 in MCM-41 matrix an antiferromagnetic transition is suppressed, a long-range order is absent and the magnetic system shows spin-glass-like characteristics [108].

Diffuse scattering on the amorphous matrix at low momentum transfer in the simplest case can be approximated as the sum of Debye-like functions:

$$I(Q) \sim \sum_{i=1}^{N} A_i \frac{sin(R_iQ)}{R_iQ} + AQ + B, \tag{3}$$

where R_i - Interatomic, the characteristic distance between the nearest neighbor, next-nearest atoms, etc., Q - momentum transfer, A and B - the variable parameters.

A fit shows that the calculated distance between the nearest atoms in the matrix close to the well-known in the literature distances of Si-O and O-O in silicate tetrahedra SiO_4 [109]. Temperature measurements show that these distances are practically constant, i.e., the size of the cavities in the glass matrix is independent on temperature [64]. The latter circumstance, as will be shown below, has a significant effect on the physical properties of the embedded material, if it has non-zero thermal expansion.

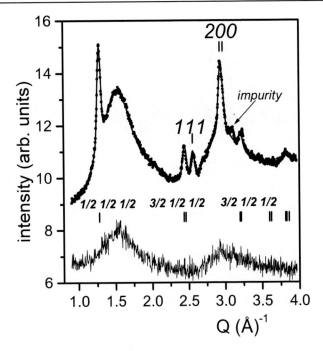

Figure 21. Neutron diffraction pattern of nanostructured CoO embedded in a porous glass: (1) the observed profile and (2) the calculated curve. Neutron diffraction pattern of porous glass without CoO is shown below in a reduced scale (from [58]).

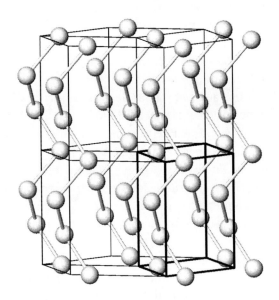

Figure 22. Hexagonal crystal structure of Se. A unit cell is shown by the solid line. Hexagonal axis is directed vertically (from [62].)

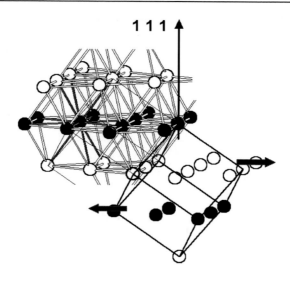

Figure 23. Magnetic structure of MnO.

4.3.2. An example of spontaneous crystallization of amorphous Se in nanopores

The crystal structure of trigonal Se consists of rigid, weakly interconnected helical chains (Figure 22). Therefore, the crystalline Se is easily transformed into amorphous state, and vice versa, keeping the short fragments of spirals (sometimes eight-membered ring) as a structural units [110]. These processes are greatly accelerated in the case of nanostructured Se and become spontaneous [62]. Profile analysis shows that Se in the nanopores, exists not only in crystalline but also in an amorphous state.

With temperature increasing, amorphous Se in the pores begins to crystallize in trigonal structure. Such formation of crystalline Se from the amorphous phase starts at about 320 K and reaches a maximum at 380-400 K. When a quantity of crystalline fraction drops sharply due to the beginning of the melting process.

Coexistence of crystalline and amorphous Se is not surprising due to the phase instability [111]. Such coexistence of disordered clusters of crystalline Se and crystalline Se was observed in the cavities of zeolite $AlPO_4$-5 using Raman spectroscopy [112]. The phase transition temperature in these experiments coincides with our observations, i.e., the phase transformation does not depend on the topology of the matrix.

The melting-freezing phase transitions of are quite complex, here the crystallized (or partially crystallized) and amorphous, molten fraction coexist. Transitions are smeared by temperature, show hysteresis, so that they can not be characterized by only one parameter, as the melting point. Moreover, the melting at the particle surface, which borders with the matrix wall, is different from melting in the "core". Obviously, the matrix and its interaction with the nanoparticle play an important role. Physics of the melting- freezing transitions in the "restricted geometry" is intensively studying and is far from a complete understanding, see, for example, work on In [113], Ga [114] and Hg [55] nanostructured into a porous glass.

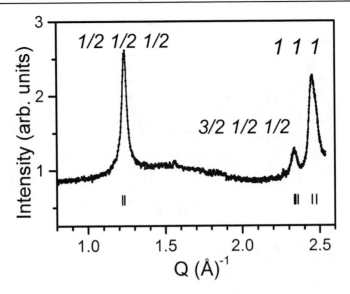

Figure 24. Observed (circles) and calculated (solid line) neutron diffraction pattern of MnO within the matrix of SBA-15 with a diameter of channel 47 Å (from [115].)

5. Magnetic ordering inside and on the surface of nanoparticles

5.1. Magnetic order and magnetic moment in oxides MnO and CoO within a porous glass

In all studied cases in oxides of manganese and cobalt confined to porous glass the Bragg reflections with half-integer indices were observed below the Neel temperature $((T_N)$, which indicate the emergence of correlated magnetic order in the embedded nanoparticles [57, 115, 58]. Observed system of the magnetic reflections corresponds to the antiferromagnetic order of type II in the face-centered cubic (fcc) lattice with wave vector $\mathbf{k} = [\frac{1}{2}\frac{1}{2}\frac{1}{2}]$, i.e., similar to the bulk [36, 38] (Figure 23).

A typical neutron diffraction pattern measured in the magnetically ordered phase at the wavelength of incident neutrons of 2.43 Å, is shown for CoO in Figure 21. Neutron diffraction pattern of MnO within the mesoporous matrix with a channel diameter of 47 Å, measured at 10 K at the wavelength of incident neutrons 4.73 Å, is shown in Figure 24. Because of the long wavelength only two magnetic reflections: $\frac{1}{2}\frac{1}{2}\frac{1}{2}$, $\frac{3}{2}\frac{1}{2}\frac{1}{2}$ and one nuclear reflection 111 can be observed. Measurements at this wavelength were used to measure the temperature dependence of the magnetic moment. Moreover, large intensity at this wave length allows accurately to determine the structural distortion caused by magnetostriction effect, which leads to the "splitting" the reflections in the ordered state [57].

From the profile analysis of neutron diffraction patterns it follow that in all cases of MnO within different porous matrices the magnetic moments lie in the plane (111). In CoO inside a porous glass the moments make an angle of 9.5(3) degrees with the plane (111). This is consistent with published data for the bulk [38, 41]. For MnO the measured average ordered magnetic moment of 3.84(4) μ_B/ion appeared to be significantly less than the moment of the free ion Mn^{2+} 5 μ_B/ion, and smaller the moment in the bulk MnO of

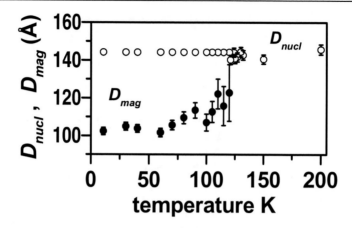

Figure 25. Temperature dependencies of the volume-averaged diameters D_{mag} and (solid circles) and D_{nucl} (open circles) for the magnetic and nuclear regions, respectively, for embedded MnO (from [57].)

4.892 μ_B/ion [116]. In CoO the average magnetic moment at 10 K was found to be 2.92(2) μ_B/ion [58], which is also much smaller than the moment of 3.80(1) μ_B/ion in the bulk [41].

Since the magnetic and nuclear reflections are separated in neutron diffraction it is easy to calculate the volume averaged diameters D_{mag} and D_{nucl} for magnetic and nuclear regions, respectively (Figure 25). It is seen, the size of magnetic cluster appeared to be much smaller than the size of the nanoparticle and this difference increases with temperature decreasing and magnetic moment increasing.

In the diffraction experiment the disordered spins on the surface do not contribute to coherent magnetic Bragg scattering. Therefore, the observed decrease of the magnetic moment and the smaller area with magnetic order, in respect with the size of nanoparticle, results from the random "canting" spins on the surface, i.e., the formation of a layer of disordered spins near the pore walls. A similar phenomenon was observed in all experiments with magnetic nanoparticles, regardless of the topology of porous media.

Assuming the observed decrease in the MnO moment is due to complete absence of magnetic order in the surface layer, the volume of this layer should be about 23 % of the volume of the nanoparticle, whereas the estimation of this layer from the difference in the diameters is of 56 %. This is a rough estimate, because the shape of the nanoparticle is actually agglomerate, even though the nanoparticle is isotropic in the sense of powder diffraction. However, it shows that spin disorder in the surface layer is incomplete and, apparently, decreases to the centre of the nanoparticle core. Anyway, the fact that the magnetic order exists in the region, less than the nanoparticle, evidences in strong magnetic inhomogeneity over the volume.

It is natural to assume the separation of magnetic area onto smaller domains. Although, it is accepted that magnetic particles with a diameter of less than 100 nm are single-domain [117], however the domain formation may be due to "bottle-necks" or other defects, naturally existed in a porous media. Indeed, the length of magnetic correlations should be close to the average distance between defects, which is smaller than the nanoparticle size.

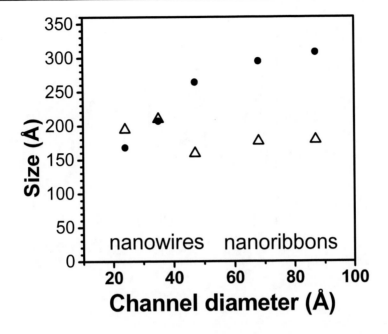

Figure 26. Lengths of the magnetic domains and the lengths of MnO nanoparticles in the channels of MCM-41 matrices with different diameters. The experimental error - order of the size of the symbol.

From the broadening of the magnetic peaks the size of magnetic domains was estimated for channel type matrices. As discussed above, the smallest size of diffracting object contributes to the "base" of the diffraction peak, while the peak width is determined mainly by the largest size. Therefore, from the peak broadening, we can only estimate the length of magnetic domains. These values are shown in Figure 26 together with lengths of nanoparticles for the MCM-41 matrices with different channel diameters. It is seen that the length of magnetic domains in the channels with different diameters are the same - about 180 AA, smaller than "nanoribbons" length. In the case of "nanowires" domain lengths coincide with the length of nanoparticles. Note, the magnetic domains formation was observed for different type of nanosystems. Recently, unusual domain fragmentation was observed in the "core-shell" nanosystems [4].

The decrease of averaged magnetic moment in magnetic nanoparticles is well-known phenomenon (see reviews [117, 118]). For the first time the surface spin disorder has been discovered in the Mössbauer experiments with ferrimagnetic γ-Fe_2O_3 [119, 120, 96]. Known numerical calculations of the magnetic moment in the small magnetic clusters, are indicating a decrease in magnetic moment [121]. ESR experiments with antiferromagnet MnO confined within a porous glass and channel type matrices demonstrate the strong signal due to disordered "surface" spins is observed below the magnetic transition, in contrast with the bulk [86].

The appearance of spin disorder and magnetic inhomogeneity in nanoparticles are fundamental phenomena, which leads to a number of unusual effects, which will be discussed below.

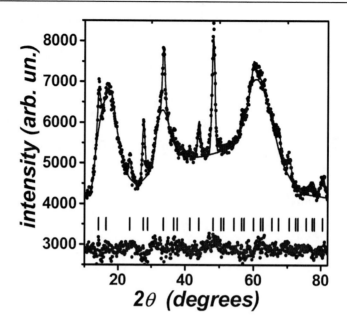

Figure 27. The neutron diffraction pattern of maghemite nanostructured in a porous glass. Observed profile (open circles) and calculated profile (solid line). Below the difference between the measured and calculated profiles is shown. Strong diffuse scattering comes from the matrix of amorphous SiO_2 (from [59]).

5.2. Magnetic moments in different crystallographic positions in maghemite γ-Fe_2O_3.

In the "restricted geometry" a noticeable part of spins exist in the conditions of the local violation of symmetry of the environment that leads to a decrease in the average magnetic moment/ion. In multi-lattice magnets this effect also depends on the geometry of a particular crystallographic sublattice. Such an example was found in maghemite - two-sublattice ferrimagnetic iron oxide γ-Fe_2O_3 with the spinel structure [59].

Profile analysis of the neutron patterns (Figure 27) showed that the magnetic structure of nanostructured oxide corresponds to the Néel model, with parallel magnetic moments belonging to each of two crystallographic sites: the tetrahedral (A) and octahedral (B), while the moments belonging to different sites are antiparallel. Néel temperature for maghemite is $T_N = 948$ K. This is an estimate, since at 573 K maghemite irreversibly transforms into hematite [51].

As expected, the magnetic moments of 3.9(1) μ_B/ion and 1.6(1)/ion μ_B for positions A and B, respectively, were measured, that is significantly smaller than μ_B/ion and 4.4(1) μ_B/ion in the bulk [122].

Surprisingly, the magnetic moments at the positions A and B are quite different. Since in both positions there is an ion Fe^3, which has no orbital angular momentum, the crystal field, which is different for the two types of positions, should not affect the magnetic moment.

About the difference between magnetic moments in the positions A and B had been reported in an earlier neutron diffraction work on $(Mn^{2+})[Fe^{3+}]_2O_4$ [123]. Because Mn^{2+}

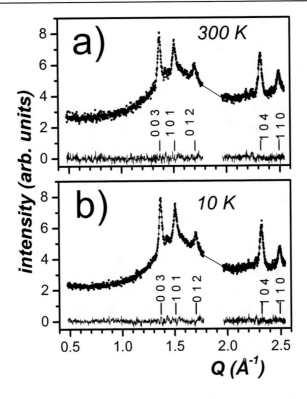

Figure 28. Neutron diffraction patterns of hematite within a porous glass at 300 K (a) and at 10 K (b). The solid line at the bottom corresponds to the difference between calculated and measured profiles. Dotted line shows an interval with parasitic reflections from cryostat (from [60]).

and Fe^{3+} ions are identical, this spinel is similar to maghemite $(Fe^{3+})[Fe^{3+}_{5/6}]_2O_4$. The paper reported that at 300 K, the magnetic moments at the positions A and B were 4.33 $m\mu_B$/ion and 3.78 μ_B/ ion, while at saturation at 4.2 K, the moments were equal to 4.6 μ_B/ion, which is close to the free ion value.

Let us consider the nearest environment of the magnetic ions at two positions. The magnetic ion in position A has 4 neighbours in the same positions with the same moment direction and 12 neighbours in positions B with the opposite spin direction. The ion in the octahedral position B has 6 neighbours with the same spin direction in positions B and 6 neighbours with the opposite spin direction in positions A. For a usual sample with the similar spin values in positions A and B, taking into account the spin directions, the exchange integrals in positions A and B are proportional to $J_A = -12J_{AB} + 4J_{AA}$ and $J_B = -6J_{BA} + 6J_{BB}$, respectively. Substituting the exchange integrals J_{AB}/k_B = -28.1, $J_{AB}/k_B = -28.1$; J_{AA}/k_B = -21.0 and J_{BB}/k_B = -8.6 proposed for maghemite in [124], one can obtain J_A/J_B = 2.19 for the ratio of the exchange integrals. Thus, the moment in position B is bound two times more weakly by the exchange interaction. The more weakly bound spin in position B should be evidently more disordered due to the breaking of local symmetry, and, as a consequence, its mean value is lower. The similar effect had been recently observed in the shell of γ-Mn_2O_3 in the core/shell system: MnO/γ-Mn_2O_3 [3].

5.3. Coexistence of two magnetic phases due to a difference of the anisotropy on the surface and in the core of hematite nanoparticles α-Fe$_2$O$_3$.

Local symmetry breaking in the spin environment on the surface, as compared with the relative ordering in the nanoparticle core leads to different local magnetic anisotropy on the surface and in the core. This explains unusual magnetic state observed in nanostructured hematite - iron oxide α-Fe$_2$O$_3$ [60].

The bulk hematite has the corundum crystal structure and presents a two-sublattice antiferromagnet with T_N = of 950 K [37]. At the temperature $T_m \sim 260$ K hematite undergoes a spin-reorientation transition, known as the "Morin transition". Below the T_m the moments in two magnetic sublattices are exactly antiparallel and aligned along the rhombohedral [111] axis (c-axis in the hexagonal setting), (AF phase). Above T_m the moments lie in the basal plane (111) with a slight canting resulting in a weak net moment originated from Dzyaloshinskii-Moriya anisotropic superexchange interaction [125] (WF phase). The spin flip is related to a competition of two terms with different temperature dependence: the magnetic dipolar interaction and the single-ion anisotropy arising from higher order spin-orbital effects, that leads to the different sign of the anisotropy constant [126] and, as result, to different spin orientations above and below Morin transition.

It is known that the stoichiometry of the nanostructured ferric oxides could substantially differ from the bulk [59]. The refinement of the x-ray pattern measured at 300 K show that the observed peak broadening is due to a size effect only without any contributions from inner stresses. No systematic peak broadening, which could be related to the anisotropic shape of the nanoparticle, was observed, as fasten happens for oxides nanoparticles fabricated by the ball-milling method [127].

The observed profile is well described by the crystal structure of corundum (space group $R\bar{3}c$) with the unit cell and the positional parameters: $a = 5.03788(5)$ Å, $c = 13.790(3)$ Å, z_{Fe} =0.35500(3) and y_O = 0.3101(3), respectively. Observed unit cell parameters turn out to be larger than those in the bulk: $a = 5.0317(1)$ Å, $c = 13.737(1)$ Å, that is consistent with the data reported for the hematite nanoparticles [127]. The measured unit cell parameter and the mean diameter of the embedded nanoparticle exactly correspond to the known empirical dependence [128]. The independent refining of the occupation factors for Fe and O ions shows the Fe occupation factor of 0.951(2) while the corresponding factor for oxygen is close to 1. Analysis of the x-ray patterns does not show any noticeable impurities.

The indexing of the magnetic reflections corresponds to the well-known antiferromagnetic ordering of the Néel type [37, 129]. The refined magnetic moment of 3.33(5) μ_B at 10 K coincides in the limits of one standard error with the moment measured at 300 K. The reduction of the averaged magnetic moments in the "restricted geometry" with respect to the free-ion value of 5 μ_B and the moment in the bulk of 4.9 μ_B [130] is not surprising and is explained by spin disorder on nanoparticle surface.

The intensity of the magnetic scattering strongly depends on the relative direction of the magnetic moment respect to the scattering vector. In particular, the intensity of the 003 magnetic reflection (the 111 reflection in the rhombohedral axis) depends on the angle θ that the sublattice magnetization makes with the rhombohedral axis, being a maximum for the $\theta = 90$ degree (WF phase) and zero for $\theta = 0$ (AF phase).

However, it turns out that the magnetic contributions into the neutron diffraction pat-

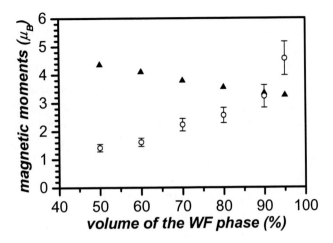

Figure 29. Magnetic moment of Fe in the AF phase (open circles) and in the WF phase (solid triangles, error bar coincides with the symbol size) calculated from the intensities of the magnetic reflection vs the volume of the WF phase. (From [60].)

terns measured at 300 K and at 10 K are similar (Figure 28a and 28b). It means that at least down to 10 K there is no any phase transition in confined nanoparticles.

Analysis of the observed intensities of the magnetic reflections shows that they substantially differ from the intensities, which correspond to the single WF or to the single AF phases. The observed patterns can be equally well described by two models, which are indistinguishable in the frame of the neutron powder diffraction. The first model assumes that the resulting moment tilts from the rhombohedral axis by an angle of 72 degrees. The alternative model assumes two magnetic phases: in one phase the magnetic moments are aligned along the rhombohedral axis, as in the bulk below the Morin transition (AF phase), and in the other phase the magnetic moments confined to the perpendicular plane, as in the bulk above the Morin transition (WF phase).

Indeed, the deviation of the magnetic moments from the rhombohedral axis was observed by the neutron diffraction in the experiments with the bulk hematite under applied pressure [131]. The observed spin deviation was explained by an increase of the higher order anisotropy terms in the free energy derived from symmetry considerations. The high-resolution x-ray powder diffraction under applied pressure confirms a progressive distortion of the hematite crystal structure signifying the increasing asymmetry of the FeO_6 octahedra, which culminated in a structural phase transition [132]. However, in all experiments with applied pressure the unit cell parameters appear to be notably smaller than in the confined nanoparticles. It means that effect of the higher order terms, which could provide the deviation of the magnetic moment is unlike for the considered case of nanoparticles.

In all reported cases of the free hematite nanoparticles [130, 133, 134] or the nanoparticles confined within a silica gel [135] with the diameters lower than ~ 20 *nm* the Morin transition is absent and the observed magnetic structure always corresponds to the single WF phase. The spin deviation from the rhombohedral axis was shown by Mössbauer spectroscopy for the only case of the strongly interacting hematite nanoparticles with 8 nm diameter. For the same particles but coated with the phosphate layer the magnetic structure corresponded to the WF phase [136].

Mössbauer spectroscopy on the free spherical hematite nanoparticles showed that the constant of the magnetic anisotropy increases by a factor of 10, when the nanoparticle size decreases from 25 *nm* to 6 *nm* [130]. Since the ratio of the surface to the total volume is inversely to a radius, such increase evidences a large positive contribution from the surface. Note that the positive sign of the anisotropy constant ensures the predominance of the AF phase at the nanoparticle surface [130]. Since the WF phase corresponds to a negative sign of the magnetic anisotropy constant the positive contribution from the surface means the strong heterogeneity of the magnetic anisotropy over the particle.

Obviously, that in the case of the tilted magnetic moments the magnetic ion environment should be uniform throughout the sample, otherwise every moment will have the particular direction like in a spin-glass system. Because the magnetic anisotropy is strongly inhomogeneous across the volume of confined nanoparticle an assumption of the co-existing magnetic phases with the different directions of the average magnetic moment "quenched" by the different local anisotropy looks more plausible. Note that for the first time the two-phase model had been supposed as the explanation of the observed small residual intensity of the 003 reflection below the Morin transition, which "could equally well arise from regions in the crystal where due to imperfections, the local anisotropy constant does not change sign" [129].

In accordance with all reported cases of the nanoparticles with diameters smaller than 20 *nm* we do not see any changes in the neutron diffraction patterns at 10-300 K temperature interval, so the Morin transition is absent. However, instead of the single WF phase, the observed magnetic intensities correspond to the two-phase (WF+AF) model.

A coexistence of the WF and the AF phases below the Morin transition, while above the transition the WF phase exists, is a known phenomenon for free spherical nanoparticles with a diameter larger than 20 *nm*. The amount of the AF phase increases with decreasing temperature, while the amount of the WF phase decreases, that reflects the strong temperature dependence of the anisotropy constant. And only for the smallest nanoparticles of 18 *nm* the Morin transition disappears and the single WF phase is observed [127].

From the diffraction experiments it is impossible to determine the magnetic moment independently, since the intensity of the magnetic scattering is proportional to the volume of the scattering object and to the magnetic moment squared. Nevertheless some conclusions could be drawn.

In Figure 29 the volume averaged values of the magnetic moments in the AF phase and the WF phase are shown as a function of the volume of the WF phase, calculated from the intensities of the magnetic reflections at 10 K. It is seen that the magnetic moment in the AF phase are smaller than in the WF phase. It is well known that the spin disordering at the surface is larger than in the core and consequently the mean magnetic moment at the surface should be smaller.

Because of the complex agglomerate shape of confined nanoparticle it is difficult to ascertain that the regions with the different local anisotropy are strictly localized. Nevertheless, the assumption that the AF phase dominates at the surface looks reasonable. Moreover it can explain why the AF phase is observed in confined nanoparticles with the shape of the agglomerate, and is not observed in the spherical particles, which has the smaller surface at the same average diameter.

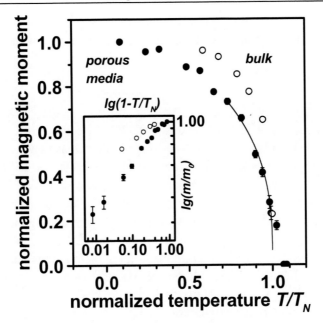

Figure 30. Temperature dependence of the scaled magnetic moment of MnO embedded in a porous glass (solid circles) and in the bulk MnO (open circles). The solid line corresponds to a fit with a power law. The moment dependencies on a logarithmic scale are shown in the inset (from [57]).

6.   Magnetic phase transitions in the "restricted geometry"

6.1.   Phase transition in nanostructured MnO and CoO

The discontinuous first order magnetic transition the bulk MnO was first explained quantitatively in the framework of the bilinear exchange interaction [137, 138, 139] as a consequence of difference in the exchange integrals between the nearest spins in the ferromagnetic layer and between nearest spins from different layers. Consideration of this issue within the phenomenological group-theoretical approach has shown that in MnO with 8-component order parameter[2] the magnetic transition should be of first order, while in CoO with magnetic moment directed at a certain angle and reducible representation described by 12 basis functions, the transition should be of second order, as indeed is observed in the experiments [43].

First neutron diffraction studies of nanostructured MnO showed that in the "restricted geometry" the magnetic transition changes from discontinuous to continuous [57]. In Figure 30 the temperature dependence of the magnetic moment measured for the nanoparticles and the bulk sample MnO [37] is shown.

Unfortunately, due to the quasielastic critical scattering it is impossible to determine exactly T_N from the temperature dependence of intensity, whereas the direct measurement of quasi-elastic scattering is strongly hindered due to weak diffraction signal. The weak

---

[2]In the magnetic structure of MnO with four-rays star of the wave vector $\mathbf{k} = [\frac{1}{2}\frac{1}{2}\frac{1}{2}]$ and the magnetic moment in the plane (111) the reducible representation is 8-dimensional.

traces of this scattering are observed up to some degrees above T_N K, that "smears" the transition. Nevertheless, an approximation of the observed dependence of the magnetic moment $m(T)$ by a power law:

$$m(T) \sim (1 - T/T_N)^\beta \qquad (4)$$

at $T/T_N ¿ 0.7$, gives $T_N = 122.0(2)$ and $\beta = 0.34(2)$. Because of the critical scattering the values obtained should be regarded as a lower limit. Measured critical exponent β is close to the values 0.362(4) and 0.326(4), obtained by computer modelling for a finite-size systems in the 3-dimensional Ising model and Heisenberg, respectively [140].

Continuous magnetic phase transition instead of the expected first order transition had been previously observed in thin films of MnO [141, 142]. Changing the nature of the transition had been attributed to a decrease in the dimension of the spin order parameter from the 8-component parameter for the bulk to the 2-component parameter in the film, where, due to strong uniaxial anisotropy, should be only one wave vector. A similar phenomenon was observed in single crystal MnO at external pressure [143], when there is only one wave vector exists and the transition becomes continuous.

In the case of porous glass with elongated pores, anisotropic interaction with the pore walls, in principle, can cause internal stresses and reduce the dimension of the spin order parameter. However, the MnO was synthesized in the pores actually *in situ*, and it is difficult to expect strong stress, as in the case of films. Indeed, profile analysis did not show any significant internal stresses. Moreover, the continuous transition was observed in MnO embedded in different porous media. Therefore, the observed change in the nature of the magnetic transition is fundamental and related to the finite size of the system.

As expected, in CoO nanostructured into a porous glass a continuous magnetic transition was observed [58] (see figure 31). Approximation of the temperature dependence of the magnetic moment with a power law in the temperature range $0.73 \div 1$ gives the value of the critical exponent $\beta = 0.31(2)$. Since there is a strong correlation between β and T_N, the search for a minimum of χ^2-functional was performed with special cautions.

The measured value of β exceeds the value of 0.25(2), measured by neutron diffraction for the bulk. CoO is traditionally regarded as a classic example of compounds with a large orbital angular momentum. Therefore, its magnetic behaviour should follow the predictions of three-dimensional Ising model with the critical exponent $\beta = 0.312$ (or 5/16), that was shown by the heat capacity measurements [144]. Birefringence measurements [145] give similar values of the critical exponent $\beta = 0.29(2)$. However, neutron diffraction on a single crystal gives lower value of the critical exponent $\beta = 0.25(2)$ [44].

It should mark a small value of the critical exponent in CoO, that it is not typical of ordinary second-order transition. In this regard, one should note the discovered decrease in the cubic symmetry to monoclinic in the bulk CoO [42]. It is possible that the magnetic transition is discontinuous, but with very little jump, which is difficult to observe experimentally. The observed critical exponent $\beta = 0.31(2)$ in nanostructured CoO is larger, compared to the bulk value $\beta = 0.25(2)$, that it is not surprising and correspond to well-known size-effect of "smearing" of the phase transition [146].

It should be bring to mind that in CoO and MnO orbital momentum contribution are different: whereas in Mn^2 orbital momentum is absent in the Co^2 it is significant and can

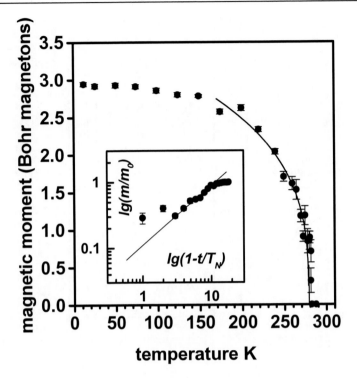

Figure 31. Temperature dependence of the magnetic moment (1) and its fitting with a power law (2) in nanostructured CoO. The inset shows the log-log dependence. The clearly pronounced deviation of two points closest to T_N is attributed to the contribution from quasi-elastic scattering (from [58]).

reach values of spin momentum [147, 148]. Obviously, in the case of CoO effects of random anisotropy on the spin disorder must be stronger than in MnO.

Strictly speaking, the singularity at the phase transitions is possible only in the thermodynamic limit, when the system is infinite [149]. If the system is finite in all directions, then the singular behaviour is impossible, as confirmed by computer simulation of phase transitions, in particularly for nanostructured systems in aerogels [150]. There is extensive literature, which shows that the continuity and "smearing" of the phase transition ("rounding") in "restricted geometry" is a common phenomenon that results from the limiting of magnetic fluctuations by the nanoparticle size [146, 151, 152]. Note that the "smearing" of the ferroelectric phase transition was observed experimentally in $NaNO_2$ embedded within a porous glass [64, 153, 154].

6.2. Evolution of magnetic phase transition in MnO, nanostructured within the channels of MCM matrices

As in the case of a porous glass, MnO nanostructured within channel type matrices has the bulk magnetic structure. The average magnetic moment of Mn does not depend on the diameter of the channel within experimental error and equal 3.98(5) μ_B/ion, which is close to the value of 3.84(4) μ_B/ion, measured for MnO in a porous glass. As well as in

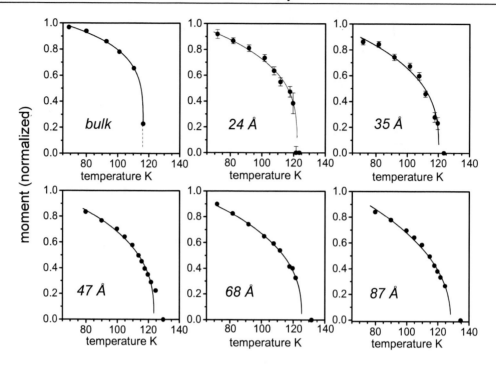

Figure 32. Temperature dependence of the normalized magnetic moment for MnO confined to the channels of different diameter. The solid line corresponds to a fit with a power law. Errors (estimated standard deviation), if not shown, do not exceed the symbol size (from [115]).

all other cases, a decrease of the magnetic moment is due to spin disorder on the surface. Temperature dependencies of the magnetic moment in MnO nanoparticles embedded in a matrices with different channel diameters are shown in Figure 32. For comparison the corresponding dependence for the bulk is also shown [37].

It is seen that the magnetic transition becomes continuous with the enhanced transition temperature T_N, respect to the bulk value $T_N = 117$ K. With decreasing diameter the phase transition character changes. In Figures 33a and 33b the dependencies of β and T_N, calculated in the approximation of a power law $m \sim (1 - T/T_N)^\beta$ on the channel diameter are displayed.

In the bulk as well as in confined MnO the magnetic transition is accompanied by a rhombohedral crystal distortion, which lifts the frustration in the first coordination sphere and stabilizes the antiferromagnetic structure. The similar distortion had been already observed in MnO confined to a porous glass [57]. From profile refinements of neutron diffraction patterns the value of crystal structure distortion measured at the lowest temperature for all studied samples appear to be the same, so it cannot be responsible for observed transformation of transition.

Since the lengths of magnetic domains in the channels of different diameter are similar (Figure 26) the observed evolution of the phase transition should be attributed to decrease of the magnetic domain diameter with channel diameter decreasing. It results in the increasing of anisotropy and in the changing of the dimensionality of the magnetic system towards to

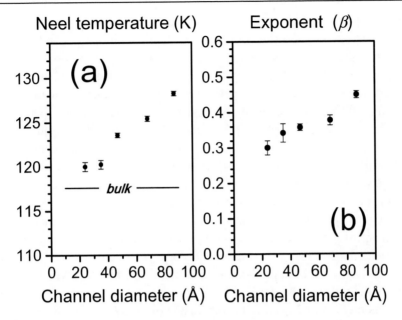

Figure 33. Dependencies of Néel temperature T_N (a) and exponent β (b) on channel diameters from the fitting with a power law (from [115]).

quasi-one-dimensional case.

Surprisingly, the measured critical exponent is changing linearly with the channel diameter, i.e. with the ratio of the volume/surface of nanoparticle. Supposing that at the surface the magnetic bonds are broken, this ratio is proportional to the number of magnetic bonds in the magneto-ordered core. Therefore the increasing the exponent β with channel diameter towards to the mean-field limit of 0.5 (Figure 33b) apparently reflects a simple fact that more and more interactions come into an action [155].

Nanoparticles confined to the large channels are expected to behave as constrained 3D-systems. However with the channel diameter decreasing one expects a crossover to one-dimensional behaviour. In this case the magnetic fluctuations should destroy the long-range magnetic order and T_N should go to zero. However, in our case T_N does not extrapolate to zero with the channel diameter decreasing (Figure 33a). Probably it is connected with high anisotropy of the embedded nanoparticles.

6.3. Temperature of magnetic transition

For MnO, nanostructured within the channel-type matrices or within a porous glass, Néel temperature depends on the size of the void, however, it always remains above the transition temperature in the bulk as well as in MnO within a porous glass. The only case described in the literature, when transition temperature in nanoparticles is higher than in the bulk, is a case of free ferromagnetic nanoparticles $MnFe_2O_4$ with the size 7.5-24.4 nm [156]. In this case the particle size increases the transition temperature decreases, approaching the Curie temperature of the bulk, while in our case T_N increases with channel diameter.

The nanoparticles embedded in the channels of MCM matrices are nanoribbons or nanowires. In spite of this topological difference T_N appears above the bulk value for all samples. The nanoparticles of MnO within a porous glass show the enhanced T_N as well. Moreover, the same phenomenon was recently observed in core/shell systems with MnO core [4]. Obviously, that topology of the nanoparticle does not play dominant role.

Neutron diffraction experiments carried out on epitaxial anisotropic thin films of MnO grown on different substrates showed T_N dependence on the substrate. The Néel temperature may be strongly enhanced above that of the bulk as well as depressed, however it does not depend on the film thickness [141, 142]. Note that a T_N enhancement is not present in MnF_2 films [157]. It looks that T_N depends on the non-stoichiometry or/and distortion of the surface between MnO and substrate and obviously strongly depends on anisotropy.

The origin of T_N enhancement had been proposed to arise from size and surface effects, where the broken symmetry at the surface and the concomitant local disorder may lead to (i) enhancement of the exchange interaction between surface atoms with respect to bulk exchange interactions [158]; (ii) variations in the crystal field resulting in high-low spin transitions [159] and (iii) the appearance of new degrees of freedom which interact with the antiferromagnetic order parameter [115].

The common effect on lowering the dimension of the system is reduction of the transition temperature. When approaching a magnetic transition from above the magnetic correlation length is limited by the size of the nanoparticle [160]. Indeed, reduction of magnetic transition temperature is shown by numerical calculations [161, 162], as well as experimentally observed in magnetic nanowires Ni[160], CuO nanoparticles [163], thin layers of CoO[48] and in other cases (see review [164]). For CoO nanostructured within a porous glass $T_N = 278.0(5)$ K smaller than $T_N = 289.0(1)$ K in the bulk [58, 44].

The boundary effect could be another factor decreases the transition temperature. One should expect that in the vicinity of nanoparticle surface, the absolute values of the exchange constants decrease due to disorder. In the phenomenological theory of a finite crystal this effect it is described by the term of a positive surface energy in the thermodynamic potential [165, 166]. The disorder is accompanied by the decrease of the mean moment and the local transition temperatures at the surface that leads to the decrease of the transition temperature of the entire nanoparticle.

It is well known that the enhanced transition temperature can be initiated by internal stresses due to structural defects. In particular, the influence of external pressure on the transition temperature for the bulk CoO was experimentally measured [167]. It was reported the observation of enhanced T_N in epitaxial films NiF_2 with a thickness of 38-60 nm, which was attributed to internal stresses [168]. However, in the case of confined nanoparticles of CoO and MnO, no significant internal stresses were detected.

7.   Structural distortions and magnetism.

Form of diffraction reflections from nanostructured MnO and CoO in all the porous media below Néel temperature indicates the structural distortions corresponding to the rhombohedral and the tetragonal distortion of a cubic lattice, respectively, as in the bulk.

For MnO within a porous glass the temperature dependence of the unit cell parameter a and the rhombohedral distortion angle α, calculated from neutron diffraction data, are shown in Figure 34. The measured dependence of the unit cell parameter is in a good

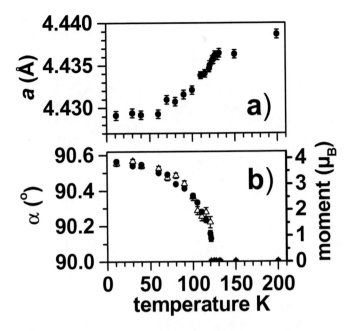

Figure 34. Temperature dependencies of (a) the unit cell parameter a, (b) the angle of rhombohedral distortion (open triangles) and the magnetic moment (solid circles) for the embedded MnO (from [57]).

agreement with the known dependence for the bulk [80]. As in the bulk, for confined MnO, an anomaly in the temperature dependence of the unit cell at the magnetic transition, due to spin correlations in the second coordination sphere [80, 139] is clearly visible.

Magnetoelastic energy in the case of the rhombohedral distortion can be written as follows [169]

$$E = b_1 \varepsilon S^2 + \frac{1}{2} b_2 \varepsilon^2. \tag{5}$$

Here, b_1, b_2 - some coefficients, S - spin, and ε - angle which determines the structural distortion. Minimizing this energy immediately leads to dependence of $\varepsilon \sim S^2$. For the bulk MnO it was stated in an earlier paper [170] and observed experimentally [80]. However, in the case of MnO, confined to a porous glass, the angle of the rhombohedral distortion is directly proportional to the magnetic moment (Figure 34b). Such dependence is not observed in the case of highly anisotropic nanoparticles in the channel type matrices. The reason for this phenomenon is still unclear.

7.1. Unusual low-temperature transition in MnO, nanostructured within the channels of MCM-41 matrix

In MnO, nanostructured within the channel type matrices, the diffraction signal in the neutron scattering is several times weaker than that in a porous glass. Therefore, the structural distortions were investigated by high-resolution X-ray diffraction at synchrotron source [171].

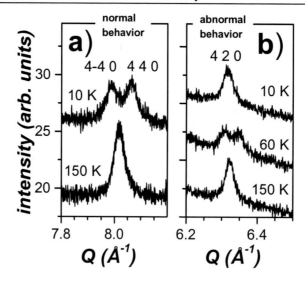

Figure 35. a) Normal behavior of the x-ray diffraction patterns of MnO within MCM with 87 Å the channel diameter above (150 K) and below (10 K) the magnetic transition; b) Abnormal behavior of the patterns of MnO within MCM matrix with 35 Å channel diameter (from [171]).

As expected, below the magnetic transition, the diffraction reflections split due to the rhombohedral distortion. In Figure 35a, a fragment of x-ray pattern measured for matrix with 87 Å channel diameter is shown above (150 K) and below (10 K) the magnetic transition. Unexpectedly, for two samples: matrix with 35 Å channel diameter and with 68 Å channel diameter the structural distortion disappeared below about 40 K (Figure 35b). So, the crystal structure of MnO became a cubic as well it is above the magnetic transition.

To check the reproducibility of the discovered phenomenon two new samples with MnO, using the same MCM matrices with 35 Å and 24 Å channel diameters, for which abnormal and normal behaviour had been observed, respectively, were prepared. Because of improved technology we succeeded in filling the matrix with a more quantity of crystalline MnO.

It turns out that the diffraction lineshape from the new samples has so called "saw-tooth" profile indicative of "nanoribbons" [81] (see insert in Figure 36). Note, in the first sample with a smaller quantity of crystalline MnO, the diffraction peaks were symmetrical that corresponds to "nanowires".

The complicated lineshape limits the profile analysis to high angles where the "saw-tooth" profile is not strongly expressed. The fragment of a profile refinement of the x-ray pattern for the new sample of MCM with 35 Å channel diameter is shown in Figure 36. No non-stoichiometry for all studied samples was detected.

In Figure 37a the temperature dependencies of the angle of rhombohedral distortion and the unit cell parameter are shown for old and new samples with 35 Å channel diameter. For comparison, the dependence for the sample with normal behaviour with 87 Å channel diameter, is shown too. The "reentrant" effect is observed at cooling as well as at warming.

Both MCM samples with 35 Å channel diameter, new and old show an abnormal be-

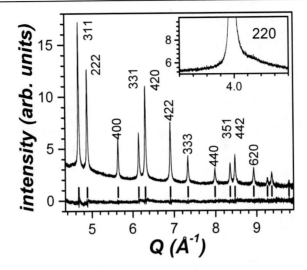

Figure 36. The x-ray diffraction patterns of MnO within MCM with 35 Å channel diameter at 10 K. In the bottom the difference of calculated and observed profiles is shown. Bars mark the reflection position. In the insert the enlarged fragment of the pattern is shown (from [171]).

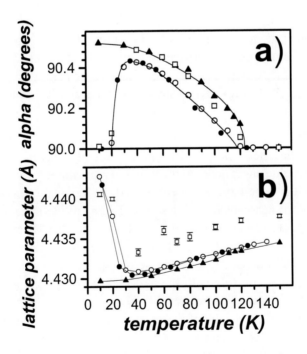

Figure 37. Temperature dependencies of the angle of rhombohedral distortion (alpha) (a) and the lattice parameter (b). Open circles corresponds to new sample with 35 Å channel diameter at warming, closed circles - at cooling; open squares corresponds to old sample at warming; solid triangles corresponds to matrix with 87 Å channel diameter (normal behaviour). Errors (e.s.d. - estimated standard deviation), if not shown, do not exceed the symbol size (from [171]).

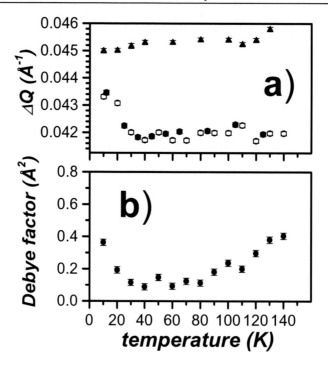

Figure 38. (a) Temperature dependences of the FWHM for reflection 200 for matrix with 35 Å channel diameter with abnormal behaviour at warming (open circles,) and cooling (closed circles); solid triangles correspond to matrix with 24 Å channel diameter and normal behaviour. (b) Temperature dependence of the Debye parameter for matrix with 35 Å channel diameter. Errors (e.s.d.), if not shown, do not exceed the symbol size (from [171]).

haviour, namely, a sharp decrease of the distortion angle below 40 K. The both samples with 24 Å channel diameters, old and new, show a normal behaviour, namely, a continuous increase of the distortion angle down to the lowest temperature.

In Figure 37b it is seen that the disappearing structural distortions in the samples with abnormal behaviour is accompanied by a sharp increase of the lattice parameter. The difference in the absolute values of the lattice constants for two samples, old and new, is readily explained by the different form of the embedded nanoparticles: "nanoribbon" and "nanowire". It is well known that in the case of two-dimensional lattice the maxima of diffraction peaks do not coincide with the nodes in the reciprocal space that leads to an underestimation of the unit cell parameter [81, 77].

Together with the sharp change of the structure parameters a change in FWHM of the diffraction peaks, which takes place together with the disappearing of the structural distortion, was found. In Figure 38a FWHM temperature dependence (in the units of reciprocal space) is shown for the reflection 200, which does not split at the rhombohedral distortion. The disappearing of the structural distortions is accompanied with an increase of the Debye-Waller parameter, which is proportional to the mean square displacement of atoms (Figure 38b).

The difference in FWHM for matrices with 35 and 24 Å channel diameters (Figure 38a) is readily explained by the difference in the nanoparticle shape. The nanoparticles

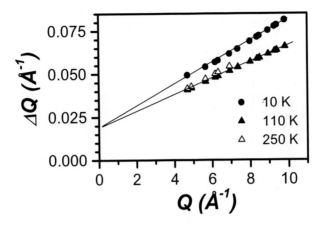

Figure 39. Q-dependencies of the peak broadening for matrix with 35 Å channel diameter and abnormal behaviour for temperatures 250, 140, and 10 K. Errors (e.s.d.) do not exceed the symbol size (from [171]).

within the thinner channels are smaller so the corresponding FWHM is larger. Because the instrumental line width is about of some percents from the measured peak width, the observed FWHM coincides with the peak broadening.

It had been pointed out [172], that the magnetic structure in MnO with spins perpendicular to the triad axis is inconsistent with rhombohedral symmetry so the true symmetry should be lower. However, a special search carried out for the bulk [173] did not show any deviation from the trigonal symmetry below the magnetic transition. In experiments with nanostructured MnO we did not detect any deviations from the trigonal symmetry neither from the cubic symmetry at low temperatures as well.

As it had been discussed above the peak broadening originates from size effect or inner stresses. In the first case broadening $\Delta Q(Q)$ does not depend on the reciprocal lattice vector Q, while in the second case it should be proportional to Q. In Figure 39 the dependencies $\Delta Q(Q)$ for some temperatures are shown. They shown for high Q only, since the effects of "saw-tooth" profile at low Q are too strong.

The experimental peak broadening has temperature and Q-independent contribution, which is attributed to the size effect, and a linear, temperature dependent, contribution whose slope is proportional to the inner stresses. In Figure 39 it is seen that additional stresses should be associated with the low temperature transition.

The observation of sharp changes in the structure parameters and the appearing of the inner stresses strongly evidences that we deal with specific low temperature phase transition, a new original phenomenon observed in confinement only.

Unfortunately, due to the large wavelength in neutron diffraction experiments, only one nuclear 111 reflection, and it is impossible to make any quantitative estimates of the peak broadening at low temperatures, similar to the above analysis based on x-ray data. In the neutron scattering one can only state the distortion of the nuclear 111 reflection, which corresponds to the appearance of inner and possible anisotropic stresses.

In neutron diffraction measurements, the antiferromagnetic order in all studied samples

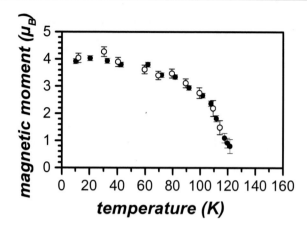

Figure 40. Temperature dependence of the magnetic moment in matrix with 35 Åchannel diameter calculated from the intensity of the magnetic reflection $\frac{1}{2}\frac{1}{2}\frac{1}{2}$. Closed and open circles belong to two sets of data measured in an interval of about two years (from [171]).

was observed [115]. As an example, the temperature dependencies of the magnetic moment for sample with abnormal behaviour is shown in Figure 40. Surprisingly, there is no any change in the temperature dependence of the ordered magnetic moment accompanied with the low temperature transition.

In the conventional theory the antiferromagnetic state in MnO is stabilized by the structural distortions. The theory predicts the angle of rhombohedral distortion to be proportional to the square of the magnetic moment [139, 170]. However, in the case of nanoparticles the high anisotropy and the inner stresses add new terms in the free energy, which could drastically change the energy balance. Today we do not know the real mechanism, which triggers the discovered phase transition. Investigations of this surprising phenomenon should be continued by other methods, in particularly, by the measurements of antiferromagnetic resonance, susceptibility and others.

7.2. Structural phase transition in the absence of a regular atomic lattice in the matrix MCM-48

In the case of very small nanoparticles of MnO within the matrix MCM-48, where a regular atomic lattice is absent, there is only one structural phase transition at the Néel temperature $\sim$ 120 K, without any additional low-temperature transitions, just the same as in the bulk [101]. In Figure 41 two profiles of reflection 220, measured at 130 K and 10 K, above and below the phase transition are shown, which clearly demonstrate the "splitting" of this reflection, while reflection 200 does not change with temperature. It corresponds to the rhombohedral (rhombohedral) structure distortion, as in the bulk.

As shown above, the observed profiles are well described by an exponential law. Using this fact, from the "splitting" of the 220 reflection the angle of the rhombohedral distortion at different temperatures was calculated (see Figure 42a). It is evident that the structural distortion in MnO within the matrix MCM-48 is the same as in the bulk: rhombohedral distortion appears below the Néel temperature and gradually increases with decreasing tem-

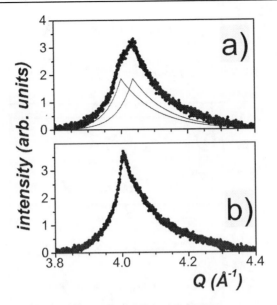

Figure 41. Fragments of x-ray diffraction patterns around 220 reflection from MnO within the matrix MCM-48 at 10 K (a) and 130 K (b). Two profiles in thin lines correspond to deconvolution of the resulting profile. Thick solid line through the experimental points is a fit with an exponential law (see text) (from [101]).

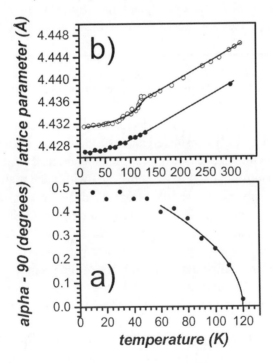

Figure 42. Temperature dependence of the angle of the rhombohedral distortion (alpha) (a) and lattice parameter (b). Open circles corresponds to the bulk MnO. Errors (e.s.d.), if not shown, do not exceed the symbol size. Solid line in (a) is a fit with a power law, in (b) solid lines are the guides for an eye (from [101]).

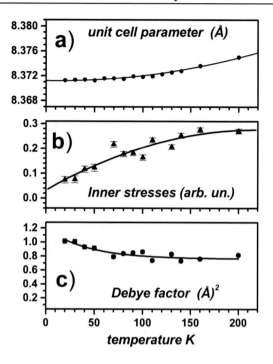

Figure 43. Temperature dependence of (a) the unit cell parameter, (b) internal stresses, and (c) the Debye-Waller factor for nanostructured maghemite. If the error bars are not shown, the corresponding e.s.d. does not exceed the symbol size. The solid lines are the guides for an eye (from [59]).

perature. In these calculations we used the unit cell parameter, calculated from the position of reflection 200 (Figure 42b).

As in the bulk, the lattice parameter decreases with decreasing temperature. As discussed above, the difference in absolute values of the lattice constant of bulk and nanostructured MnO sample due to quasi-two form of nanoparticles.

In the case of anisotropic nanoparticles of MnO within the matrix of channel type, no anomaly at the Néel temperature, resulted from the difference in magnetic correlations in the second coordination sphere [174, 80], were not observed. At the same time in isotropic nanoparticles MnO, within the porous glass, this anomaly exists [57] that means that the spin correlations are sensitive to the shape of the nanoparticles.

Approximation of the temperature dependence of the rhombohedral angle by a power law $(1 - T/T_N)^\beta$ gives the next value: $\beta = 0.49(6)$ and $T_N = 119.7(4)$ K (Figure 42a). The latter is close to T_N, measured for MnO in MCM-41 with a system of parallel channels and the same diameter.

Assuming that the rhombohedral angle is proportional to the square of the magnetic moment observed for the matrix with a system of parallel channels with the same diameter, the corresponding exponent in the temperature dependence of the moment is equal to 0.25(3) that coincides with the value observed in this matrix [115]. It is evident that the absence of a regular atomic lattice does not affect the magnetic transition.

8. Atomic vibrations in confined nanoparticles.

Thermal motion of atoms is one of the fundamental properties of solids. However, little is known about the atomic vibrations when size of the system is comparable to the character-istic length of atomic interactions. There are some fundamental restrictions, in particularly; a wavelength of phonons in confined nanoparticles cannot be greater than the nanoparticle size that leads to modification of a spectrum of thermal fluctuations. Moreover, the last should depend on the surface atoms, which are in the condition of the local disorder. For confined nanoparticles the interactions with matrix walls play important role as well.

It is well known that the scattering of the thermal neutrons can pass over with either a change of the incident neutron energy - inelastic scattering, and without an energy change - elastic scattering. With the help of inelastic scattering on a single crystal the energy of lattice excitation (phonon) and its momentum can be measured. In the case of powder sample because of the power averaging the information about the momentum direction is lost. Thus, one can only measure the phonon density of states as a function of phonon en-ergy. Using elastic neutron scattering one can also get information about atomic vibrations. However, due to limitations of the powder diffraction this information is not complete as in the studies on single crystals; however, in a nano-structured state we deal with polycrys-tals only. Therefore, the measurements of Bragg reflections or diffuse scattering or phonon density measurements remain the main diffraction methods.

For a quantitative description of atomic motion the thermal or Debye-Waller factor $T_k(\mathbf{Q})$ is introduced, which is the Fourier transform of the quantity $p_k(\mathbf{u})$ - the probability density to detect k-th atom at a distance $\mathbf{u}$ from its equilibrium position:

$$T_k(\mathbf{Q}) = \frac{1}{(2\pi)^3} \int p_k(\mathbf{u}) \exp\left[i\mathbf{Q} \cdot \mathbf{u}\right] d^3\mathbf{u}. \tag{6}$$

Here

$$p_k(\mathbf{u}) = \int T_k(\mathbf{Q}) \exp\left[-i\mathbf{Q} \cdot \mathbf{u}\right] d^3\mathbf{Q}. \tag{7}$$

Thermal motion of atoms changes the conditions of interference, which affects the in-tensity of Bragg reflections:

$$I_{Bragg}(\mathbf{Q}) \sim \left|\sum_{k=1}^{n} f_k T_k(\mathbf{Q}) \exp\left[i\mathbf{Q} \cdot \mathbf{r}_k\right]\right|^2, \tag{8}$$

here the summation is over all atoms of the unit cell, f_k scattering power (scattering length) of k-th atom, which in the case of scattering of x-ray radiation includes the form factor, vector $\mathbf{r}_k$ determines the equilibrium position of the atom; $\mathbf{Q}$ - momentum transfer.

Thus, by measuring the intensity of Bragg reflections as a function of $\mathbf{Q}$, one can deter-mine the thermal factor.

The character of atomic motion is determined by the interatomic potential V($\mathbf{u}$). In the harmonic approximation, the potential is described by a parabolic dependence. In this case, one can easily show that the probability to detect an atom at a distance $\mathbf{u}$ corresponds to the normal distribution and thermal factor can be written as [175]:

$$T_k(\mathbf{Q}) = \exp(-\frac{1}{2}\mathbf{Q}^T\mathbf{B}\mathbf{Q}) \tag{9}$$

here $\mathbf{B}$ - third-rank tensor (corresponds to an ellipse of thermal vibrations), which determines the averaged deviation of atom from its equilibrium position:

$$\mathbf{B} = \left\langle \mathbf{u} \cdot \mathbf{u}^T \right\rangle. \tag{10}$$

In the case of isotropic vibrations the equation 9 reduces to simpler notation:

$$T_k(\mathbf{Q}) = \exp(-\frac{1}{2}\langle(\mathbf{Q} \cdot \mathbf{u}_k)^2\rangle) \tag{11}$$

The crystal symmetry imposes restrictions on the elements of $\mathbf{B}$. It is obvious that since $\langle u^2 \rangle \geq 0$ the determinant of $\mathbf{B}$ must be positive definite. It is generally accepted to describe the atomic displacements in terms of anisotropic thermal parameters $beta_{ij}$, which are refined variables and associated with the mean square displacement amplitude unitary transformation [175].

In many cases, parabolic (harmonic) approximation adequately describes the thermal factor $T(\mathbf{Q})$. However, if the interatomic interaction potential significantly deviates from the quadratic law, then there are anharmonic terms, in this case it is necessary to consider a more accurate model.

Effect of "restricted geometry" on the atomic vibrations is manifested in many different ways. For example, measurements of the temperature dependence of the lattice parameters proved that the thermal expansion coefficient (TEC) in the embedded nanostructured compounds is different from those in the bulk. For example, neutron diffraction studies of the thermal evolution of structure of CuO nanoparticles with the size of 146(5) Å synthesized within porous glass shown that TEC is about $2 \cdot 10^{-5}$ K^{-1} and temperature independent in interval of 1.5-250 K, whereas for the bulk CuO TEC is an order of larger [176]. It is contrast with giant negative TEC found in the nanoparticle fabricated by mechanical ball milling and means that TEC strongly depends on the preparation procedure [177].

In some cases, the effects caused by the interaction of nanoparticle with the walls of the matrix were found. For example, in studies of maghemite in porous glass [59] it was showed that the inner stresses, calculated from the width of the diffraction lines, decreases with temperature decreases (Figure 43b). Matrix of amorphous silica has very low TEC. For commercial porous glass vycor-7930 of Corning TEC is $\sim 7.5 \cdot 10^{-7}K^{-1}$, whereas for nanostructured maghemite TEC is equal of $\sim 2.3 \cdot 10^{-6}K^{-1}$ at room temperature. Since the oxide "compresses" faster with decreasing temperature the internal stresses caused by the interaction with the pore walls, disappear. This explains the unusual behaviour of the Debye-Waller factor, which is proportional to the mean square amplitude of thermal vibrations (Figure 43c). Instead of the expected "freezing" of the thermal vibrations as the temperature decreases their amplitude increases.

It should be noted that the overall increase of the Debye-Waller in "limited geometry" was observed for the oxides MnO and CoO embedded in porous glass. Such a strong increase in the Debye-Waller factor, reflects the presence of a large number of defects, which is typical for nanostructured compounds and, in turn, leads to a large static component factor. It is also known about the strong effect of "restricted geometry" on amplitude of atomic vibrations in confined ferroelectrics [64, 178].

Most pronounced effect of "restricted geometry" on the thermal vibrations manifested in the nanoparticles of low-melting metals, since they demonstrate high amplitude of atomic vibrations. As examples, some the results of the studies of confined nanoparticles of lead [63] and selenium [62] inside a porous glass are presented below.

8.1. Atomic vibrations in nanostructured lead within porous glass.

It is well known that an accurate assessment of the integral intensity of the diffraction reflection the contribution to thermal diffuse scattering (TDS), which usually takes the form of a broad peak under the Bragg reflection, is very important. However, in our case it is practically impossible to calculate TDS from a diffraction pattern due to the non-linear background from the amorphous matrix. At the same time, since the diffraction signal is weak, the contribution of TDS is of the order of statistical errors and can be neglected, at least up to temperatures around melting point. At higher temperatures, TDS can be taken into account by the appropriate correction of the background.

Since the confined Pb nanoparticles (see section 3.2) are elongated along [111] direction (see Figure 6), it is natural to seek the appropriate anisotropy of the mean-square displacement taking into consideration the off-diagonal terms: $beta_{12} = \beta_{13} = \beta_{23}$ in addition to diagonal terms: $\beta_{11} = \beta_{22} = \beta_{33}$ in the tensor of thermal factors. The condition of equality of the off-diagonal terms corresponds to orthorhombic symmetry with the anisotropy axis along the [111] direction [175].

The values of the off-diagonal elements at high temperatures, calculated by the profile analysis of neutron diffraction, differ significantly from zero. The mean-square displacement along and perpendicular to the [111] axis, which coincide with a pore axis, calculated from anisotropic thermal parameters, is shown in Figure 44a). It is seen that the atomic vibrations perpendicular to the axis has changed only slightly at high temperatures, in contrast to the vibrations along the axes of the pores.

This behaviour can be explained by the fact that the pore walls limit the atomic motion, whereas atoms vibrate more freely along the pore axis. Note that due to low statistics, to distinguish the atomic motion along and perpendicular to the axis of a pore is possible only at high temperatures, when atomic displacements are maximal. Observed experimental results are consistent with the theoretical predictions, that the thermal vibrations of atoms located on the surface of nanoparticles are limited, if the matrix, in our case SiO_2, has a higher melting point than the embedded particle [179]

In Figure 44b an isotropic mean-square displacements in the "restricted geometry" and in the bulk Pb [180, 181, 182], measured with diffraction method, are displayed. The anomalous behaviour of the mean-square displacement in nanostructured samples at about the Debye temperature ~ 105 K ia clearly seen. This anomaly is also observed in the temperature dependence of the lattice parameter.

The interatomic potential in nanostructured objects and in the bulk differ drastically, that leads to differences in the atomic motion. Therefore, neither the model of one particle potential used in work [181] for the description of the mean-square displacement in the

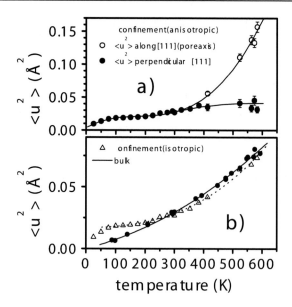

Figure 44. (a) Anisotropic mean-square displacement vs temperature for confined Pb along (open circles) and perpendicular (solid circles) to the [111] direction. (b) Isotropic mean-square displacement vs temperature for confined (triangles) and the bulk Pb (circles). The errors do not exceed the size of the symbols, if not shown (from [63]).

bulk lead, nor the numerical calculations based on the Lennard-Jones potential [183] can be applied in the case of the confined nanoparticles.

To describe the mean-square displacement a power-series expansion of $\langle u^2 \rangle$ can be used: following [184] the mean-square displacement for cubic crystals may be written as follows:

$$\langle u^2 \rangle_{bulk} = \frac{3 \cdot \hbar^2}{m \cdot k_B \cdot \Theta_D} \cdot T + \alpha_2 \cdot T^2 + \alpha_3 \cdot T^3, \qquad (12)$$

where α_2 and α_3 are isotropic, anharmonic contributions due to the cubic and quartic terms in the interatomic potential, respectively, m - the atomic mass, k_B - the Boltzmann constant, and Θ_D - the Debye temperature.

Setting the Debye temperature to 105 K, the best fit of the experimental data reported for the bulk [181] is shown by the solid line in Figure 44. The similar fitting for confinement at temperatures above 70 K gives surprising result - the linear term turns out to be close to zero. In Figure 44 the corresponding curve is shown by the dashed line.

Linear term in equation 12 describes a well known effect of "softening" of the normal vibration frequencies due to thermal expansion, known as Grüneisen effect. The thermal expansion of the mean distance between atoms increases, the strength of interatomic interaction reduces, leading to a "softening" of the normal vibration frequencies. This phenomenon is characterized by a dimensionless Grüneisen constant, which is defined as the average change in the oscillation frequency divided by the volume change: $d\omega/\omega = -\gamma_G \cdot dV/V$, where V - volume [185].

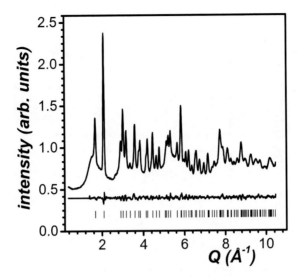

Figure 45. Observed (at the top) and difference (at the bottom) neutron diffraction patterns of the nanostructured Se embedded in a porous glass at 300 K. Strong diffuse scattering origins mainly from amorphous porous glass. Vertical bars mark the reflection positions (from [62]).

So, the softening of the vibration due to the thermal expansion is very small in confined nanoparticles in contrast with the bulk, where this effect was demonstrated. Because the long-wavelength vibrations are strongly reduced or absent in the nanostructured particles, obviously the observed difference in the temperature behaviour for the bulk and confinement reflects the fundamental difference in their vibration spectra.

8.2.   Atomic vibrations in nanostructured selenium.

A typical neutron diffraction pattern of nanostructured selenium in a porous glass with average pore diameter of 70 Å, is shown in Figure 45. Internal stresses are found only below 100 K. Above this temperature, the broadening of diffraction peaks is due to the size effect only. From the peak broadening the average nanoparticle diameter is defined as 183(6) Å. Starting with about 400 K, the size of nanoparticles increases sharply, showing the melting process.

Se in nanostructured porous glass crystallizes in trigonal structure, as in the bulk with close unit cell parameters. Due to a strongly anisotropic shape of the unit cell is natural to expect an anisotropic shape for the embedded nanoparticles. However, because of low statistical accuracy, no systematic broadening of the reflections was found.

From the temperature dependence of the Bragg reflections the anisotropic thermal factors of β_{ij} and the quantity of so-called the root-mean-square (rms), which characterizes the average amplitude of atomic vibrations, were calculated. In Figure 46 the temperature dependencies of the isotropic and anisotropic parts of the rms amplitudes are shown.

As expected, the anisotropic averaged amplitude of atomic vibrations (rms) (Figure 46b) in the basal plane and in the perpendicular direction are different, that coincides with

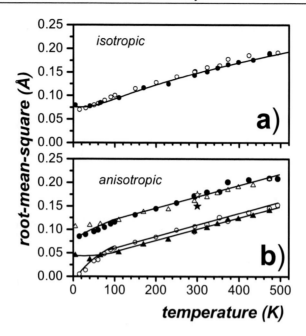

Figure 46. (a) Isotropic root-mean-square displacement versus temperature for confined Se (open circles) and the bulk Se (solid circles). Solid line is a fit by the Einstein law. b) Anisotropic root-mean-square displacement versus temperature: in the basal plane for confined Se (solid circles) and for the bulk (open triangles); along the hexagonal axis c for confined Se (open circles) and for the bulk (solid triangles). By the open and solid stars the rms amplitudes in the basal plane and along the hexagonal axis, respectively, from a single crystal experiments, are shown [88]. The error is about the size of the symbols. Lines are the guide for the eye (from [62]).

measurements on single crystals of Se [88], shown for comparison in Figure 46.

To date, there is no theory of atomic vibrations in the "restricted geometry". The usual approximation of the temperature dependence of the isotropic part of the mean-square displacement for the bulk $\langle u^2 \rangle$ is described in terms of Debye theory [186, 175]:

$$\langle u^2 \rangle = \frac{3\hbar^2 T}{m k_B \Theta_D^2} [F(\frac{\Theta_D}{T}) + \frac{\Theta_D}{4T}] \tag{13}$$

where m, $\hbar$? k_B, T - the atomic mass, the Planck and the Boltzmann constants and temperature, respectively, and F(x) - Debye function:

$$F(x) = \frac{1}{x} \int_0^x \frac{x dx}{e^x - 1} \tag{14}$$

It leads to the characteristic Debye temperature $\Theta_D = 315(10)$ K, that much above 135 K - the value given for the bulk [187]. Variation of the static constants does not improve the quality of convergence. However, the best results are obtained using a the model of independent vibrational modes of Einstein [175]:

$$\langle u^2 \rangle = \frac{3\hbar^2}{mk_B\Theta_E}\left[\frac{1}{exp(\Theta_E/T)-1}+\frac{1}{2}\right] \tag{15}$$

Approximation in the frame of this model with a single independent parameter Θ_E, the characteristic Einstein temperature, gives much better convergence with $Theta_E = 161(5)$ K, a corresponding fit is shown by thick line in Figure 46a). So, in the "restricted geometry" the description of the atomic motion in confined Se by the model of independent oscillators is more adequate.

The Einstein model assumes an equality of frequencies of all vibrations, i.e., it approximates only the optical mode. In the Debye model there is no difference between optical and acoustic branches, so it applies only to cubic monatomic compounds [169, 188]. Selenium can not be considered as monatomic compound, because its atoms are in different crystallographic positions. Moreover, due to anisotropy, atomic bonds between different atoms are different.

Anisotropy of the atomic motion in Se is intrinsic feature, which reflects the high anisotropy of the crystal structure, which consists of loosely coupled, rigid helical chains along the hexagonal axis. Indeed, the interaction between Se atoms in helical chains along the hexagonal axis is much stronger than the interaction between the chains. Therefore, the average amplitude in the basal plane, which corresponds to interaction between the chains, is more than in the perpendicular direction.An unexpected sharp drop in the amplitude of vibrations - the "freezing" of the atomic motion along the chains, while the atomic motion in the perpendicular plane (the motion of chains) preserves (Figure 46b), was observed in the "restricted geometry". In addition, it appears that a sharp decrease in the amplitude is accompanied by the appearance of internal stresses and increase in the average size of the crystallized nanoparticles (Figure 47).

Note, the unit cell of Se at low temperatures demonstrates the noticeable deformation, namely, with decreasing temperature the hexagonal prism length increases while the basal edge decreases. Obviously such deformation should favour a change of the nanoparticle shape.

We did not detect any change in the background at low temperatures, i.e. the total amount of amorphous and crystallized Se constant. Therefore, the observed increase in the average size of nanoparticles should be attributed to a change in a shape of nanoparticle, which becomes anisotropic. Unfortunately, because of low statistical precision anisotropy of nanoparticle shape is not possible to register directly through the systematic broadening of the peaks.

It is known that the thermal expansion of the silica matrix is insignificant compared that of the embedded Se. Moreover, Se in a porous glass crystallizes into adjacent pores in the form of agglomerates. Therefore, the deformation of nanoparticles, "trapped" in the pores of the matrix, can cause the internal stresses. So the interaction of the matrix wall and nanoparticle leads to a significant change in atomic motion of the last. Therefore the observed "freezing" of the thermal vibration in the confined nanoparticles one should attribute to this specific effect. Observed "blocking" of nanoparticles in the pores due to the substantial difference in thermal expansion, appears to be a general phenomenon in the confined geometry, which leads to a substantial modification of the atomic motion.

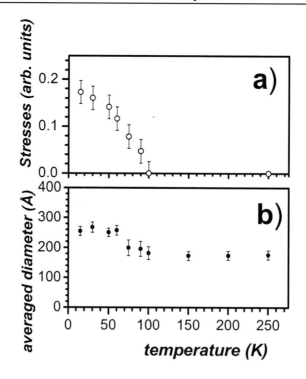

Figure 47. Temperature dependencies of the inner stresses (a) and the volume averaged nanoparticle size (b) (from [62]).

9. Acknowledgments

I wish to thank my colleagues, who participated in these researches over the years for fruitful collaboration: I. Mirebeau, S. Vakhrushev, Yu. Kumzerov, D. A. Kurdyukov and V. Sakhnenko.

References

[1] Bratkovsky A. M. "Spintronic effects in metallic, semiconductor, metal-oxide and metal-semiconductor heterostructures." *Reports on progress in physics*, **71**, (2008), 026502.

[2] Shen J, Kirschenr J. "Tailoring magnetism in artificially structured materials: the new frontier." *Surface science*, **500**, (2002), 300–322.

[3] I.V. Golosovsky, G. Salazar-Alvarez, A. López-Ortega, M. A. González, J. Sort, M. Estrader, S. Suriñach, M. D. Baró and J. Nogués. "Magnetic Proximity Effect Features in Antiferromagnetic/Ferrimagnetic Core-Shell Nanoparticles." *Phys. Rev. Lett.*, **102**, (2009), 247201.

[4] López-Ortega, D. Tobia, E. Winkler, I. V. Golosovsky, G. Salazar-Alvarez, S. Estradé, M. Estrader, J. Sort, M. A. González, S. Suriñach, J. Arbiol, F. Peiró, R.

D. Zysler, M. D. Baró and J. Nogués. "Size dependent passivation shell and magnetic properties in antiferromagnetic /ferrimagnetic, core/shell MnO nanoparticles." *Journal of the American Chemical Society*, **132**, (2010), 9398–9407.

[5] C. T. Kresge, M. E. Leonowicz, W. J. Roth, J. C. Vartuli, J. S. Beck. "Ordered mesoporous molecular sieves syntehesized by a liquid-crystal template mechanism." *Letters to Nature*, **359**, (1992), 710–712.

[6] M. Grün, I. Lauer and K. Unger. "The synthesis of micrometer- and submicrometer-size spheres of ordered mesoporous oxide MCM-41." *Adv. Mater.*, **9**, (1997), 254–256.

[7] J. Nogués, J. Sort, V. Langlais, V. Skumryev, S. Suriñach, J. S. Muñoz, M. D. Baró. "Exchange bias in nanostructures." *Physics Reports*, **422**, (2005), 65117.

[8] V. Skumryev, S. Stoyanov, Y. Zhang, G. Hadjipanayis, D. Givord and J. Nogues. "Beating the superparamagnetic limit with exchange bias." *Nature*, **423**, (2003), 850–853.

[9] J. Weston, A. Butera, D. Otte and J. Barnard. "Nanostructured magnetic networks: a materials comparison." *J. Magn. Magn. Mater.*, 515–519.

[10] W. Goldburg, F. Aliev and X-L. Wu. "Behavior of liquid crystals and fluids in porous media." *Physica A*, **213A**, (1995), 61–70.

[11] K. Morishige and Y. Ogisu. "Liquidsolid and solidsolid phase transitions of oxygen in a single cylindrical pore." *J. Chem. Phys.*, **114**, (2001), 7166–7173.

[12] D. Wallacher, R. Ackerman, P. Huber, M. Enderle and K. Knorr. "Diffraction study of solid oxygen embedded in porous glasses." *Phys. Rev. B*, **64**, (2001), 185203–1–9.

[13] M. Calbi, M. Cole, S. Gatica, M. Bojan, G. Stan. "Condensed phases of gases inside nanotube bundles." *Rev. Mod. Phys.*, **73**, (2001), 857–865.

[14] R. Trasca, M. Calbi and M. Cole. "Lattice model of gas condensation within nanopores." *Phys. Rev. E*, **65**, (2002), 061607–1–9.

[15] M. R. Fitzsimmons, S. D. Bader, J. A. Borchers, G. P. Felcher, J. K. Furdyna, A. Hoffmann, J. B. Kortright, I. K. Schuller, T. C. Schulthess, S. K. Sinha, M. F. Toney, D. Weller, S. Wolf. "Neutron scattering studies of nanomagnetism and artificially structured materials." *J. Magn. Magn. Mater.*, **271**, (2004), 103146.

[16] Feygenson, M., Schweika, W., Ioffe, A., Vakhrushev, S.B., Brückel, T. "Magnetic phase transition in confined MnO nanoparticles studied by polarized neutron scattering." *Physical Review B*, **81**, (2010), 2064423.

[17] S. Polarz and B. Smarsly. "Nanoporous Materials." *Journ. Nanosci. and Nanotech.*, **2**, (2002), 581–612.

[18] P. Levitz, G. Ehret, S. K. Sinha and J. M. Drake. "Porous vycor glass: The microstructure as probed by electron microscopy direct energy transfer, small-angle scattering and molecular adsorption." *J. Chem. Phys.*, **95**, (1991), 6151–6160.

[19] D. Enke, F. Janovsky, W. Schwieger. "Porous glasses in the 21st century a short review." *Micro. Mesopor. Mater.*, **60**, (2003), 19–30.

[20] Yu. A. Kumzerov. *Nanostructured films and coatings ed. by G. M. Chow, I. Ovidko and T. Tsakalakos* (Kluwer Academic Publishers, Dordrecht, 2000). Nato Science Series, 3, High Technology, v.78.

[21] M. A. Aleksashkina, B. I. Venzel and L. G. Svatovskaya. "Application of Porous Glasses as Matrices for Nanocomposites." *Glass Physics and Chemistry*, **31**(3), (2005), 269–274.

[22] D. Morineau, G. Dossen, C. Alba-Simionesco, P. Llewellyn. "Glass transition, freezing and melting of liquids confined in the mesoporous silicate MCM-41." *Philosophical Magazine B*, **79**, (1999), 1845–1855.

[23] D. Zhao, J. Feng, Q. Huo, N. Melosh, G. H. Fredrickson, B. F. Chmelka, G. D. Stucky. "Triblock Copolymer Syntheses of Mesoporous Silica with Periodic 50 to 300 Angstrom Pores." *Science*, **279**, (1998), 548–552.

[24] M. Weyland, P. A. Midgley and R. E. Dunin-Borkowski. The Electron Microscopy group, Department of Material Science and Metallurgy, University of Cambridge, UK, http://www-hrem.msm.cam.ac.uk/research.

[25] M. Kaneda, T. Tsubakiyama, A. Carlsson, Y. Sakamoto, T. Ohsuna, O. Terasaki, S. H. Joo and R. Ryoo. "Structural Study of Mesoporous MCM-48 and Carbon Networks Synthesized in the Spaces of MCM-48 by Electron Crystallography." *J. Phys. Chem. B*, **106**, (2002), 1256–1266.

[26] Bozhi Tian, Xiaoying Liu, L. A. Solovyov, Zheng Liu, Haifeng Yang, Zhendong Zhang, Songhai Xie, Fuqiang Zhang, Bo Tu, Chengzhong Yu, Osamu Terasaki and Dongyuan Zhao. "Facile synthesis and characterisation of novel mesoporous ans mesorelief oxides with gyroidal structures." *J. Am. Chem. Soc.*, **126**, (2004), 865–874.

[27] K. Schumacher, P. I. Ravikovitch, A. Du Chesne, A. V. Neimark and K. K. Unger. "Characterization of MCM-48 Materials." *Langmuir*, **16**, (2000), 4648–4654.

[28] M. Wohlgemuth, N. Yufa, J. Hoffman and E. L. Thomas. "Triply Periodic Bicontinuous Cubic Microdomain Morphologies by Symmetries." *Macromolecules*, **34**, (2001), 6083–6089.

[29] J. Harting, M. J. Harvey, J. Chin, P V. Coveney. "Detection and tracking of defects in the gyroid mesophase." *Computer Physics Communications*, **165**, (2005), 97109.

[30] S. Borisov, T. Hansen, Y. Kumzerov, A. Naberezhnov, V. Simkin, O. Smirnov, A. Sotnikov, M. Tovar and S. Vakhrushev. "Neutron diffraction study of $NaNO_2$ ferroelectric nanowires." *Physica B*, **350**, (2004), e1119e1121.

[31] S. K. Gordeev. *Advanced composite materials, ed. by M. Prelas, A. Benedictus and L. Lin, Diamond based composites* (Kluwer Academic Publishers, Netherlands, Dordrecht, 1997).

[32] D. J. Sellmyer, M. Zheng and R. Skomski. "Magnetism of Fe, Co and Ni nanowires in self-assembled arrays." *J. Phys.: Condens. Matter*, **13**, (2001), R433R460.

[33] A. W. Zhao, G. W. Meng, L. D. Zhang, T. Gao, S. H. Sun, Y. T. Pang. "Electrochemical synthesis of ordered CdTe nanowire arrays." *Appl. Phys. A*. Published online: 8 January 2003.

[34] S. Mazumder, D. Sen and A. K. Patra. "Characterization of porous materials by small-angle scattering." *PRAMANA, Indian Academy of Sciences*, **63**(1), (2004), 165–173.

[35] L. Sun, C. L. Chien and P. C. Searson. "Fabrication of nanoporous single crystal mica template for electrochemical deposition of nanowire arrays." *J. Mater. Sciences*, **35**, (2000), 1097–1103.

[36] C. G. Shull, J. S. Smart. "Detection of antiferromagnetism by neutron diffraction." *Phys. Rev.*, **76**, (1949), 1256–1257.

[37] C. G. Shull, W. A. Strauser and E. O. Wollan. "Neutron diffraction by paramagnetic and antiferromagnetic substances." *Phys. Rev.*, **83**, (1951), 333–345.

[38] W. L. Roth. "Magnetic structure of MnO, FeO, CoO and NiO." *Phys. Rev.*, **110**, (1958), 1333–1341.

[39] H. Shaked, J. Faber, Jr. and R. L. Hitterman. "Low-temperature magnetic structure of MnO: a high-resolution neutron-diffraction study." *Phys. Rev. B*, **38**, (1988), 11901–11903.

[40] H. Fjellvag, F. Gronvold, S. Stolen. "On the crystallographic and magnetic structures of nearly stoichiometric iron monoxide." *J. Solid State Chem.*, **124**, (1996), 52–57.

[41] D. Herrmann-Ronzaud, P. Barlet and J. Rossat-Mignod. "Equivalent type-II magnetic structures: CoO, a collinear antiferromagnet." *J. Phys. C*, **11**, (1978), 2123–2137.

[42] W. Jauch, M. Reehuis, H. J. Bleif, F. Kubanek and P. Pattison. "Crystallographic symmetry and magnetic structure of CoO." *Phys. Rev. B*, **64**, (2001), 052102–1–3.

[43] D. Mukamel and S. Krinsky. "Physical realisation of n¿4-component vector models." *Phys. Rev. B*, **13**, (1976), 5065–5077.

[44] M. D. Rechtin, S. C. Moss and B. L. Averbach. "Influence of lattice contraction on long-range order in CoO near T_N." *Phys. Rev. Lett.*, **24**, (1970), 1485–1489.

[45] P. J. van der Zaag, Y. Ijiri, J. A. Borchers, L. F. Feiner, R. M. Wolf, J. M. Gaines, R.W. Erwin and M. A. Verheijen. "Difference between blocking and Néel temperatures in the exchange biased Fe_3O_4/CoO." *Phys. Rev. Lett.*, **84**, (2000), 6102–6105.

[46] M. J. Carey, A. E. Berkowitz, J. A. Borchers and R. W. Erwin. "Strong interlayer coupling in CoO/NiO antiferromagnetic superlattices." *Phys. Rev. B*, **47**, (1993), 99529955.

[47] Y. J. Tang, D. J. Smith, B. L. Zink, F. Hellman and A. E. Berkowitz. "Finite size effects on the moment and ordering temperature in antiferromagnetic CoO layers." *Phys. Rev. B*, **67**, (2003), 054408–1–7.

[48] T. Ambrose and C. L. Chien. "Finite-Size Effects and Uncompensated Magnetization in Thin Antiferromagnetic CoO Layers." *Phys. Rev. Lett.*, **76**, (1996), 1743–1746.

[49] E. N. Abarra, K. Takano, F. Hellman and A. E. Berkowitz. "Thermodynamic Measurements of Magnetic Ordering in Antiferromagnetic Superlattices." *Phys. Rev. Lett.*, **77**, (1996), 3451–3454.

[50] M. Gich, A. Roig, C. Frontera, E. Molins, J. Sort, M. Popovici, G. Chouteau, D. Martin y Marero and J. Nogués,. "Large coercivity and low-temperature magnetic reorientation in ϵ-Fe_2O_3 nanoparticles." *Journal of Applied Physics*, **98**, (2005), 044307–1–6.

[51] R. Dronskowski. "The Little Maghemite Story: A Classic Functional Material." *Adv. Funct. Mater.*, **11**, (2001), 27–29.

[52] E. V. Charnaya, Cheng Tien, W. Wang, M. K. Lee, D. Michel, D. Yaskov, S. Y. Sun,and Yu. A. Kumzerov. "X-ray studies of the melting and freezing phase transitions for gallium in a porous glass." *Phys. Rev. B*, **72**, (2005), 035406.

[53] P. Thompson, D. Cox and B. Hastings. "Rietveld refinement of Debye-Scherrer synchrotron x-ray data from Al_2O_3." *J. Appl. Cryst.*, **20**, (1987), 79–83.

[54] J. I. Langford. "A rapid method for analysing the breadths of diffraction and spectral lines using the Voight function." *J. Appl. Crystallogr.*, **11**, (1978), 10–14.

[55] Yu. A. Kumzerov, A. A. Nabereznov, S. B. Vakhrushev and B. N. Savenko. "Freezing and melting of mercury in porous glass." *Phys. Rev. B*, **52**, (1995), 4772–4774.

[56] I. V. Golosovsky et al. unpublished.

[57] I. V. Golosovsky, I. Mirebeau, G. André, D. A. Kurdyukov, Yu. A. Kumzerov and S. B. Vakhrushev. "Magnetic ordering and phase transition in MnO embedded in a porous glass." *Phys. Rev. Lett.*, **86**, (2001), 5783–5786.

[58] I. V. Golosovsky, I. Mirebeau, G. André, M. Tovar, D. M. Tobbens, D. A. Kurdyukov, and Yu. A. Kumzerov . "Magnetic Phase Transition in a Nanostructured Antiferromagnet CoO Embedded in Porous Glass." *Physics of the Solid State*, **48**, (2006), 2130–2133.

[59] I. V. Golosovsky, M. Tovar, U. Hoffman, I. Mirebeau, F. Fauth, D. A. Kurdyukov and Yu. A. Kumzerov. "Diffraction Studies of the Crystalline and Magnetic Structures of γ-Fe_2O_3 Iron Oxide Nanostructured in Porous Glass." *JETP Letters*, **83**(7), (2006), 298–301.

[60] I. V. Golosovsky, I. Mirebeau, F. Fauth, D. A. Kurdyukov, Yu. A. Kumzerov. "Magnetic structure of hematite nanostructured in a porous glass." *Solid State Communications*, **141**, (2007), 178–182.

[61] I. Golosovsky, D. Többens and M. Tovar. "Magnetic phase transitions in the antiferromagnetic oxides embedded in a porous glass." *BENSC Experimental report 2001, HMI, Berlin*, 64.

[62] I. V. Golosovsky, O. P. Smirnov, R. G. Delaplane, A. Wannberg, Y. A. Kibalin, A. A. Naberezhnov and S. B. Vakhrushev. "Atomic motion in Se nanoparticles embedded into a porous glass matrix." *Eur. Phys. J. B*, **54**, (2006), 1434–6028.

[63] I. V. Golosovsky, R. G. Delaplane, A. A. Naberezhnov and Y. A. Kumzerov. "Thermal motion in lead confined within a porous glass." *Phys. Rev. B*, **69**, (2004), 132301–1–4.

[64] A. V. Fokin, Yu. A. Kumzerov, N. M. Okuneva, A. A. Naberezhnov, S. B. Vakhrushev, I. V. Golosovsky and A. I. Kurbakov. "Temperature Evolution of Sodium Nitrite Structure in a Restricted Geometry." *Phys. Rev. Lett.*, **89**, (2002), 175503–1–4.

[65] B. Dorner, I. Golosovsky, Yu. Kumzerov, D. Kurdukov, A. Naberezhnov, A. Sotnikov, S. Vakhrushev. "Structure of KD_2PO_4 embedded in a porous glass." *Ferroelectrics*, **286**, (2003), 213–219.

[66] H. M. Rietveld. "Line profiles of neutron powder-diffraction peaks for structure refinement." *Acta Cryst.*, **22**, (1967), 151–152.

[67] J. Rodriguez-Carvajal. "The computer code Fullprof, 2003-2006, (http://www-llb cea fr/fullweb/powder.htm)." *Physica B*, **192**, (1993), 55.

[68] R. W. James. *The optical prinsiples of the diffraction of x-rays* (Ox bow Press, Woodbridge, Connecticut, 1982).

[69] C. Rottman, M. Wortis, J. C. Heyraud and J. J. Metois. "Equilibrium shapes of small lead crystals: observation of Pokrovsky-Talapov critical behavior." *Phys. Rev. Lett.*, **52**, (1984), 1009–1012.

[70] D. Reznik, C. H. Olk, D. A. Neumann and J. R. D. Copley. "X-ray powder diffraction from carbon nanotubes and nanoparticles." *Phys. Rev. B*, **52**, (1995), 116–125.

[71] J. K. Kjems, L. Passell and H. Taub, J. G. Dash and A. D. Novaco. "Neutron scattering study of nitrogen adsorbed on basal-plane-oriented graphite." *Phys. Rev. B*, **13**, (1976), 1446–1462.

[72] P. W. Stephens, P. A. Heiney, R. J. Birgeneau, P. M. Horn, D. E. Moncton and G. S. Brown. "High-resolution x-ray scattering study of the commensurate-incommensurate transition of monolayer Kr on graphite." *Phys. Rev. B*, **29**, (1984), 3512–3532.

[73] P. A. Heiney, P. W. Stephens and R. J. Birgeneau, P. M. Horn and D. E. Moncton. "X-scattering study of the structure and freezing transition of monolayer xenon on graphite." *Phys. Rev. B*, **28**, (1983), 6416–6434.

[74] D. Yang and R. F. Frindt. "Powder x-ray diffraction of two-dimensional materials." *J. Appl. Phys.*, **79**, (1996), 2377–2385.

[75] Ph. Lambin and A. A. Lucas. "Quantitative theory of diffraction by carbon nanotubes." *Phys. Rev. B*, **56**, (1997), 35713574.

[76] J. L. Schlenker and B. K. Peterson. "Computed X-ray Powder Diffraction Patterns for Ultrasmall Zeolite Crystals." *Appl. Cryst.*, **29**, (1996), 178–185.

[77] B. E. Warren. "X-ray diffraction in random layer lattices." *Phys. Rev.*, **59**, (1941), 693–698.

[78] S. Rols, R. Almairac, L. Henrard, E. Anglaret and J.-L. Sauvajol. "Diffraction by finite-size crystalline bundles of single wall nanotubes." *Eur. Phys. J. B*, **10**, (1999), 263–270.

[79] S. V. Cherepanova and S. V. Tsybulya. "Simulation of X-ray powder diffraction patterns for one-dimensionally disordered crystals." *Materials Science Forum*, **443-444**, (2004), 87–90.

[80] B. Morosin. "Exchange striction effects in MnO and MnS." *Phys. Rev. B*, **1**, (1970), 236–243.

[81] I. V. Golosovsky, I. Mirebeau, E. Elkaim, D. A. Kurdyukov and Y. A. Kumzerov. "Structure of MnO nanoparticles embedded into channel-type matrices." *Eur. Phys. J. B*, **47**, (2005), 55–62.

[82] P. Debye. "X-ray dispersal." *Annalen der Physik*, **46**(6), (1915), 809. Leipzig.

[83] A. Cervellino, C. Giannini and A. Guagliardi. "Determination of nanoparticle structure type, size and strain distribution from X-ray data for monatomic f.c.c.-derived non-crystallographic nanoclusters." *J. Appl. Cryst.*, **36**, (2003), 1148–1158.

[84] J. I. Langford and D. Loüer. "Diffraction line profiles and Scherrer constants for materials with cylindrical crystallites." *J. Appl. Cryst.*, **15**, (1982), 20–26.

[85] J. I. Langford and D. Loüer. "Powder diffraction." *Rep. Prog. Phys.*, **59**, (1996), 131–234.

[86] I. V. Golosovsky, D. Arčon, Z. Jagličič, P. Cevc, V. P. Sakhnenko, D. A. Kurdyukov and Y. A. Kumzerov. "ESR of MnO embedded into the silica nanoporous matrices with different topology." *Phys. Rev. B*, **72**, (2005), 144410–1–6.

[87] H. Pearson. *Pearson's handbook of crystallographic data for intermetallic phases* (ASM, Metals Park Metals Park, Ohio, 1985).

[88] P. Cherin and P. Unger. "The Crystal Structure of Trigonal Selenium." *Inorganic Chemistry*, **6**, (1967), 1589–1591.

[89] K.E. Murphy, M. B. Altman and B. Wunderlich. "The monoclinic-to-trigonal transformation in selenium." *J. Appl. Phys.*, **48**, (1977), 4122–4131.

[90] R. Nelmes, G. Meyer, J. Tibballs. "The crystal structure of tetragonal KH_2PO_4 and KD_2PO_4 as a function of temperature." *J. Phys. C: Solid State Phys.*, **15**, (1982), 59–75.

[91] R. Nelmes. "The structure of ammonium hydrogen sulfate in its ferroelectric phase and the ferroelectric." *Phys. Stat. Sol. (b)*, **52**, (1972), K89–K91.

[92] C. Falah, L. Smiri-Dogguy, A. Driss, T. Jouini. "Preparation et Determination Structurale de la Forme Haute Temperature de KH_2PO_4." *Journ. Solid State Chemistry*, **141**, (1998), 486–491.

[93] K. Itoh, T. Matsubayashi, E. Nakamura and H. Motegi. "X-Ray Study of High-Temperature Phase Transitions in KH_2PO_4." *J. Phys. Soc. Japan*, **39**, (1975), 843–844.

[94] M. Mathew and W. Wong-Ng. "Crystal structure of a new monoclinic form of potassium dihydrogen phosphate containing orthophosphacidium ion, $H_4PO_4^{+1}$." *Journ. Solid State Chem.*, **114**, (1995), 219–223.

[95] A. R. Ubbelohde. *Melting and Crystal Structure* (Clarendon, Oxford, 1965).

[96] M. P. Morales, C. J. Serna, F. Bødker and S. Mørup. "Spin canting due to structural disorder in maghemite." *J. Phys.: Condens. Matter*, **9**, (1997), 5461–5467.

[97] A. N. Shmakov, G. N. Kryukova, S. V. Tsybulya, A. L. Chuvilin and L. P. Solovyeva. "Vacancy Ordering in γ-Fe_2O_3: Synchrotron X-ray Powder Diffraction and High-Resolution Electron Microscopy Studies." *J. Appl. Cryst.*, **28**, (1995), 141–145.

[98] A. F. Wells. *Structural inorganic chemistry* (Clarendon Press, Oxford, 1984).

[99] F. Walz. "The Verwey transition a topical review." *J. Phys.: Condens. Matter*, **14**, (2002), R285–R340.

[100] L. A. Solovyov, V. I. Zaikovskii, A. N. Shmakov, O. V. Belousov and Ryong Ryoo. "Framework Characterization of Mesostructured Carbon CMK-1 by X-ray Powder Diffraction and Electron Microscopy." *J. Phys. Chem. B*, **106**, (2002), 12198–12202.

[101] I. V. Golosovsky, I. Mirebeau, F. Fauth, M. Mazaj, D. A. Kurdyukov and Yu. A. Kumzerov. "High-resolution x-ray diffraction study of MnO nanostructured within a MCM-48 silica matrix with a gyroidal system of channels." *Phys. Rev. B.*, **74**, (2006), 15440-1–5.

[102] N. D. Mermin. "Crystalline order in two dimensions." *Phys. Rev.*, **176**, (1968), 250–254.

[103] J. Als-Nielsen, J. D. Litster, R. J. Birgeneau, M. Kaplan, C. R. Safinya, A. Lindegaard-Andersen and S. Mathiesen. "Observation of algebraic decay of positional order in a smectic liquid crystal." *Phys. Rev. B*, **22**, (1980), 312–320.

[104] P. Dutta and S. K. Sinha. "Analitic form for the static structure factor for a finite two-dimensional harmonic lattice." *Phys. Rev. Lett.*, **47**, (1981), 50–53.

[105] Y. Imry and L. Gunther. "Fluctuations and physical properties of the two-dimensional crystal lattice." *Phys. Rev. B*, **3**, (1971), 3939–3945.

[106] V. Petkov, Yong Peng, G. Williams, Baohua Huang, D. Tomalia and Yang Ren. "Structure of gold nanoparticles suspended in water studied by x-ray diffraction and computer simulation." *Phys. Rev. B*, **72**, (2005), 195402-1–7.

[107] S. Sako and K. Ohshima. "Antiferromagnetic transition temperature of MnO ultrafine particle." *J. Phys. Soc. Japan*, **64**, (1995), 944–950.

[108] Y. Hayakawa, S. Kohiki, M. Sato, Y. Sonda, T. Babasaki, H. Deguchi, A. Hidaka, H. Shimooka and S. Takahashi. "Magnetism of diluted Co_3O_4 nanocrystals." *Physica E*, **9**, (2001), 250–252.

[109] I. Naray-Szabo. *Krystalykemia* (Academiai Kiado, Budapest, 1969).

[110] P. Jóvári, R. G. Delaplane, L. Pusztai. "Structural models of amorphous selenium." *Phys. Rev. B*, **67**, (2003), 172201-1–4.

[111] K. Matsuishi, K. Nogi, J. Ohmori, S. Onari and T. Arai. "Structural phase stability of Se clusters in zeolites." *Z. Phys. D: At., Mol. Clusters*, **40**, (1997), 530–533.

[112] I. L. Li and Z. K. Tang. "Phase transition in confined Se nanoclusters." *J. Appl. Phys.*, **95**, (2004), 6364–6367.

[113] K. M. Unruh, T. E. Huber, C. A. Huber. "Melting and freezing of indium metal in porous glasses." *Phys. Rev. B*, **48**, (1993), 9021–9027.

[114] E. V. Charnaya, P. G. Plotnikov, D. Michel, C. Tien, B. F. Borisov, I. G. Sorina, E. I. Martynova. "Acoustic studies of melting and freezing for mercury embedded into Vycor glass." *Physica B*, **299**(1), (2001), 56–63.

[115] I. V. Golosovsky, I. Mirebeau, V. P. Sakhnenko, D. A. Kurdyukov and Y. A. Kumzerov. "Evolution of the magnetic phase transition in MnO confined to channel type matrices. Neutron diffraction study." *Phys. Rev. B*, **72**, (2005), 144409-1–5.

[116] M. Bonfante, B. Hennion, F. Moussa and G. Pepy. "Spin waves in MnO at 4.2 K." *Solid State Commun.*, **10**, (1972), 553–556.

[117] R. H. Kodama. "Magnetic nanoparticles - Condens. Matter." *J. Magn. Magn. Mater.*, **200**, (1999), 359–372.

[118] A. E. Berkowitz and K. Takano. "Exchange anisotropy a review." *J. Magn. Magn. Mater.*, **200**, (1999), 552–570.

[119] J. M. D. Coey. "Noncollinear spin arrangement in ultrafine ferrimagnetic crystallites." *Phys. Rev. Lett.*, **27**, (1971), 1140–1142.

[120] C., J. Serna, F. Bødker, S. Mørup, M. P. Morales, F. Sandiumenge,and S. Veintesmillas-Verdaguer. "Spin frustration in maghemite nanoparticles." *Solid State Comm.*, **118**, (2001), 437–440.

[121] J. Merikosky, J. Timonen and M. Manninen and P. Jena. "Ferromagnetism in small clusters." *Phys. Rev. Lett.*, **66**, (1991), 938–941.

[122] C. Greaves. "A Powder Neutron Diffraction Investigation of Vacancy Ordering and Covalence in γ-Fe$_2$O$_3$." *J. Solid State Chem.*, **49**, (1983), 325–333.

[123] J. Hastings and L. Corliss. "Neutron diffraction study of manganese ferrite." *Phys. Rev.*, **104**, (1956), 328–331.

[124] R. H. Kodama and A. E. Berkowitz. "Atomic-scale magnetic modelling of oxide nanoparticles." *Phys. Rev. B*, **59**, (1999), 6321–6336.

[125] T. Moriya. "Anisotropic superexchange interaction and weak ferromagnetism." *Phys. Rev.*, **120**, (1960), 91–98.

[126] J. O. Artman, J. C. Murphy and S. Foner. "Magnetic Anisotropy in Antiferromagnetic Corundum-Type Sesquioxides." *Phys. Rev.*, **138**, (1965), A912–A917.

[127] S. J. Stewart, R. A. Borzi, E. D. Cabanillas, G. Punte, R. C. Mercader. "Effects of milling-induced disorder on the lattice parameters and magnetic properties of hematite." *J. Magn. Magn. Mater.*, **260**, (2003), 447–454.

[128] D. Schroeer and R. C. Nininger Jr. "Morin transition in $\alpha-$Fe$_2$O$_3$ microcrystals." *Phys. Rev. Lett.*, **19**, (1967), 632–634.

[129] R. Nathans, S. J. Pickart, H. A. Alperin, P. J. Brown. "Polarised-neutrons study of hematite." *Phys. Rev.*, **136**, (1964), A1641–A1647.

[130] F. Bødker, M. F. Hansen, C. Koch, K. Lefmann, S. Mørup. "Magnetic properties of hematite nanoparticles." *Phys. Rev. B*, **61**, (2000), 6826–6838.

[131] R. Worlton, D. L. Decker. "Neutron diffraction study of the magnetic structure of hematite to 41 kbar." *Phys. Rev.*, **171**, (1968), 596–599.

[132] G. Kh. Rozenberg, L. S. Dubrovinsky, M. P. Pasternak, O. Naaman, T. Le Bihan and R. Ahuja. "High-pressure structural studies of hematite Fe$_2$O$_3$." *Phys. Rev. B*, **65**, (2002), 064112–1–8.

[133] C. Frandsen, S. Mørup. "Inter-particle interactions in composites of antiferromagnetic nanoparticles." *J. Magn. Magn. Mater.*, **266**, (2003), 36–48.

[134] R. A. Borzi, S. J. Stewart, G. Punte, R. C. Mercader, M. Vasquez-Mansilla, R. D. Zysler, E. D. Cabanillas. "Magnetic interactions in hematite small particles obtained by ball milling." *J. Magn. Magn. Mater.*, **205**, (1999), 234–240.

[135] W. Kündig, H. Bömmel, G. Constabaris, R. H. Lindquist. "Some Properties of Supported Small α-Fe$_2$O$_3$, Particles Determined with the Mössbauer Effect." *Phys. Rev.*, **142**, (1966), 327–333.

[136] C. Frandsen, S. Mørup. "Spin Rotation in α-Fe$_2$O$_3$ Nanoparticles by Interparticle Interactions." *Phys. Rev. Lett.*, **94**, (2005), 027202-1–4.

[137] M. E. Lines. "Green functions in the theory of antiferromagnetism." *Phys. Rev.*, **135**(5A), (1964), 1336–1346.

[138] M. E. Lines. "Antiferromagnetism in the Face-Centered Cubic Lattice. I. The random-phase Green's function approximation." *Phys. Rev.*, **139**(4A), (1965), A1304–A1312.

[139] M. E. Lines and E. D. Jones. "Antiferromagnetism in the Face-Centered Cubic Lattice. II. Magnetic Properties of MnO." *Phys. Rev.*, **139**(4A), (1965), A1313–A1327.

[140] D. P. Landau. "Computer simulation studies of magnetic phase transitions." *J. Magn. Magn. Mater.*, **200**, (1999), 231–247.

[141] W. Neubeck, C. Vettier, D. Mannix, N. Bernhoeft, A. Hiess, L. Ranno and D. Givord. "Magnetic neutron diffraction of MnO thin films." *Annual Reports of Institute Laue Langevin 1998, Grenoble*.

[142] W. Neubeck, L. Ranno, M. B. Hunt, C. Vettier and D. Givord. "Epitaxial MnO thin films grown by pulsed laser deposition." *Appl. Surf. Sci.*, **138-139**, (1999), 195–198.

[143] D. Bloch, D. Hermann-Ronzaud, C. Vettier, W. B. Yelon and R. Alben. "Stress-induced tricritical phase transition in manganese oxide." *Phys. Rev. Lett.*, **35**, (1975), 963–967.

[144] M. B. Salomon. "Specific heat of CoO near T_N: anisotropy effect." *Phys. Rev. B*, **2**, (1970), 214–219.

[145] K. H. Germann, K. Maier and E. Strauss. "Linear magnetic birefringence in transition metal oxides: CoO." *Phys. Stat. Sol. b*, **61**, (1974), 449.

[146] Y. Imry. "Finite-size rounding of a first-order phase transition." *Phys. Rev. B*, **21**, (1980), 20422043.

[147] W. Neubeck, C. Vettier, K.-B. Lee and F. de Bergevin. "K-edge resonant x-ray magnetic scattering from CoO." *Phys. Rev. B*, **60**, (1999), R9912–R9915.

[148] C. F. J. Flipse, C. B. Rouwelaar and F. M. F. de Groot. "Magnetic properties of CoO nanoparticles." *Eur. Phys. J. D*, **9**, (1999), 479–481.

[149] G. G. Cabrera. "First order phase transitions in a solid with finite size." *Inter. J. Mod. Phys. B*, **4**, (1990), 1671.

[150] K. Uzelac, A. Hasmy and R. Jullien. "Numerical study of phase transitions in the pores of an aerogel." *Phys. Rev. Lett.*, **74**, (1995), 422–425.

[151] M. S. S. Challa, D. P. Landau and K. Binder. "Finite-size effects at temperature-driven first-order transitions." *Phys. Rev. B*, **34**, (1986), 18411852.

[152] K. Binder and D. P. Landau. "Finite-size scaling at first-order phase transition." *Phys. Rev. B*, **30**, (1984), 1477–1485.

[153] A. Beskrovny, I. Golosovsky, A. Fokin, Yu. Kumzerov, A. Kurbakov, A. Naberezhnov, S. Vakhrushev. "Structure evolution and formation of a pre-melted state in $NaNO_2$ confined within porous glass." *Appl. Phys. A*, **74**, (2002), S1001–S1003.

[154] I. Golosovsky, V. Dvornikov, T. Hansen, A. FokinII, E. Koroleva, L. Korotkov, A. Naberezhnov and M. Tovar. "Structure and Conductivity of Nanostructured Sodium Nitrite." *Solid State Phenomena*, **115**, (2006), 221–228.

[155] H. Stanly. *Introduction to phase transitions and critical phenomena* (Clarendon Press, Oxford, 1971).

[156] Z. X. Tang, C. M. Sorensen, and K. J. Klabunde and G. C. Hadjipanayis. "Size-Dependent Curie Temperature in Nanoscale $MnFe_2O_4$ Particles." *Phys. Rev. Lett.*, **67**, (1991), 3602.

[157] I. V. Golosovsky, N. S. Sokolov, A. Gukasov, A. Bataille, M. Boehm, and J. Nogués. "Size-dependent magnetic behaviour and spin-wave gap in MnF2 epitaxial films with orthorhombic crystal structure." *Journ. of Magn. Mag. Mat.*, **322**, (2010), 668–671.

[158] I. V. Golosovsky, I. Mirebeau, F. Fauth, M. Mazaj, D. A. Kurdyukov and Yu. A. Kumzerov. "Size and anisotropy effects on magnetic properties of antiferromagnetic nanoparticles." *J. Magn. Magn. Mater.*, **322**, (2010), 234–237.

[159] Morales M.A., Skomski R., Fritz S., Shelburne G., Shiled j.E., Yin M., O'Brien S., Leslie-Pelecky D.L. "Surface anisotropy and magnetic freezing of MnO nanoparticles." *Phys. Rev. B*, **75**, (2007), 134423.

[160] L. Sun, P. C. Searson and C. L. Chien. "Finite-size effect in nickel nanowires arrays." *Phys. Rev. B*, **61**, (2000), R6463–R6466.

[161] T. MacFarland and G. T. Barkema, J. F. Marko. "Equilibrium phase transitions in a porous medium." *Phys. Rev. B*, **53**, (1996), 148–158.

[162] P. V. Hendriksen, S. Linderoth, P.-A. Lingård. "Finite-size modifications of the magnetic properties of clusters." *Phys. Rev. B*, **48**, (1993), 7259–7273.

[163] A. Punnoose, H. Magnone, and M. S. Seehra. "Bulk to nanoscale magnetism and exchange bias in CuO nanoparticles." *Phys. Rev. B*, **64**, (2001), 174420-1–7.

[164] X. Batlle and A. Labarta. "Finite-size effects in fine particles: magnetic and transport properties." *J. Phys. D: Appl. Phys.*, **35**, (2002), R15–R42.

[165] M. I. Kaganov and A. N. Omelyanchuk. "Phenomenological theory of phase transition in a thin ferromagnetic plate." *Sov. Phys. JETP*, **34**, (1972), 895.

[166] D. L. Mills. "Surface Effects in Magnetic Crystals near the Ordering Temperature." *Phys. Rev. B*, **3**, (1971), 3887–389.

[167] D. Bloch. "Contribution to the study of the magnetic properties of solids under hydrostatic pressure." *Annales de physique, 14-th series*, **1**, (1966), 93–125.

[168] Hongtao Shi and D. Lederman, K. V. ODonovan and J. A. Borchers. "Exchange bias and enhancement of the Néel temperature in thin NiF_2 films." *Phys. Rev. B*, **69**, (2004), 214416–1–9.

[169] Kittel, Charles. *Introduction to solid state physics* (Wiley, New York., 1967).

[170] D. S. Rodbell and J. Owen. "Sublattice magnetisation and lattice distortions in MnO and NiO." *J. Appl. Phys.*, **35**, (1964), 1002–1003.

[171] I. V. Golosovsky, I. Mirebeau, F. Fauth, D. A. Kurdyukov and Yu. A. Kumzerov. "Low-temperature phase transition in nanostructured MnO embedded within the channels of MCM-41-type matrices." *Phys. Rev. B.*, **74**, (2006), 054433–1–5.

[172] A. K. Cheetham and D. A. Hope. "Magnetic ordering and exchange effects in the antiferromagnetic solid solution $Mn_xN_{1-x}O$." *Phys. Rev. B*, **27**, (1983), 6964–6967.

[173] W. Jauch and M. Reehuis. "Electron density distribution in paramagnetic and antiferromagnetic MnO: a x-ray diffraction study." *Phys. Rev. B*, **67**, (2003), 184420–1–7.

[174] D. S. Rodbell, J. Owen and P. E. Lawrence. "Low-temperature lattice distortions in MnO and EuTe." *J. Appl. Phys.*, **36**, (1965), 666–667.

[175] B. T. Willis and A. W. Pryor. *Thermal vibration in crystallography* (Cambridge University Press, 1974).

[176] I. V. Golosovsky, A. A. Naberezhnov, D. A. Kurdyukov, I. Mirebeau, G. André. "Temperature evolution of the structure in CuO nanoparticles within a porous glass." *Kristallographia (in Russian)* , **55**, (2010), 919–923.

[177] X.G. Zheng, H. Kubozono, H. Yamada, K. Kato, Y. Ishiwata and C.N. Xu. " ." *Nature Nanotechnology*, **3**, (2008), 724.

[178] S. V. Pan'kova, V. V. Poborchii and V. G. Solov'ev. "The giant dielectric constant of opal containing sodium nitrate nanoparticles." *J. Phys.: Condens. Matter*, **8**, (1996), L203–L206.

[179] R. Kofman, P. Cheyssac, A. Aouaj, Y. Lereah, G. Deutscher, T. Ben-David, J. M. Penisson and A. Bourret. "Surface melting enhanced by curvature effects." *Surf. Science*, **303**, (1994), 231–246.

[180] J. Lisher. "The Debye-Waller factor of lead from 296 to 550 K." *Acta Cryst.*, **A32**, (1976), 506–509.

[181] M. Merisalo, M. Lehmann and F. Larsen. "Temperature dependence of thermal vibration in lead as determined from short-wavelength neutron diffraction data." *Acta Cryst.*, **A40**, (1984), 127–133.

[182] S. L. Schuster and J.W. Weymouth. "Study of thermal diffuse x-ray sacttering from lead single crystals." *Phys. Rev. B*, **3**, (1971), 4143–4153.

[183] G. A. Heiser, R. C. Shukla and E. R. Cowly. "Average square atomic displacement: a comparison of the lattice-dynamics, molecular-dynamics and Monte Carlo results." *Phys. Rev. B.*, **33**, (1986), 2158–2162.

[184] J. T. Day, J. G. Mullen and R. C. Shukla. "Anharmonic contribution to the Debye-Waller factor for copper, silver and lead." *Phys. Rev. B*, **52**, (1995), 168–176.

[185] D. J. Pastin. "Formulation of the Gruneisen Parameter for Monatomic Cubic Crystals." *Phys. Rev.*, **138**, (1965), A767–A770.

[186] P. Debye. "Interference of x-rays and heat movement." *Annalen der Physik*, **43**(1), (1914), 49–95. Leipzig.

[187] *International Tables for X-ray Crystallography* (The Kynoch Press, Birmingham, 1962).

[188] R. D. Horning and J.-L. Staudenmann. "The Debye-Waller factor for polyatomic solids. Relationships between x-ray and specific-heat Debye temperatures. The Debye-Einstein model." *Acta Cryst.*, **A44**, (1988), 136–142.

In: Neutron Scattering Methods and Studies
Editor: Michael J. Lyons

ISBN: 978-1-61122-521-1
© 2011 Nova Science Publishers, Inc.

Chapter 8

NEUTRON SCATTERING FROM SOLUTIONS OF BIOLOGICAL MACROMOLECULES*

M. W. Roessle and D. I. Svergun

EMBL-Outstation Hamburg, Building 25A, Notkestrasse 85
D-22603 Hamburg, Germany

ABSTRACT

Small-angle scattering (SAS) of X-rays and neutrons is a powerful method for the analysis of biological macromolecules in solution. The method allows one to study low resolution structure of native particles in nearly physiological environments and to analyze dynamic processes like complex formation, assembly and folding. Thanks to recent progress in SAS instrumentation and novel analysis methods, which substantially improved the resolution and reliability of the structural models, the method has attracted renewed interest of biologists and biochemists. In this chapter devoted to biological neutron scattering (SANS) we cover basics of the neutron scattering theory, SANS instrumentation and modern methods used in data analysis and modeling. Differences between X-ray and neutron scattering and advantages/shortcomings of the two techniques are demonstrated. The major advantage of SANS, *i.e.* possibility of contrast variation and specific labeling due to isotope hydrogen/deuterium exchange is comprehensively discussed. Recent applications of SANS to proteins and nucleoprotein complexes are reviewed and a test case, contrast variation analysis of a GroEL/GroES chaperonin system using modern data interpretation methods, is presented in detail.

INTRODUCTION

The scattering of X-rays at small angles (*i.e.* close to the primary beam) was found to provide low resolution structural information (between one and a few hundred nm) the late nineteen thirties by A. Guinier [1]. The SAXS method became an important tool for the study

---

* A version of this chapter was also published in *Methods in Protein Structure and Stability Analysis: Conformational Stability, Size, Shape and Surface of Protein Molecules*, edited by Vladimir N. Uversky and Eugene A. Permyakov, Nova Science Publishers. It was submitted for appropriate modifications in an effort to encourage wider dissemination of research.

of biological macromolecules in solution tool in the sixties, allowing one to assess the overall particle in the absence of crystals. Whereas SAXS could be used at laboratory X-ray devices, practical applications of SANS were hampered by the necessity of having a neutron source. In the seventies, when SAS has increasingly started to move to large scale facilities (synchrotrons for X-rays and research reactors for neutrons), major advantages of SANS compared to SAXS were realized. These were absence of radiation damage and possibility of contrast variation by solvent exchange (H_2O/D_2O) [2] or specific deuteration [3]. Since then, SANS has become a technique complementary to SAXS and actively used in biological research.

SAXS and SANS, although the scattering mechanisms of X-rays and neutrons are very different, can be described with similar mathematical formalisms, and the scattering patterns provide in principle similar information. The experiments typically need homogeneous dilute solutions of macromolecules but do not require specific sample preparation. This relative simplicity of sample preparation results in low information content of the scattering data in the absence of crystalline order. Only overall particle parameters (e.g. volume, radius of gyration) of the macromolecules can be directly determined from the experimental data, whereas the analysis in terms of three-dimensional models is ambiguous. In this sense, SANS can be superior to SAXS as additional information is gained in neutron scattering by hydrogen/deuterium exchange, potentially yielding more spectacular results, especially for complex particles (a classical example is protein triangulation in the small ribosomal subunit [4].). It should however be noted that SANS studies typically require more time and effort than SAXS (an extreme case, the above triangulation study, took more than 15 years!).

Recent progress in SAXS/SANS data analysis methods allows reliable *ab initio* shape and domain structure determination and detailed modelling of macromolecular complexes using rigid body refinement. This progress was accompanied by further advances in instrumentation and biological sample preparation procedures. For X-rays, high brilliance third generation synchrotrons have been built, for neutrons, modern pulsed spallation sources and high yield production of deuterated material are of major importance. In the present chapter, after a brief account of SANS theory and instrumentation, novel data analysis methods for the interpretation of the scattering patterns from macromolecular solutions will be presented. Special attention will be paid to the use of hydrogen/deuterium exchange for contrast variation in SANS, and recent applications of contrast variation for biological studies in solution will be briefly reviewed. A test case will be considered, where modern analysis methods in combination with specific deuteration will be employed to characterize the structure of a large macromolecular complex, a GroEL/GroES chaperonin system.

BASICS OF SMALL ANGLE NEUTRON SCATTERING

Properties of Neutrons

The x-ray photons are scattered by electrons of the atom whereas the neutrons are scattered by the atomic nuclei. The physical mechanisms of elastic neutron and x-ray scattering are fundamentally different, but can be mathematically described by the same formalism. The neutron wavelength is given by de Broglie's relationship

$$\lambda = h/mv \tag{1}$$

where the neutrons mass $m_n = 1.675 \times 10^{-27}$ kg, h is the Planck constant $h = 6.626 \times 10^{-34}$ Js and v is the neutron velocity in m/s.

The spectral range of neutrons used for scattering experiments depends on the neutron source. Neutrons are produced by means of nuclear reactions such as fission or spallation as shown in Figure 1. In the classical fission reaction a thermal neutron is absorbed by a ^{235}Uranium nucleus, which is split into fission fragments. From the 2 to 5 produced neutrons in this reaction an average of 1 neutron is useable, whereas the others are needed to keep the chain reaction going on. Steady state high flux reactors, such as at the Institute Laue-Langenvin (Grenoble/France) or at the Oak Ridge National Laboratory (Oak Ridge/USA) reach a neutron flux of about 1.5×10^{15} n cm^{-2} s^{-1}.

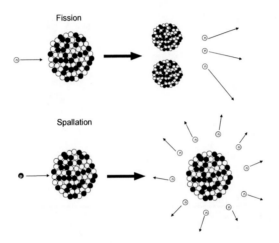

Figure 1. Fission and Spallation.

The other possibility for neutron production is spallation sources. In the spallation process an accelerated high energy proton (typical energy ~ 1GeV) hits a high-target nucleus. Beside particles such as protons, pions *etc* about 20 neutrons per proton are emitted during this reaction. These neutrons have a comparable spectrum to those produced by fission, but a few neutrons carry the full proton energy up to many 100MeV. Spallation sources such as ISIS (Appleton/United Kingdom) or SINQ at the Paul-Scherrer Institute (Villingen/Switzerland) using tantalum or lead-bismuth targets and produce neutron fluxes, which are comparable to high flux nuclear reactors. Table 1 shows parameters for neutrons produced by fission or spallation [5].

Table 1. Typical parameters of neutrons used for scattering experiments

		cold	thermal	hot
Energy,	meV	1	25	1000
Temperature,	K	12	290	12000
Wavelength,	nm	0.9	0.18	0.029
Velocity,	m/s	440	2200	14000

The hot neutrons emerging from fission or spallation have a too high energy (*e.g.* too small wavelength) to be used for scattering experiments on biological macromolecules. In order to slow down the particles the so called "cold sources" have to be used, where the hot neutrons are moderated by inelastic collisions with light atoms such as hydrogen, deuterium or beryllium. The resulting neutron spectrum after the cold source is well described by a Maxwel distribution with a maximum depending on the temperature of the cold source [6]; a typical spectrum is displayed in Figure 2.

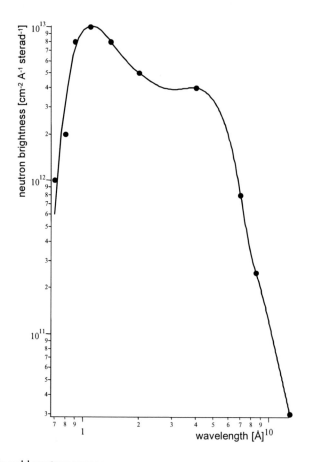

Figure 2. Spectra of a cold neutron source.

Interaction of Neutrons with Matter

The strength of interaction of neutrons with matter is proportional to the scattering cross-section, *i.e.* the number of neutrons per solid angle scattered by the nuclei. This cross-section is related to the scattering length b of a nucleus by

$$d\sigma/d\Omega = b^2 \tag{2}$$

In the elastic case with no energy transfer between the neutron and the nucleus b is a real number and the total scattering length of the nucleus can be written as

$$\sigma = 4\pi\, b^2 \tag{3}$$

The total scattering length consists of a spin dependent component and a contribution from the nuclear potential

$$\sigma_{tot} = \sigma_{spin} + \sigma_{pot} \tag{4}$$

The potential component is always structurally related, whereas the spin scattering contains for randomly oriented spins no structural information and produces a flat background only. Spin contrast variation techniques using polarized neutrons and oriented spins in strong magnetic fields can give further advantages for structural analysis [7,8]. Table 2 presents a comparison of atomic scattering lengths for neutrons and x-rays [9].

Table 2. Neutron and x-ray scattering density length of biological relevant elements

atomic number	Element or Isotope	b_{coh} for neutrons 10^{-12} cm	f_x for X-rays 10^{-12} cm
1	H	-0.374	0.28
1	D	0.667	0.28
6	C	0.665	1.69
6	^{13}C	0.600	1.69
7	N	0.940	1.97
8	O	0.580	2.25
8	^{17}O	0.578	2.25
12	Mg	0.530	3.38
15	P	0.510	4.23
16	S	0.285	4.50
19	K	0.370	5.30

From the table one can derive the following:

(i) The b-factor does not increase with the atomic number as for x-rays
(ii) Neutrons are more sensitive to light atoms such as hydrogen or deuterium than x-rays
(iii) There is a large difference in neutron scattering between the biologically relevant atom hydrogen and its isotope deuterium

The peculiarities of neutrons are actively used for structural analysis and ensure complementarities between x-ray and neutron studies.

Beamlines for Small Angle Neutron Scattering

The main components of a classical small angle neutron scattering experimental setup are presented in Figure 3 [10].

Flight tubes guide the moderated neutrons towards the sample, which can be 10m to 100m away from the cold source in the reactor vessel. These guides taking advantage of the low total reflection angles of the neutrons are usually built from boron glass mirror plates with a rectangular cross section and the inner surface coated by a thin layer of metal (e.g. Ni ~100nm). In order to minimize absorption by air these guides are evacuated and the mirror planes have to be highly polished up to a low surface roughness in the order of about 10% of the metal layer thickness [11]. SANS is an elastic scattering technique and in a classical experiment monochromatic neutron beam is needed. Using the de Broglie equation neutrons with a wavelength of 10Å travel at about 400m/s – a rather low number compared to the velocity of light c=300 000 km/s. This low velocity permits the neutron wavelength selection by mechanical choppers made of disks of neutron absorbing material with transparent aperture slits, which rotate with high speed around the axis parallel to the beam. Modern velocity selectors consist of multiple disks or a helical arrangement of flight path channels. The wavelength (i.e. velocity) is selected by changing the rotation speed of the chopper. The wavelength resolution of such monochromator is about $\Delta\lambda/\lambda{\approx}5\text{-}10\%$.

The monochromatic neutrons are guided by flight tubes to the sample, whereas the guides are also acting as collimating devices, and have to be aligned to provide desired beam size and on the sample (typically about 10 x 10mm^2).

The scattered neutrons are detected by area detectors usually mounted on movable chariots in evacuated detector tubes. This permits one to select the angular range by changing the sample-detector distance. Standard gas-filled SANS detectors register neutrons indirectly by a nuclear reaction such as

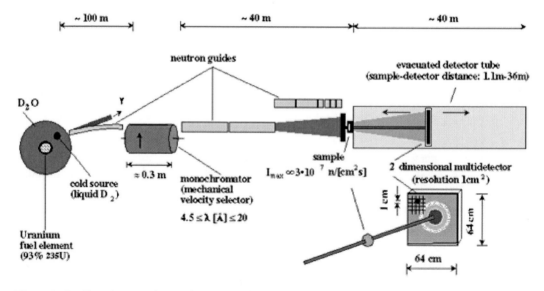

Figure 3. Small angle scattering station D11 at the ILL.

n + ^{3}He → ^{3}Tritium + proton

whereas the proton can be detected by a gas filled proportional counter. In scintillation detectors the neutrons are transformed into light, which can be detected by CCD or image plates.

SANS beamlines on spallation sources have a different design to that of reactor based beamlines. The neutron flux of a spallation source is pulsed, based on the time structure of the proton beam hitting the target. To account for this, SANS experiments can be operated in a time-of-flight (TOF) mode. As pointed out before the neutron velocity depends on the wavelength and thermal neutrons move relativly slow. The TOF method uses this property to discriminate the neutrons arriving at the detector in time frames representing different wavelengths within the pulse. The disperse time T along a flight path L can be determined by variation of the de Broglie relationship

$$T = \frac{m}{h} L \lambda \tag{5}$$

and the wavelength can be estimated as

$$\lambda = \frac{h}{m} \frac{T}{L} \tag{6}$$

for time T, L the distance from the moderator tank to the detector, h the Planck konstant and m the neutrons mass [12,13].

Spallation sources such as ISIS operate at a proton pulse frequency of about 50Hz. At such repetition rates one has to ensure that the fastest neutrons produced by the proton pulse do not overtake the slower ones from the previous pulse. The maximal flight path from the source to the detector is typically in the range of a few dozen meters and the wavelength bandwidth can be estimated by

$$\lambda_{max} - \lambda_{min} = \frac{h}{m} \left(\frac{T_1}{L_D} - \frac{T_0}{L_D} \right) = \frac{h}{m} \frac{T_P}{L_D} \tag{7}$$

where T_P is the repetition time of the pulse and L_D the moderator-detector distance. Using the values for spallation sources above the wavelength bandwidth is about 3Å. From neutron wavelength distribution in Figure 2 one can assume a wavelength cutoff at 10Å to take advantage of the full neutron flux. A mechanical disk chopper is required to catch also these slower neutrons. The chopper cuts out the faster neutrons of the subsequent pulse and increases the effective T_P value and allows λ_{max} comparable to the maximal wavelength of the cold source.

The data recording and reduction on the pulsed sources are more complicated than for the steady state reactors since the neutrons with different wavelength λ have to be separated in different time channels.

Scattering of Thermal Neutrons by Disordered Systems

If an object is illuminated by a monochromatic plane wave with wavevector $k_0 = |k_0| = 2\pi/\lambda$, atoms within the object interacting with the incident radiation become sources of spherical waves. In the elastic scattering case (i.e. without energy transfer) the modulus of the scattered wave $k_1 = |\vec{k}_1| = k_0$ remains unchanged (Figure 4).

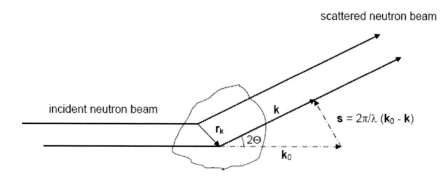

Figure 4. Schematic drawing of elastic scattering. The incident neutron beam is scattered by two different scattering centers under the angle 2Θ. The scattered beams interfere because of the resulting phase shift due to the momentum transfers.

The amplitude of the wave scattered by each atom is described by its scattering length, b.

$$A_k = b_k e^{-i\vec{s}\cdot\vec{r}} \tag{8}$$

where $\vec{r}$ describes the coordinate of the scattering center and $\vec{s}$ the scattering vector

$$\vec{s} = \left(\vec{k} - \vec{k}_0\right) \tag{9}$$

The total scattering amplitude of a particle is derived by summation over all scattering centers:

$$A(\vec{s}) = \sum_k A_k(\vec{s}) = \sum_k b_k e^{-i\vec{s}\cdot\vec{r}} \tag{10}$$

Introducing the scattering length density $\rho(\vec{r})$ in the particle volume V

$$\rho(\vec{r}) = \frac{1}{V}\sum_k b_k \delta(\vec{r}-\vec{r}_k) \tag{11}$$

the total scattering amplitude is given by:

$$A(\vec{s}) = \int\limits_V b_k\, e^{-i\vec{s}\cdot\vec{r}}\, dV \qquad (12)$$

The amplitude $A(\vec{s})$ is not accessible in a scattering experiment, but the scattering intensity

$$I(\vec{s}) = A(\vec{s}) \times A(\vec{s}) \qquad (13)$$

which is proportional to the number of scattered neutrons.
Applying the convolution theorem to (13) results in:

$$I(\vec{s}) = \int\limits_v \rho^*(\vec{r})^2\, e^{-i\vec{s}\cdot\vec{r}}\, dV \qquad (14)$$

where $p^*(\vec{r})^2$ denotes the autocorrelation function of the particle, i.e. self correlation function of the scattering length density $\rho(\vec{r})$.

The scattering from ensembles of identical particles depends strongly on the distribution of these particles in the irradiated volume. In the crystal molecules have defined correlated orientations and are regularly distributed in space. The intensity function in this case is a discrete three-dimensional function $I(\vec{s}_{hkl})$ corresponding to the density distribution of the reciprocal lattice of the single unit cell [14]. For solution scattering positions and orientations of the molecules are uncorrelated and the particles are therefore considered to be randomly distributed. Consequently, the intensity from the entire ensemble is proportional to the scattering from a single particle averaged over all orientations $I(s) = I(\vec{s})$. Here, s denotes the modulus of the scattering vector (or momentum transfer), $s = 4\pi\,\lambda^{-1}\,sin(\theta)$ where 2θ is the scattering angle.

From this theoretical consideration one can derive two limiting conditions for small angle scattering as a method for analysis of single macromolecules:

(i) The ensemble of particles has to be mono-disperse i.e. all particles should be identical
(ii) The particles of the ensemble must not interact and the positions and orientations have to be random

Especially if (ii) is not fulfilled, the scattering intensity of interacting particles will contain an additional interference term depended on the interparticle interactions and can be useful for investigations of the interaction potentials, which, however, would impede the single molecule structure analysis.

The I(s) function can be written now as

$$I(s) = \int\limits_0^\infty \rho^*(r)r^2\,\frac{sin(sr)}{sr}\, dr \qquad (15)$$

taking the transformation

$$\left\langle e^{-i\vec{s}\cdot\vec{r}} \right\rangle_\Omega = \frac{\sin(sr)}{sr} \tag{16}$$

and

$$\left\langle \rho^*(\vec{r})^2 \right\rangle_\Omega = \rho^*(r)r^2$$

for $\vec{s}$ and $\vec{r}$ over all possible orientations Ω into account. This equation (15) was first derived in its discrete form by Debye [15] and is valid for a diluted, mono-disperse solution of independent particles

As pointed out above the scattering lengths of atoms are different and the difference between the scattering density of the particle and the density of the environment has to be taken into account. The average excess density of the particle is

$$\Delta\rho = \left\langle \Delta\rho(\vec{r}) \right\rangle = \left\langle \rho(\vec{r}) - \rho_S \right\rangle \tag{17}$$

where ρ_s is the scattering density of the solvent. The particle density can be represented as

$$\rho(\vec{r}) = \Delta\rho\rho_C(\vec{r}) - \rho_F(\vec{r}) \tag{18}$$

where $\rho_C(r)$ is the shape function equal to 1 inside the particle and 0 outside, and scattering length density fluctuations are written as $\rho_F(\vec{r}) = \rho(\vec{r}) - \left\langle \rho(\vec{r}) \right\rangle$. Introducing this expression into equation (12), the scattering amplitude contains two terms

$$A(\vec{s}) = \Delta\rho A_C(\vec{s}) + A_F(\vec{s}) \tag{19}$$

and the spherically averaged scattering intensity is composed of three basic scattering functions:

$$I(s) = (\Delta\rho)^2 I_C(s) + 2\Delta\rho I_{CF}(s) + I_F(s) \tag{20}$$

where $I_C(s)$ is the scattering from the particle shape i.e. from the excluded volume filled with an unitary density, $I_F(s)$ emerges from the density fluctuations around the average value and $I_{CF}(s)$ is a cross-term. This equation is of general value and the contributions from the overall shape and internal structure of particles can be separated using measurements in solutions with different solvent density (i.e. for different $\Delta\rho$).

The above equations hold for an ideal dilute solution of non-interacting particles. For a specimen, which exhibits interactions between the particles, the scattering intensity changes to:

$I_S(s) = I(s) \times S(s)$ (21)

The interference term S(s) characterizes the interaction between the particles and is called structure factor. Separation of the two terms for semi-dilute solutions is possible by measurements at different concentrations or/and in different solvent conditions (pH, ionic strength, etc). For systems of particles differing in size and/or shape, the total scattering intensity is given by the weight average of the scattering from the different types of particles.

DATA ANALYSIS AND INTERPRETATION

Overall Parameters Derived from SANS

Following equation (16) the scattering intensity is the Fourier transformation of the spherically averaged autocorrelation function $\rho^*(r)$. This function of excess scattering density is obviously equal to zero for distances exceeding the maximum particle diameter D_{max}. For small angle scattering the function $p(r) = r^2 \Delta\rho^*(r)$ is often used, which corresponds to the distribution of distances between volume elements inside the particle weighted by the excess density distribution $\Delta\rho^*(r)$. Using the p(r) function the scattering intensity can be written as

$$I(s)=4\pi \int_0^{D_{max}} p(r)\frac{\sin(sr)}{sr}dr$$ (22)

and the distance distribution function is computed by the inverse Fourier transformation

$$p(r)=\sqrt{\frac{1}{2\pi^2}}\int_0^{\infty} I(s)s^2 \frac{\sin(sr)}{sr}ds$$ (23)

For s=0 the I(s) function (22) is reduced to

$$I(s = 0)=4\pi \int_0^{\infty} p(r)dr=(\Delta\rho)^2 V^2$$ (24)

and the total scattering intensity I(0) is proportional to the area of the p(r) function, i.e. to the total scattering length of the particle. Knowing the chemical composition of the particle, its molecular mass can be estimated from the I(0) value.

The second moment of this distance distribution function p(r) is called in analogy to classical mechanics "radius of gyration" R_g. As the p(r) function is equal to zero for distances larger than the maximum particle diameter D_{max}, R_g is expressed as

$$R_g^2 = \frac{1}{2} \frac{\int_0^{D_{max}} p(r)r^2\,dr}{\int_0^{D_{max}} p(r)\,dr} \tag{25}$$

The value of the R_g can also be computed directly from the scattering function by the "Guinier approximation" [1,16], which is derived by expanding the $sin(sr)/sr$ term in (22) in a McLaurin series leading to the equation:

$$I(s) = 4\pi \left\{ \int_0^{D_{max}} p(r)\,dr - \frac{s^2}{3!} \int_0^{D_{max}} p(r)r^2\,dr + \ldots \right\} \tag{26}$$

Using the definitions for R_g and $I(0)$ the series expansion can be reduced to

$$I(s) = I(0) \left\{ 1 - \frac{R_g^2}{3} s^2 + \ldots \right\} \tag{27}$$

and for $sR_g < 1.3$ the inner part of the scattering function can be described as a Gaussian function

$$I(s) = I(0)e^{-\frac{R_g^2}{3} s^2} \tag{28}$$

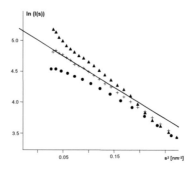

Figure 5. Guinier plots of proteins in solution: Symbols (▲) interacting particles, (●) attractive interaction (protein aggregation) and (+) monodisperse non interacting particle.

This approximation derived by Guinier in the late thirties has long been the most important tool in the analysis of scattering from isotropic systems and continues to be very useful at the first stage of data analysis. The Guinier plot $ln(I(s))$ versus s^2 is in the case of an ideal monodisperse systems a linear function, whose intercept gives $I(0)$ and the slope yields R_g. Deviations from the linearity of the Guinier plot indicate attractive or repulsive interparticle interactions leading to interference effects. In Figure 5, Guinier plots of interacting systems as well as of an ideal solution of molecules are shown.

The Guinier approximation holds for particles of arbitrary size and the R_g characterize the shape of a particle. For simple geometrical bodies the R_g can be calculated [17] analytically as shown in Table 3.

Table 3. Radii of gyration of simple geometrical bodies

body	R_g^2
sphere (radius R)	$(3/5) R^2$
hollow sphere (inner radius R_1, outer radiusR_2)	$(3/5) (R_2^5 - R_1^5)/(R_2^3 - R_1^3)$
ellipsoid (semi-axes a, b, c)	$(a^2 + b^2 + c^2)/5$
elliptic cylinder (semi-axes a, b; height h)	$(a^2 + b^2)/4 + h^2$
hollow cylinder (radii R_1, R_2, height h)	$(R_1^2 + R_2^2)/2 + h^2/12$

For elongated biological macromolecules such as actin, myosin and chromatin filaments, which may be hundreds nm long, it may not be possible to record reliable data in the Guinier region ($s \leq 1.3/R_g$). For these very elongated particles, the radius of gyration of the cross-section R_c can be derived using a similar representation plotting $sI(s)$ versus s^2, and for flattened particles, the radius of gyration of the thickness R_t is computed from the plot of $s^2I(s)$ versus s^2:

$$s \cdot I(s) \cong I_c(0)e^{-\frac{R_c^2}{2}s^2} \tag{29}$$

or

$$s^2 \cdot I(s) \cong I_t(0)e^{-R_t^2 s^2} \tag{30}$$

The cross-sectional or thickness information is additional data to the overall parameters of the particle, but less structural information can be obtained by SANS from such particles than for more isometric ones.

The so-called "Kratky plot" of $s^2I(s)$ versus s [18,19] allows fast analysis of the protein folding state. A linear increase at high s-values indicates in this representation an unfolded protein. This random chain like behavior changes for more compact particles. In this case the $s^2I(s)$ function decays at high s-values indicating a globular (folded) protein state. Figure 6 shows typical Kratky plots of completely random chain like behavior for the unfolded protein changing to a completely folded compact state [20-23].

This analysis allows a rapid differentiation between the random chain and globular shape, but requires precise measurements at higher scattering angles.

For compact particles, analysis of the asymptotic behavior of scattering data for s→∞ is done using the so called Porod law [17,24]. From equation (22) one can approximately express the scattering intensity at high angles as

$$I(s) \cong 8\pi s^{-4} \rho^*(0) + o(s) \tag{31}$$

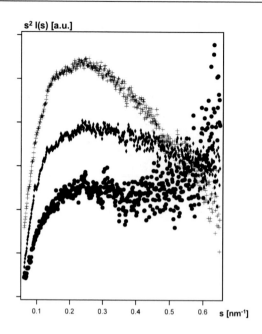

Figure 6. Kratky plots of different protein folding states. Symbols (●) indicate a completely unfolded state and (+) the folded compact state. An intermediated state is shown in the middle (◆). An offset was applied to the functions for better visualization.

where o(s) are oscillating terms of the form $sin(sD_{max})$. This s^{-4} intensity decay is known as the Porod law for higher scattering angles. From the $\rho^{*}(0)$ the particle surface S is defined by $\rho^{*}(0)= (\Delta\rho)^2 S/4$. Moreover, the total scattering from the particle Q can be derived

$$Q=\int_{0}^{\infty}I(s)s^{2}ds=2\pi^{2}\int_{V}(\Delta\rho(r))^{2}d\vec{r} \qquad (32)$$

and is called Porods invariant. In the case of homogeneous particles this invariant simplifies to $Q=2\pi^2 (\Delta\rho)^2 V$ and the excluded particle volume $V=2\pi^2 I(0) Q^{-1}$ is derived by recalling that $I(0)= (\Delta\rho)^2 V^2$. Summing up the above definitions the asymptotic behavior for s→∞ gives a specific particle surface

$$\frac{S}{V}=\pi\frac{\lim_{s\to\infty} s^{4} I(s)}{Q} \qquad (33)$$

The Porod approximation for larger angles holds for homogenous particles, and internal inhomogeneities make this analysis more difficult. In the case of large (MM>40 kDa) macromolecules, which can be measured at high contrasts (for instance protein in pure D_2O) the inhomogeneities can be taken into account by simply subtracting a constant term from the experimental data. A linear plot of $s^4I(s)$ against s^4 coordinates: $s^4I(s) \approx Bs^4 + A$, allows one to estimate A and B. Subtraction of the constant B from I(s) yields an approximation to the scattering of the corresponding homogeneous body. Alternatively, although more labor consuming, one can measure a full contrast variation series with several scattering data sets at different D_2O concentrations in solution and retrieve the shape scattering function $I_C(s)$.

The Pair Distance Distribution Function p(r)

Computation of the *p(r)* function can in principle be done by the inverse Fourier transformation (23). The necessary integration limits of this mathematical procedure (s=0 and s→∞) are however not reachable in a SANS experiment. Moreover the measured intensity *I(s)* is recorded at a number N of discrete data points *(s_i)*, with statistical and systematical errors. The main source of systematical errors in a SANS experiment are beam polychromaticity and smearing of the scattering function by the large geometrical dimensions of the neutron beam. These factors impede the straight forward *p(r)* computation. To overcome these problems indirect transformation (IFT) methods have been developed and advanced over the years [25-27]. The IFT method first introduced by O. Glatter [25] starts by representing the *p(r)* function as a linear combination of k orthogonal functions φ_k.

$$p(r) = \sum_k c_k \varphi_k(r) \tag{34}$$

This approximation holds for $0 < r < Dmax$ with coefficients ci. The Fourier transformation results in

$$I_{approx}(s) = \sum_k c_k \int_0^{D_{max}} \varphi_k(r) \frac{\sin(sr)}{sr} dr = \sum_k c_k \phi_k(s_i) \tag{35}$$

where the $\phi_k(s_i)$ represent the Fourier transformed functions φ_k smeared by the instrument resolution function if required. The coefficients c_k are calculated by fitting the experimental scattering function to minimize the functional

$$\Phi_\alpha(c_k) = \sum_i^N \left[\frac{I_{exp}(s_i) - \sum_k c_k \phi_k(s_i)}{\sigma(s_i)} \right]^2 + \alpha \int_0^{D_{max}} \left[\frac{dp}{dr} \right]^2 dr \tag{36}$$

The regularization multiplier $\alpha \geq 0$ maintains the balance between the goodness of fit to the data (first summand) and the smoothness of the *p(r)* function (second summand).

The information content of the *p(r)* and the *I(s)* function is identical, but the real-space *p(r)* is often more intuitive than the *I(s)* function. Moreover, parameters such as *I(0)* and R_g are readily calculated from the *p(r)* function (Eqs (24-25)), which in some cases eases the strict acceptance range of the Guinier approximation. Typical representatives of the IFT methods are the program *IFT* [25] and *GNOM* [28].

The Contrast Variation Technique

The differences in the scattering lengths b of different atoms (see Table 2) are used for the so-called contrast variation method. Here, the significant difference in the b values between hydrogen and its isotope deuteron is especially important for biological applications. This difference can be used in two ways. A classical contrast variation in mixtures of H_2O/D_2O allows one to effectively study multi component systems like protein-nucleic acid complexes. Using this technique the ratio of H_2O/D_2O in the solvent can be selected in such a way, that the solvent scattering density coincides with the averaged scattering density of one of the components. The ratio is called "matching point", and $\Delta\rho=0$ for this component. Note that in practice measurements at higher D_2O concentration are more accurate as these solvents produce less incoherent spin scattering from hydrogen and thus better signal to noise ratios.

Further, extremely valuable information can be obtained by selective labeling, where a portion of the structure is specifically per-deuterated. This can be done by growing cell cultures in heavy water medium. Under this conditions not only the exchangeable hydrogens (which belong to NH, NH_2, NH_3, OH and SH groups and exchange easily with the hydrogens/deuteriums in the solvent) and the non-exchangeable carbon bound H-atoms are replaced by deuterons. The per-deuteration of proteins allows one to effectively see contrast variation e.g. on protein-protein complexes. Figure 7 presents the scattering densities of deuterated, protonated proteins, nucleic acids and lipids as functions of the D_2O buffer content.

The scattering density of purely protonated (~native) protein is matched out in a mixture of 40% D_2O in H_2O, whereas for the nucleic acids the contrast matching point is in 70% H_2O. Hydrogen and deuterium are chemically very similar and exchange of hydrogen by deuterium usually does not compromise the protein function generally. Hence, deuterated recombinant proteins can be obtained by growing for instance bacteria such as *Escherichia coli* in D_2O media [29]. The yield of protein is lower compared to the non deuterated media, but modern bacteria fermentation techniques often allow to harvest sufficient material for structural investigations.

This classical contrast variation allows one to investigate selectively labeled protein in a complex in the bound *in situ* state. Additional information about the distances between labeled components can be obtained if such a complex is measured under different contrast conditions. Equation (24) describes the dependency of the scattering intensity at the zero angle (s=0) as a function of contrast $\Delta\rho$. If one assumes a completely homogenous particle the scattering intensity at the matching points equals to zero, however structural inhomogeneities inside the particle produce additional scattering intensity and the radius of gyration R_g can be written as

$$R_g^2 = R_C^2 + \frac{\alpha}{\Delta\rho} - \frac{\beta}{(\Delta\rho)^2}$$

$$\alpha = \frac{1}{V} \int \Delta\rho(r) r^2 \, dr \qquad \text{and}$$

(37)

$$\beta = \frac{1}{V^2} \int (\Delta \rho_1)^2 (\Delta \rho_2)^2 \, r_1^2 \, r_2^2 \, dr_1 \, dr_2$$

where R_C is the the radius of gyration of the particle shape at infinite contrast. For the ideal case of a homogenous particle both α and β are equal to zero. The sign of the α-term depends on the distribution of inhomogeneïties (α is positive if the molecule has higher scattering density in its outer part than at the inner part and vice versa), whereas β describes the displacement of the scattering density center and is always positive. Both parameters can be derived from the so-called "Stuhrmann plot" [30] (see Figure 8 (A)) where the radius of gyration R_g is measured for different contrasts and plotted versus $(1/\Delta \rho)^2$. One should note that the estimate of β is highly dependent on the data quality at low contrasts, which are in many cases difficult to obtain.

In the case of a two component protein system measured at different contrasts the center to center distance d between the subunits is calculated from the center displacement parameter β following the equation

$$d = \left(\frac{\beta}{\Delta \rho_1^2} \right)^{\frac{1}{2}} + \left(\frac{\beta}{\Delta \rho_2^2} \right)^{\frac{1}{2}} \qquad (38)$$

where the $\Delta \rho_1$ ($\Delta \rho_2$) is the contrast of the first (second) subunit under contrast matching conditions of the second (first) subunit [31]. Alternatively, d can be determined in analogy to the parallel axis theorem of classical mechanics

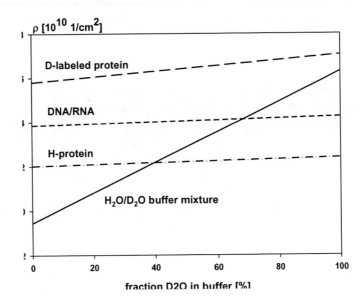

Figure 7. Scattering length density of biological macromolecules as a function of D_2O concentration in the solvent. The matching point can be determined by the crossing point of the solvent and the molecule (H-protein at 40% and RNA/DNA at 70%). Note that the density slightly increases with D_2O content due to H/D substitution of hydrogens in exchangeable groups.

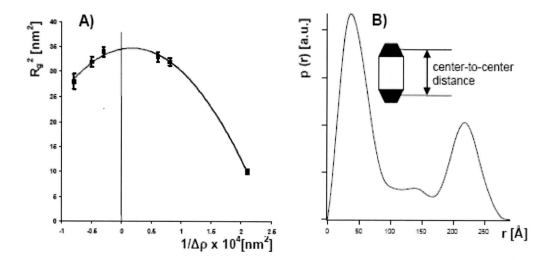

Figure 8. Center to center distance evaluation: A) Stuhrmann plot for obtaining the α and β parameters by a parabolic fit and B) direct distance estimation from *p(r)*.

$$R_g^2 = \omega_1 R_1^2 + (1-\omega_1)R_2^2 + \omega_1(1-\omega_1)d^2 \tag{39}$$

where R_1 and R_2 are the radii of gyration of the components and $\omega_1 = \Delta\rho_1 V_1$ is the fraction of scattering density of the first component. The above approach was successfully used for the estimation of intermolecular distances in the 30S ribosomal subunit and a complete 3D map of the ribosomal proteins could be obtained by this triangulation method [4,31-33]. The mapping of the ribosomal proteins of the larger 50S subunit could not be achieved by classical contrast variation only and the spin contrast variation using polarized neutrons was applied [34,35].

A direct determination of the center to center distance based on the *p(r)* is possible in special cases only, for example if two independent subunits in a trimeric assembly can be selectively labeled and reassembled with the entire complex. The entire scattering intensity of such an assembly can be written following equation (13)

$$\begin{aligned} I(s) &= (\Delta\rho_1 A_1(\vec{s}) + \Delta\rho_2 A_2(\vec{s}) + \Delta\rho_3 A_3(\vec{s}))^2 \\ &= \Delta\rho_1^2 I_1(s) + \Delta\rho_2^2 I_2(s) + \Delta\rho_3^2 I_3(s) + 2\Delta\rho_1\Delta\rho_2 I_1(s)I_2(s) \\ &\quad + 2\Delta\rho_1\Delta\rho_3 I_1(s)I_3(s) + 2\Delta\rho_2\Delta\rho_3 I_2(s)I_3(s) \end{aligned} \tag{40}$$

with $\Delta\rho_i I_i(s)$ as scattering contributions of the different complex components and the mixed expressions are cross-terms between the components. Under contrast matching conditions for the first component, its contrast $\Delta\rho_1$ equals to zero, and only the labeled proteins contribute to the scattering pattern

$$I(s) = \Delta\rho_2^2 I_2(s) + \Delta\rho_3^2 I_3(s) + 2\Delta\rho_2\Delta\rho_3 I_2(s)I_3(s) \tag{41}$$

and the Fourier transformation results in

$$p_C(r) = p_2(r) + p_3(r) + p_{23}(r)$$ (42)

The $p_C(r)$ function of such an assembly reveals the distribution functions of the subunits and of the inference between the two subunits. Figure 8 (B) displays a $p(r)$ of a trimeric protein complex with two deuterium labeled proteins measured in 40% D_2O. The first peak at 30Å indicates the contribution of the intrasubunit distances whereas the second peak at 220Å denotes to the center to center distance of the two subunits. The analysis as in Figure 8 (B) is possible for well separated subunits showing little overlap between the intra- and intersubunit vectors. Note that equations (41-42) neglect density fluctuations within the components, which is true if the difference in scattering density between the subunits is much larger than the differences created by the fluctuations.

Calculation of SANS from Atomic Models and Rigid Body Refinement

In many cases one would like to compare SANS data with the scattering from high resolution protein models from protein crystallography or NMR. In principle the scattering intensity $I(s)$ can be calculated by equation (22). This equation however represents the scattering intensity in vacuum and does not take any solvent effects into account. The scattering amplitude of a particle in solution is

$$A_{PS}(\vec{s}) = A_V(\vec{s}) - \rho_S A_S(\vec{s}) + \Delta\rho_B A_B(\vec{s})$$ (43)

where $A_V(\vec{s})$ is the scattering amplitude for the particle in vacuum, $A_S(\vec{s})$ represents the scattering from the excluded particle volume and $A_B(\vec{s})$ is the contribution from the hydration layer. The equation takes the different density of bound solvent molecules in difference to the bulk into account, which gives a non-zero contribution of the hydration layer $\Delta\rho_B = \rho_S - \rho_B$. This effect can not be neglected since the density of the hydration layer of a protein is 5% to 20% higher than the density of the bulk solution as found first in X-ray studies [36]. Moreover, SANS investigations showed that the higher density can not be explained by an increased mobility of the peripheral side chains of the protein and should be ascribed to a different solvent density in the solvation shell [37]. This finding was later confirmed by molecular dynamics simulations [38]. The program CRYSON [37] surrounds the particle with a 3Å thick hydration layer and allows adjusting the ρ_B value. The scattering amplitudes are calculated by a series of spherical harmonics and the total spherical averaged scattering intensity $I_{PS}(s)$ is given by

$$I_{PS}(s) = \left\langle A_{PS}(\vec{s}) \times A_{PS}(\vec{s}) \right\rangle_\Omega = \left\langle \left| A_V(\vec{s}) - \rho_S A_S(\vec{s}) + \Delta\rho_B A_B(\vec{s}) \right| \right\rangle_\Omega^2$$ (44)

with spherical avaraging over the all possible orientations $\langle \ \rangle_\Omega$.

CRYSON (as well CRYSOL [36] for x-rays) are able to read in the atomic coordinates protein data base (PDB [39]) and predict the theoretical scattering function based on the calculated partial amplitudes. For comparison of the theoretical prediction with experimental data CRYSON fits the theoretical model by adjusting the excluded particle volume and the $\Delta\rho_B$ as free parameters.

The amplitudes in CRYSON are rapidly computed using spherical harmonics, and this also allows to perform fast rigid body modeling [40], where a model of the complex particle is built from high resolution structure of individual subunits to fit the experimental scattering from the complex. In the program MASSHA [41], the subunits can be interactively manipulated as rigid bodies while observing corresponding changes in the fit to the experimental data. A limited automated refinement mode is available for performing an exhaustive search in the vicinity of the current configuration.

Several approaches were also developed for automated rigid body modeling, including an 'automated constrained fit' procedure [42-44], ellipsoidal modelling [45-47], screening of randomly or systematically generated models, also against multiple contrast variation data sets [48-50]. A suite of global refinement programs based on spherical harmonics calculations, allowing to account for particle symmetry and for intersubunit contacts has recently been developed [51], and further extended to account for specific deuteration and H2O/D2O exchange in the solvent.

Ab Initio Model Building Based on SANS Data

In contrast to x-ray crystallography where 3D structures are calculated based on sets of 3D data, the reconstruction of the structure from isotropic scattering as it is in SANS is a rather under-determined problem. Recently, thanks to the increased computing power Monte-Carlo type bead modeling first introduced by [52], became a useful method for *ab inito* shape determination, and using contrast variation, also for internal structure analysis.

Let us consider a general model of a particle consisting of $K\geq1$ components (phases) with significantly different contrasts $\Delta\rho_k$. One can define a volume, which encloses the particle (*e.g.* a sphere of radius $R=D_{max}/2$) and fill this volume with N closely packed small spheres (beads or dummy atoms) of radius $r_0 << R$. Each dummy atom is assigned an index X_j indicating to which phase it belongs, where X_j may range from 0 (solvent) to K. The bead positions being fixed, the shape and structure of the dummy atom model (DAM) are described by a phase assignment (configuration) vector X with $N\approx(R/r_0)^3$ components. The scattering intensity from the DAM is

$$ I(s) = \left\langle \left[\sum_{k=1}^{K} \Delta\rho_k A_k(\vec{s}) \right]^2 \right\rangle_\Omega \tag{45} $$

where $A_k(\vec{s})$ is the scattering amplitude from the volume occupied by the k-th phase. Representing the amplitudes using the spherical harmonics $Y_{lm}(\Omega)$

$$A_k(\vec{s}) = \sum_{l=0}^{\infty} \sum_{m=-l}^{l} A_{lm}^{(k)}(\vec{s}) Y_{lm}(\Omega) \tag{46}$$

one obtains for the total scattering intensity

$$I(s) = 2\pi^2 \sum_{l=0}^{\infty} \sum_{m=-l}^{l} \left\{ \sum_{k=1}^{K} \left[\Delta\rho_k A_{lm}^{(k)}(s) \right]^2 + 2 \sum_{n>k} \Delta\rho_k A_{lm}^{(k)}(s) \Delta\rho_n \left[A_{lm}^{(n)}(s) \right]^* \right\} \tag{47}$$

Here, the partial amplitudes from the volume occupied by the k-th phase in a DAM are

$$A_{lm}^{(k)}(s) = i^l \sqrt{2/\pi} \, f(s) \sum_{j=1}^{N_k} j_l(sr_j) Y_{lm}^*(\omega_j) \tag{48}$$

where the sum runs over the dummy atoms of the k-th phase, r_j, ω_j are their polar coordinates, $j_l(x)$ the spherical Bessel function and $f(s)$ is the dummy atom formfactor. Using the above equations, one can compute the scattering curves from a multiphase DAM for an arbitrary configuration X and arbitrary contrasts $\Delta\rho_k$.

This formalism describing the scattering from a multiphase particle has a rather general value, which makes it especially interesting for contrast variation studies in neutron scattering. Let us imagine that a set of $M \geq 1$ neutron scattering curves has been measured for a nucleoprotein complex in solvents with different D_2O concentrations. In terms of a DAM, this would be a two-phase system ($K=2$), so that each bead can belong either to solvent ($X_j=0$), to protein ($X_j=1$) or to the nucleic acid ($X_j=2$). A Monte-Carlo type algorithm may thus be constructed, which starts from random phase assignments and changes them randomly to find out a configuration vector, which leads to the structure compatible with the experimental data. Most importantly, as the contrasts of protein and nucleic acid are known at different D_2O concentrations (see Figure 7), the scattering curves from the model can be computed and multiple scattering curves $I^{(i)}_{exp}(s)$, $i=1,\dots M$, fitted simultaneously to minimize the discrepancy

$$\chi^2 = \frac{1}{M} \sum_{i=1}^{M} \sum_{j=1}^{N(i)} \left[(I_{exp}^{(i)}(s_j) - I^{(i)}(s_j)) / \sigma(s_j) \right]^2, \tag{49}$$

where $N(i)$ is the number of points in the i-th curve and $\sigma(s)$ denotes the experimental errors.

For an adequate description of a structure the number of dummy atoms must, however, be large ($N \approx 10^3$). Additional constrains are therefore required to obtain meaningful low resolution models where the volumes occupied by the phases would not contain just a single dummy atom or a few atoms only and the interfacial area is not too detailed. For a quantitative estimate, a list of contacts (i.e. atoms at a distance less than $2r_0$) is defined for each dummy atom. The connectivity of a non-solvent atom is characterized by counting among its contacts the number of atoms N_e belonging to the same phase (the connectivity function [53] is displayed in Figure 9). The looseness of a given configuration X can be computed using the average connectivity of all non-solvent atoms as $P(X) = 1 - <C(N_e)>$.

To retrieve a low resolution model one can thus search for the configuration X minimizing a goal function

$$f(X) = \chi^2 + \alpha P(X) \qquad (50)$$

where $\alpha > 0$ is the weight of the looseness penalty, selected in such a way that the second term yields a significant (say, about *10* to *50* percent) contribution to the function at the end of the minimization. Since χ^2 is expected to be around *1* for a correct solution and $P(X)$ is of order of 10^{-2} for compact bodies, $\alpha \approx 10^1$ is a reasonable choice.

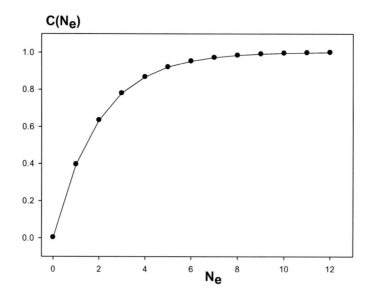

Figure 9. Penalty function as used for *ab initio* dummy atom model building. The connectivity decreases from 12 neighbors (= maximal number) C(12)= 1 and C(2)~0.5 down to C(0)~0 for an isolated dummy atom (=worst case).

The algorithm for solving the minimization problem has to cope with the large number of variables. A possible solution for this problem is provided by a so–called simulated annealing (SA) procedure [54], which starts from a random vector X. Random modifications are made to drive the system to the state with lower $f(X)$, but SA, in contrast to pure Monte-Carlo procedure, does not necessarily impede steps which increase $f(X)$. At the beginning when the system is "hot" these increasing steps will appear more often than in the more minimized "cold" configuration. During the minimization the phase assignment of only one dummy atom is changed per step resulting in only one recalculation of the scattering amplitude $A_{lm}^{(k)}(s)$.

The above approach is implemented in a computer program MONSA [53], which has successfully been used to construct models of multicomponent particles from the neutron scattering data. In particular, a map of protein-RNA distribution in the 70S *E.coli* ribosome was built [55] before the high resolution structures of ribosomal subunits have become available [56-58].

For a particular case K=1 (single component system), the method is reduced to an ab initio shape determination procedure (program DAMMIN [53]). There are also other ab initio programs available, e.g. original bead modeling program DALAI_GA based on genetic algorithm [52,59], give-n-take Monte-Carlo program *SAXS3D* [60], moving spheres modeling program *GA_STRUCT* [61].

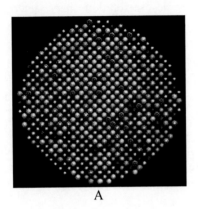
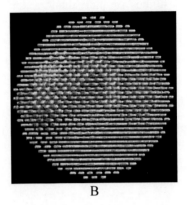

A   B

Figure 10. Assignment of different phases to a dummy model A) completely random distribution of phases for the starting conformation and B) final distribution with three different phases.

Recent developments allow one to account also for eventual particle symmetry. In particular, *DAMMIN* incorporates symmetry as a rigid constraint, so that the initial generation, randomisation and each mutation of the model are performed according to the symmetry rules. During the modeling process the phase of symmetrically related beads is changed (instead of a single bead if no symmetry is applied). This speeds up the computation significantly and selection rules for the spherical harmonics, that are used to compute the scattering amplitudes of the DAM, further shorten the computation effort. The groups with up to 19-fold symmetry axis, Pn2 (n =1,.,12) groups as well as cubic groups P23 and P432 as well as icosahedral symmetry are supported by *DAMMIN*.

A further development for *ab initio* model building is the use of dummy amino acid residues rather than beads. This approach takes advantage of the fact, that a protein is a chain of individual residues. In the amino acid backbone the C_α-atoms are separated by approximately 0.38 nm. Taking the low resolution of small angle scattering into account the backbone can be interpreted at a resolution of 0.5 nm as a necklace chain of dummy residues. Centering the dummy residues on the C_α position and the fixed distances to the neighboring residue constrains the solution. These restrictions change the penalty function and allow a more realistic model building than the bead representation for the shape determination. As initial step a random walk chain could be taken and start the minimization on that a chain. However, better results have been obtained for starting with a spherical search volume, such as used for the dummy atom modeling and condense the dummy residues inside this volume. This method is currently developed for x-rays [62], but may also become useful for neutron studies.

SELECTED APPLICATIONS

SANS is currently used to elucidate the structure of different biological macromolecules in solution, most notably, proteins, nucleoprotein and lipid-protein complexes. Below we shall briefly review some recent applications of the technique and present in more detail a case contrast variation study of the chaperonin GroE system demonstrating also the possibilities of the novel data analysis methods.

In the studies of individual proteins (e.g. for low resolution shape determination or oligomerization analysis) SANS does not usually present any advantage to SAXS, and the latter technique is employed much more often. In principle, SANS should be more useful in the study of solubilized membrane proteins, where the scattering from phospholipid or lipid detergent micelles or vesicles can be matched out in about 10 to 20% D_2O and the net scattering from the protein can be observed. The measurements at these D_2O concentrations, however, are not easy because of high hydrogen content in solution leading to strong incoherent scattering and because of low protein contrast. Oligomeric state and conformation of a large, multifunctional protein SecA was examined in small unilamellar vesicles of *E.coli* phospholipids [63]. It was found that the enzyme forms a dimer in the absence of nucleotide but dissociates into monomers in the presence of either ADP or a non-hydrolyzable ATP analog.

SANS is often employed to study reversible processes of protein folding/unfolding (thanks to the absence of radiation damage, the process can be repeated under the neutron beam many times). Photoresponsive interaction of light-sensitive surfactants, with bovine serum albumin was studied [64] . The surfactants undergo reversible photoisomerization, with the visible-light (trans) form of the surfactant being more hydrophobic than the UV-light (cis) form, allowing to control protein folding with light irradiation. SANS measurements provided information about the protein conformation in solution as a function of the surfactant concentration and indicated that the light induced unfolding is a completely reversible process. In another study [65] SANS was combined with molecular simulations in the analysis of pressure-induced denaturation of staphylococcal nuclease. Using SANS it was found that the protein retains a globular structure during the folding/unfolding transition although this structure is less compact and elongated compared to the native state. Molecular dynamics simulations of the unfolding were done by inserting water molecules into the protein interior and applying pressure. The p(r) function calculated from these simulations broadens and shows a unimodal-to-bimodal transition with increasing pressure, similar to that observed experimentally. It is suggested that the protein expands and forms two subdomains, which apart during initial stages of unfolding.

Most SANS studies are devoted to complex particles, where the major advantages of SANS come into play. For protein complexes, specific deuteration of individual proteins allows one to obtain valuable additional information compared to SAXS, and the two methods are often used together. A good example is a series SAXS and SANS studies of the cAMP-dependent protein kinase A (PKA), which serves as a prototype for understanding kinase structure-function relationship [46,66]. Analysis of scattering from samples with selectively deuterated catalytic and regulatory subunits led to the structural model of the catalytic subunit-regulatory subunit dimeric complex built of homology models of the subunits further refined using molecular dynamics and energy minimization. In a more recent paper [67], specific deuteration was employed to detect a large scale conformational change within the homodimer upon catalytic subunit binding to the PKA dimer. A SAXS/SANS combination was employed to analyze the structure of a heterotrimeric chicken skeletal muscle troponin complex consisting of a Ca2+-binding subunit, an inhibitory subunit, and a tropomyosin-binding subunit. A full set of SANS data was collected from hybrid recombinant ternary complexes, in which all possible combinations of the subunits have been deuterated, and the data were analyzed together with the SAXS curve from the complex. In all cases, Ca^{2+} bound and Ca^{2+} free solutions were measured. Starting from a model based on the human cardiac troponin crystal structure, a rigid-body Monte Carlo search yielded models of chicken skeletal muscle troponin in solution, in the presence and in the absence of regulatory calcium.

Historically, SANS is actively used for nucleoprotein complexes, exploiting both natural contrast between nucleic acids and proteins, and specific deuteration of the components. Ribosome was a classical SANS object for decades, and SANS was the first method to prove that proteins are located on the periphery of both ribosomal subunits [68,69]. A milestone in the ribosome study was the already mentioned protein triangulation in the small subunit [4]. Another example is given by the study of the 70S E. coli ribosome, which was the first practical application of the multi-phase simulated annealing bead modelling [53]. A total of 42 X-ray and neutron solution scattering curves from reconstituted ribosomes were collected, where the proteins and rRNA moieties in the ribosomal subunits were either protonated or deuterated in all possible combinations. The search volume defined by a cryo-EM model [70] was filled with densely packed beads and they were assigned either to solvent, to protein or to ribosomal RNA moieties to simultaneously fit all the scattering curves. The resulting 3 nm resolution map of RNA and protein moieties in the entire ribosome [55] is in a remarkably good agreement with the later high-resolution crystallographic models of the ribosomal subunits from other species [57,71,72]. A more recent example is given by the study of the Tn3 resolvase-crossover site synaptic complex consisting of a tetrameric protein (resolvase) and two short DNA segments [73]. By fitting SANS and SAXS data using rigid body modeling in terms of the crystallographic resolvase dimer and the DNA fragments a model of the quaternary structure of the complex was generated.

The Chaperonin System GroE from *E. coli*

Proteins of the chaperonin family are a special group of helper proteins during the protein folding or refolding process. The chaperonins were found as part of the heat shock response of bacteria cells, but are also expressed in normal growing cells. The eubacteria *E. coli* chaperonin system consists of two proteins called GroEL and GroES. The GroEL is composed of 14 identical subunits (57kDa) each arranged in two heptameric rings[74]. These rings bind back-to-back together forming a characteristic hollow cylindrical shape of P72 symmetry. The co-chaperonin GroES has a dome like structure built of seven identical subunits [75]. In the presence of ATP or ADP GroES binds to one end of the GroEL closing the hollow cylinder [76,77]. Upon this assembly reaction the GroEL undergoes large conformational changes especially at the GroES binding region in the so called apical domains. These proteins can be obtained by over expression directly from *E. coli* and can also be cultivated under D_2O conditions [78]. In the past SANS has already been used to investigate the structures of GroEL with bound substrate protein and of the asymmetric complex of the GroEL with its co-chaperonin GroES [78,79]. Below the application of modern analysis methods to a contrast variation study of the GroEL/GroES system is presented. Figure 11 displays the scattering data of GroEL, GroES and its complexes collected at the two SANS beamlines D11 and D22 at the Institute Laue-Langevin [78].

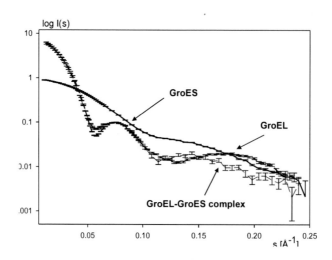

Figure 11. SANS data for GroEL, GroES and the asymmetric complex GroEL-GroES.

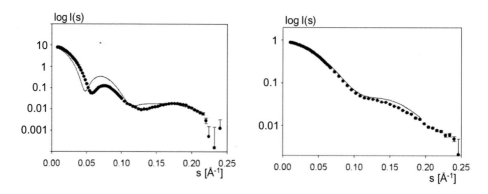

Figure 12. Comparison of experimental data (dots) with the curves computed from high resolution X-ray structures (solid lines). A) GroEL and B) GroES. Dots: experimental data; Solid line: model fit.

A comparison of the experimental SANS data with those computed from the high resolution structures determined by protein X-ray crystallography using *CRYSON* is shown in Figure 12. Deviations between the experimental and calculated data are observed, which can be attributed to missing fragments in the high resolution X-ray models. Indeed, in all three crystallographic structures: GroEL, GroES and the asymmetrical GroEL-GroES complex amino acids at the C and N terminal ends for GroEL and at a flexible loop for GroES could not be resolved.

The SANS data processed by GNOM were further used to construct particle shapes under different conditions using DAMMIN. First, the shape determination was done without symmetry and anisometry restrictions. Second, the particle was assumed to be cylindrical i.e.

the search was performed inside a cylindrical volume, but still no symmetry was applied. In the third run the search volume was kept spherical, but a P72 symmetry was used as a constrain. The obtained results are presented in the Figure 13 (note that the three models yield the same fit to the experimental data). The model without any constrains (Figure 13 A) yields only the overall particle shape, but does not represent the characteristic cylindrical apperance of the GroEL molecule. More information is seen in the structure where the search was limited to a cylindrical volume (Figure 13 B). The derived model shows a cylindrical shape, but the refolding cavity in the middle of the GroEL assembly is not modeled. In contrast, the model with the P72 symmetry (Figure 13 C) neatly reproduces the main features of the GroEL 14mer.

The inner channel of the GroEL is clearly visible and moreover beads are placed in the center of the channel dividing it into two parts. These results support the hypothesis from sequence analysis, that the residues missing in the crystal structure most likely separate the GroEL inner channel into two independent protein refolding cavities.

For the shape reconstruction of GroES, a P7 symmetry was assumed. The restored shape does not match well the x-ray structure as about 20% of the amino acid residues are missing in the high resolution model (not shown). A better agreement was achieved with the theoretical model of GroES where the missing loops have been remodeling by molecular dynamics. Superimposing this theoretical model to the ab initio SANS (see Figure 13 D) the basic structural features of the GroES molecule are well represented, although the low resolution (2.5nm) impedes a more detailed analysis of the SANS derived model.

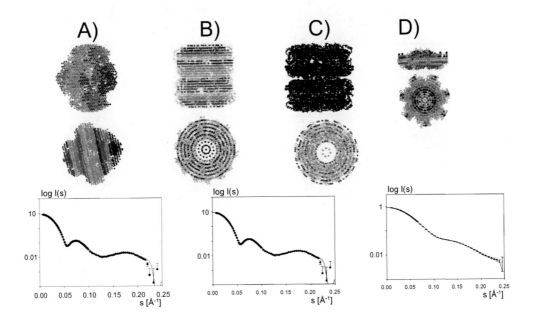

Figure 13. *Ab initio* modeling of GroEL and GroES from SANS data. A) No information about the symmetry was applied; B) No information about the symmetry, but a cylindrical search volume C) computed with in a spherical search volume and P72 symmetry. D) SANS model for GroES calculated under P7 symmetry. Dots: experimental data; Solid line: model fit.

For the investigation of the asymmetric GroEL-GroES complex the completely deuterated GroEL (e.g. with all hydrogens substituted by deuterons) was bound to the native "protonated" GroES in 40% D_2O buffer. Under these conditions the GroES matches the scattering density of the buffer and does not contribute to the scattering pattern. The deuterated GroEL remains visible and can be studied in the bound state in situ. The inverse experiment (protonated GroEL with deuterated GroES) reveals structural rearrangements of GroES upon binding to GroEL. This approach allows one to analyze structural changes of the deuterated protein upon binding its reaction partner in detail.

The two in situ structures of GroEL and GroES derived from the two above experiments are presented in Figure 14. The apical domains of the GroEL subunit are pivoted outwards and rotated in order to facilitate the binding of GroES. The conformational changes of GroES are slightly smaller, but the mobile loop regions rotate downwards to further facilitate the binding. Also in this structures a comparison with high resolution data from X-ray protein crystallography shows a good agreement within the resolution of SANS. The reconstruction of the asymmetric complex was done by interactive manipulations with the two in situ models together using the rigid body modeling program MASSHA.

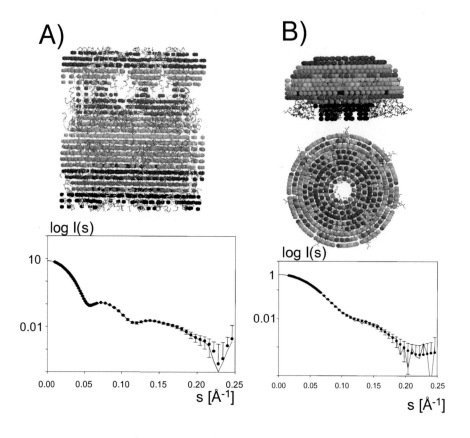

Figure 14. Dummy atom models for GroEL and GroES *in situ*. In model A) the GroEL was calculated using a P7 symmetry. The change in the upper apical domain indicate the binding of GroES. These domains are rotated outwards in order to facilitate the GroES binding. B) GroES in the bound *in situ* state. Compared to the unbound state the GroES seems to become a more dome-like structure. Dots: experimental data; Solid line: model fit.

The assembled complex has a high affinity for unfolded proteins and is an important reaction intermediate in the chaperonin driven refolding process. The first step in the refolding activity is binding of damaged proteins to the hydrophobic patches of the GroEL apical domains and the inner channel. The binding process is completely unspecific and every denatured polypeptide chain can be bound. Small proteins, up to a molecular mass of 60kDa can be completely covered in the refolding cavity of GroEL, whereas with larger proteins binding of GroES increases the cavity significantly and miss-folded protein binds on the opposite GroEL ring. This situation is shown in Figure 15 (panel B). The filled GroEL channel in the DAMMIN model points out to binding of small proteins inside the open GroEL channel.

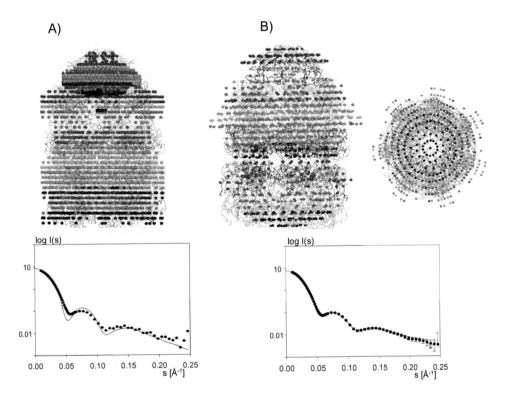

Figure 15. Model for the asymmetric GroEL-GroES complex. A) assembled from the two in situ models shown in Figure 12. B) Complex of GroEL-GroES together with bound denatured protein as substrate. The inner cavity of the ring where GroES is bound is enlarged. Dots:experimental data; Solid line: model fit.

In the proposed chaperonin refolding cycle binding and hydrolysis of ATP to the GroES bound GroEL heptameric ring will drive the system to the next step and GroES will change its position. By enlarging the refolding cavity upon this binding the miss-folded protein could be unfolded. After a second ATP binding and hydrolysis cycle the refolded protein is released. The kinetic of this ATP driven cycle can be investigated in detail by time resolved methods as described in the next section.

Time Resolved Small Angle Neutron Scattering

Compared to SAXS beamlines on third generation synchrotron sources neutron facilities provide a rather low neutron flux. For instance the high flux beamline D22 at the Institute Laue-Langevin yields about 10^7 n s^{-1} cm^{-2} on the sample, whereas high brilliance undulator SAXS beamlines such as ID02 (European Synchrotron Radiation Facility ESRF, Grenoble, France) or BioCAT (Advanced Photon Source APS, Argonne, USA) are able to provide to 10^{13} ph s^{-1} cm^{-2}. This high photon flux is successfully used for time resolution down to ms range in the studies on biological fibers as well as for solution scattering [80-82]. The limitation on neutron flux can partly be compensated by taking advantage of the contrast variation method [83]. If proteins are measured in the buffer containing D_2O instead of H_2O the contrast increases and even the lower flux can be used for time resolved SANS on protein kinetics. Recent experiments showed that time resolution in the range of 1 second is feasible using a standard stopped-flow mixing device. Such an apparatus is able to perform fast mixing of 100µl to 150µl volume necessary for a typical neutron beamsize of 10 x 10 mm^2. The scattering data can be recorded with sufficiently high precision if the experiment is repeated several times and the resulting scattering patterns are averaged. The data accuracy improvement upon averaging of ten low statistic experiments is shown in Figure 16. Here, the 1 second time frame of the scattering from GroEL in D_2O buffer features the (somewhat noisy) first side maxim, whereas the curve after the averaging yields the entire scattering pattern with rather good precision.

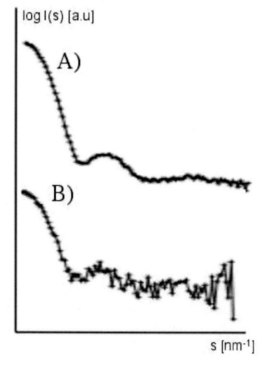

Figure 16. The statistics of time resolved SANS data (1s frames) are improved by repeating the experiment several times. A) a single 1s frame and B) its average over 10 independent 1s frames.

A serious problem of solution scattering experiments are changes of the scattering data upon aggregation. Especially in time-resolved investigations, production of such unspecific macromolecular assemblies has to be avoided. For high flux X-ray experiments aggregation can be prompted by radiation damage. This effect is practically absent for thermal neutrons, thus giving SANS an advantage over SAXS. In particular, assembly kinetics of protein complexes can be efficiently investigated using a mixture of D-labeled protein and normal H-protein. As seen in Figure 7 for such a system the H-protein has a strong negative scattering contribution in D_2O buffer, while the D-labeled protein has a positive contrast. Consequently the radius of gyration for the mixed assembly is significantly smaller than for the D-labeled protein alone and the decrease of the radius of gyration indicates binding of the two reaction partners. The kinetics of the assembly can be followed by investigating the time course of the radius of gyration and the kinetic parameters can be derived.

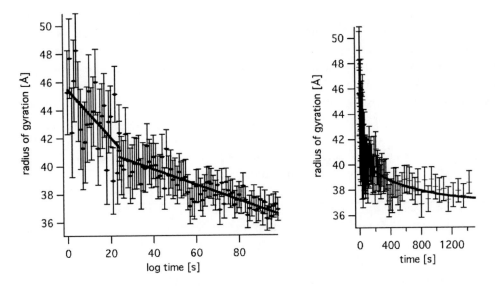

Figure 17. Time resolved measurements of the R_g. The decrease of the R_g indicate binding of native GroES to deuterated GroEL. The kinetics can be fitted to a double exponential indicating a two step kinetics.

CONCLUSIONS

During the last decade, SAS has become an increasingly important tool in studies of solutions of biological macromolecules thanks to broader accessibility of instruments on large scale facilities and novel data analysis software to a larger user community. The major instrumental progress in SAXS is related to third generation synchrotron sources, where the scattering patterns can be collected in less than a second. Several sources are currently operational (ESRF (Grenoble); Elettra (Trieste), Spring-8 (Himeji) APS (Argonne)), and yet more low emittance synchrotrons are being constructed or planned (e.g. Soleil, France; Diamond, UK; Petra, Germany). In SANS, future hopes for instrumental progress are put on the development of spallation sources (e.g. second target at ISIS, Oxford, UK or SNS in Oak Ridge, USA, currently under construction). The SANS instruments D11 and D22 built at the high flux steady-state reactor at ILL, Grenoble, continue to be the world standard, and the new instrument (D33) is currently planned.

Although the neutron sources are much weaker than synchrotrons, the power of the contrast variation technique, high contrast and lack of radiation damage make experiments with neutrons possible within minutes to hours. A critical aspect of biological SANS is the amount of material needed for the structure analysis, which is 3-5 times that required for X-rays (typically, about 10-20 mg versus 3-5 mg of purified material for a complete static SANS and SAXS, respectively). This requirement may be eased with the advent of focusing SANS using magnetic lenses [84], which should permit one to significantly reduce the beam size. Another important aspect is high yield production of perdeuterated biological samples. Here one should note recent creation of a joint EMBL-ILL Deuteration Laboratory in Grenoble, which will provide the tools and facilities required for the specific and selective isotopic-labelling of proteins, nucleic acids and lipids. As availability of perdeuterated samples is in many cases a necessary prerequisite for efficient contrast variation studies, one can expect that similar laboratories will in future be made available for users of other neutron facilities.

Methods for the analysis of static SAS patterns from monodisperse solutions and mixtures are now well developed, and many modelling programs are publicly accessible on the Web. A comprehensive program suite ATSAS for data processing, *ab initio* analysis, rigid body refinement, 3D visualization and characterization of mixtures for biological SAXS and SANS is available from www.embl-hamburg.de/ExternalInfo/Research/Sax/. Some *ab initio* shape determination programs, e.g. DALAI_GA (akilonia.cib.csic.es/ DALAI_GA2/) and SAXS3D (www.cmpharm.ucsf.edu/~walther/saxs/) can also be downloaded. A program FISH for modelling of form and structure factors is available from http://www .isis.rl.ac.uk/largescale/ LOQ/FISH/FISH_intro.htm. Useful references can also be found at the D22 home page (http://whisky.ill.fr/YellowBook/D22/D22_info/welcome.html)

SANS is often employed in combination with SAXS and other structural and biochemical methods to generate consistent models. The synergy of different methods is especially important in the study of large macromolecular complexes. The above mentioned studies (protein-RNA map in the 70S ribosome using SANS and SAXS data constrained by the overall shape obtained from cryo-EM, or analysis of chaperonins based on SANS and crystallographic models) present approaches, which should be extremely useful in future. The extremely important advantage of SANS, the possibility of selective isotopic labeling of biological complexes without compromising biological activity, will become indispensable for the analysis of complex structures in which the high resolution models from crystallography and NMR can meaningfully be fitted.

REFERENCES

[1]   Guinier, A. 1939, *Ann. Phys.* (Paris), 12, 161.
[2]   Ibel, K. and Stuhrmann, H. B. 1975, *J Mol Biol*, 93, 255.
[3]   Engelman, D. M. and Moore, P. B. 1972, *Proc Natl Acad Sci U S A*, 69, 1997.
[4]   Capel, M. S., Engelman, D. M., Freeborn, B. R., Kjeldgaard, M., Langer, J. A., Ramakrishnan, V., Schindler, D. G., Schneider, D. K., Schoenborn, B. P., Sillers, I. Y. and et al. 1987, *Science*, 238, 1403.

[5]   Scherm, R. and Fak, B. 1993, in *Neutron and Sychrotron Radiation for Condensed Matter Studies* (eds. Baruchel, J., Hodeau, J. L., Lehmann, M. S., Regnard, J. R. and Schlenker, C.) 113, Springer/Les Editions de Physique.

[6]   Ageron, P. 1989, *Nuc. Instr. Meth in Physics Research*, 284, 197.

[7]   Stuhrmann, H. B., Scharpf, O., Krumpolc, M., Niinikoski, T. O., Rieubland, M. and Rijllart, A. 1986, *Eur Biophys J*, 14, 1.

[8]   Willumeit, R., Burkhardt, N., Diedrich, G., Zhao, J., Nierhaus, K. H. and Stuhrmann, H. B. 1996, *J. Mol. Struct.*, 383, 201.

[9]   Bacon, G. E. 1975, *Neutron Diffraction.* Oxford University Press, Oxford).

[10]   Lindner, P., May, R. P. and Timmins, P. A. 1992, *Physica B,* 180-181, 967.

[11]   Rehm, C. and Agamalian, A. 2002, *Appllied Physics* A, 1483–1485.

[12]   Thiyagarajan, P., Epperson, J. E., Crawford, R. K., Carpenter, J. M., Klippert, T. E. and Wozniak, D. G. 1997, *J Appl Crystallogr,* 30, 280.

[13]   Heenan, R. K., Penfold, J. and King, S. M. 1997, *J. Appl. Cryst.*, 30, 1140.

[14]   Giacovazzo, C., Monaco, H. L., Voiterbo, D., Scordari, F., Gilli, G., Zanotti, G. and Catti, M. 1992, *Fundamentals of Crystallography* (ed. Giacovazzo, C.), Oxford University Press, New York).

[15]   Debye, P. 1915, *Ann. Physik,* 46, 809.

[16]   Guinier, A. and Fournet, G. 1955, *Small Angle Scattering of X-Rays* Wiley, New York).

[17]   Porod, G. 1951, *Kolloid-Z.*, 124, 83.

[18]   Kratky, O., Pilz, I. and Schmitz, P. J. 1966, *Colloid and Interface Science*, 21, 24.

[19]   Kratky, O. and Pilz, I. 1972, *Q Rev Biophys*, 5, 481.

[20]   Pollack, L., Tate, M. W., Darnton, N. C., Knight, J. B., Gruner, S. M., Eaton, W. A. and Austin, R. H. 1999, *Proc. Natl. Acad. Sci. USA*, 96, 10115.

[21]   Pollack, L., Tate, M. W., Finnefrock, A. C., Kalidas, C., Trotter, S., Darnton, N. C., Lurio, L., Austin, R. H., Batt, C. A., Gruner, S. M. and Mochrie, S. G. J. 2001, *Phys. Rev. Lett.*, 86, 4962.

[22]   Petrescu, A. J., Receveur, V., Calmettes, P., Durand, D., Desmadril, M., Roux, B. and Smith, J. C. 1997, *Biophys J,* 72, 335.

[23]   Roessle, M., Panine, P., Urban, V. S. and Riekel, C. 2004, *Biopolymers,* 74, 316.

[24]   Porod, G. 1982, in *Small-angle X-ray scattering* (eds. Glatter, O. and Kratky, O.) 17, Academic Press, London.

[25]   Glatter, O. 1977, *J. Appl. Cryst.*, 10, 415.

[26]   Semenyuk, A. V. and Svergun, D. I. 1991, *J. Appl. Crystallogr.*, 24, 537.

[27]   Svergun, D. I. 1992, *J. Appl. Crystallogr.*, 25, 495

[28]   Svergun, D. I. 1992, *J. Appl. Crystallogr.*, 25, 495.

[29]   Lederer, H., May, R. P., Kjems, J. K., Schaefer, W., Crespi, H. L. and Heumann, H. 1986, *Eur J Biochem*, 156, 655.

[30]   Stuhrmann, H. B. and Kirste, R. G. 1965, *Z. Phys. Chem. Frankfurt,* 56, 247.

[31]   Koch, M. H. J. and Stuhrmann, H. B. 1979, *Methods Enzymol,* 59, 670.

[32]   Serdyuk, I. N. and Grenader, A. K. 1975, *FEBS Lett*, 59, 133.

[33]   Lixin Fan, Svergun, D. I., Volkov, V. V., Aksenov, V. L., Agalarov, C. C., Selivanova, O. M., Shcherbakova, I. N., Koch, M. H. J., Gilles, R., Wiedenmann, A., May, R. and Serdyuk, I. N. 2000, *J. Appl. Crystallogr.*, 33, 515.

[34]   Willumeit, R., Diedrich, G., Forthmann, S., Beckmann, J., May, R. P., Stuhrmann, H. B. and Nierhaus, K. H. 2001, *Biochim Biophys Acta,* 1520, 7.

[35] Willumeit, R., Forthmann, S., Beckmann, J., Diedrich, G., Ratering, R., Stuhrmann, H. B. and Nierhaus, K. H. 2001, *J Mol Biol,* 305, 167.

[36] Svergun, D. I., Barberato, C. and Koch, M. H. J. 1995, *J. Appl. Crystallogr.,* 28, 768.

[37] Svergun, D. I., Richard, S., Koch, M. H. J., Sayers, Z., Kuprin, S. and Zaccai, G. 1998, *Proc Natl Acad Sci U S A*, 95, 2267.

[38] Merzel, F. and Smith, J. C. 2002, *Proc Natl Acad Sci U S A*, 99, 5378.

[39] Berman, H. M., Westbrook, J., Feng, Z., Gilliland, G., Bhat, T. N., Weissig, H., Shindyalov, I. N. and Bourne, P. E. 2000, *Nucleic Acids Res.,* 28, 235.

[40] Svergun, D. I. 1991, *J. Appl. Crystallogr.,* 24, 485.

[41] Konarev, P. V., Petoukhov, M. V. and Svergun, D. I. 2001, *J. Appl. Crystallogr.,* 34, 527.

[42] Boehm, M. K., Woof, J. M., Kerr, M. A. and Perkins, S. J. 1999, *J Mol Biol,* 286, 1421.

[43] Petoukhov, M. V. and Svergun, D. I. 2006, *Eur. Biophys.,35,* 567-576

[44] Furtado, P. B., Whitty, P. W., Robertson, A., Eaton, J. T., Almogren, A., Kerr, M. A., Woof, J. M. and Perkins, S. J. 2004, *J Mol Biol,* 338, 921.

[45] Krueger, J. K., Zhi, G., Stull, J. T. and Trewhella, J. 1998, *Biochemistry,* 37, 13997.

[46] Wall, M. E., Gallagher, S. C. and Trewhella, J. 2000, *Annu Rev Phys Chem*, 51, 355.

[47] Priddy, T. S., MacDonald, B. A., Heller, W. T., Nadeau, O. W., Trewhella, J. and Carlson, G. M. 2005, *Protein Sci*, 14, 1039.

[48] Heller, W. T., Finley, N. L., Dong, W. J., Timmins, P., Cheung, H. C., Rosevear, P. R. and Trewhella, J. 2003, *Biochemistry*, 42, 7790.

[49] Nollmann, M., Stark, W. M. and Byron, O. 2004, *Biophys J,* 86, 3060.

[50] King, W. A., Stone, D. B., Timmins, P. A., Narayanan, T., von Brasch, A. A., Mendelson, R. A. and Curmi, P. M. 2005, *J Mol Biol*, 345(4), 797.

[51] Petoukhov, M. V. and Svergun, D. I. 2005, *Biophys J.,* 89, 1237-1250

[52] Chacon, P., Moran, F., Diaz, J. F., Pantos, E. and Andreu, J. M. 1998, *Biophys J,* 74, 2760.

[53] Svergun, D. I. 1999, *Biophys J,* 76, 2879.

[54] Kirkpatrick, S., Gelatt, C. D., Jr. and Vecci, M. P. 1983, *Science,* 220, 671.

[55] Svergun, D. I. and Nierhaus, K. H. 2000, *J Biol Chem*, 275, 14432.

[56] Ban, N., Freeborn, B., Nissen, P., Penczek, P., Grassucci, R. A., Sweet, R., Frank, J., Moore, P. B. and Steitz, T. A. 1998, *Cell*, 93, 1105.

[57] Ban, N., Nissen, P., Hansen, J., Moore, P. B. and Steitz, T. A. 2000, *Science*, 289, 905.

[58] Ban, N., Nissen, P., Hansen, J., Capel, M., Moore, P. B. and Steitz, T. A. 1999, *Nature*, 400, 841.

[59] Chacon, P., Diaz, J. F., Moran, F. and Andreu, J. M. 2000, *J Mol Biol*, 299, 1289.

[60] Walther, D., Cohen, F. E. and Doniach, S. 2000, *J. Appl. Crystallogr.,* 33, 350.

[61] Heller, W. T., Abusamhadneh, E., Finley, N., Rosevear, P. R. and Trewhella, J. 2002, *Biochemistry,* 41, 15654.

[62] Svergun, D. I., Petoukhov, M. V. and Koch, M. H. J. 2001, *Biophys J,* 80, 2946.

[63] Bu, Z., Wang, L. and Kendall, D. A. 2003, *J Mol Biol,* 332, 23.

[64] Lee, C. T. J., Smith, K. A. and Hatton, T. A. 2005, *Biochemistry*, 44, 524.

[65] Paliwal, A., Asthagiri, D., Bossev, D. P. and Paulaitis, M. E. 2004, *Biophys J.,* 87, 3479.

[66] Tung, C. S., Walsh, D. A. and Trewhella, J. 2002, *J Biol Chem*, 277, 12423.

[67] Heller, W. T., Vigil, D., Brown, S., Blumenthal, D. K., Taylor, S. S. and Trewhella, J. 2004, *J Biol Chem*, 279, 19084 19090.

[68] Stuhrmann, H. B., Haas, J., Ibel, K., Wolf, B., Koch, M. H. J., Parfait, R. and Crichton, R. R. 1976, *Proc Natl Acad Sci U S A,* 73, 2379.

[69] Stuhrmann, H. B., Koch, M. H. J., Parfait, R., Haas, J., Ibel, K. and Crichton, R. R. 1978, *J Mol Biol*, 119, 203.

[70] Frank, J., Zhu, J., Penczek, P., Li, Y., Srivastava, S., Verschoor, A., Radermacher, M., Grassucci, R., Lata, R. K. and Agrawal, R. K. 1995, *Nature*, 376, 441.

[71] Schluenzen, F., Tocilj, A., Zarivach, R., Harms, J., Gluehmann, M., Janell, D., Bashan, A., Bartels, H., Agmon, I., Franceschi, F. and Yonath, A. 2000, *Cell,* 102, 615.

[72] Brodersen, D. E., Clemons, W. M., Jr., Carter, A. P., Wimberly, B. T. and Ramakrishnan, V. 2002, *J Mol Biol*, 316, 725.

[73] Nollmann, M., He, J., Byron, O. and Stark, W. M. 2004, *Mol Cell,* 16, 127.

[74] Braig, K., Otwinowski, Z., Hegde, R., Boisvert, D. C., Joachimiak, A., Horwich, A. L. and Sigler, P. B. 1994, *Nature*, 371, 578.

[75] Hunt, J. F., Weaver, A. J., Landry, S. J., Gierasch, L. and Deisenhofer, J. 1996, *Nature,* 379, 37.

[76] Xu, Z. H., Horwich, A. L. and Sigler, P. B. 1997, *Nature,* 388, 741.

[77] Sigler, P. B., Xu, Z. H., Rye, H. S., Burston, S. G., Fenton, W. A. and Horwich, A. L. 1998, *Annual Review of Biochemistry*, 67, 581.

[78] Stegmann, R., Manakova, E., Rossle, M., Heumann, H., Nieba-Axmann, S. E., Pluckthun, A., Hermann, T., May, R. P. and Wiedenmann, A. 1998, *J Struct Biol,* 121, 30.

[79] Thiyagarajan, P., Henderson, S. J. and Joachimiak, A. 1996, *Structure*, 4, 79.

[80] Woenckhaus, J., Kohling, R., Thiyagarajan, P., Littrell, K. C., Seifert, S., Royer, C. A. and Winter, R. 2001, *Biophys J,* 80, 1518.

[81] Roessle, M., Manakova, E., Laure, I., Nawroth, T., Holzinger, J., Narayanan, T., Bernstorff, S., Amenitsch, H. and Heumann, H. 2000, *J. Appl. Cryst.*, 33, 548.

[82] Ferenczi, M. A., Bershitsky, S. Y., Koubassova, N., Siththanandan, V., Helsby, W. I., Panine, P., Roessle, M., Narayanan, T. and Tsaturyan, A. K. 2005, *Structure*, 13, 131.

[83] Roessle, M., Manakova, E., Holzinger, J., Vanatalu, K., May, R. P. and Heumann, H. 2000, *Physica* B, 276, 532.

[84] Suzuki, J., Oku, T., Adachi, T., Shimizu, H. M., Hirumachi, T., Tsuchihashi, T. and Watanabe, I. 2003, *J Appl Crystallogr,* 36, 795.

In: Neutron Scattering Methods and Studies
Editor: Michael J. Lyons

ISBN: 978-1-61122-521-1
© 2011 Nova Science Publishers, Inc.

Chapter 9

ELASTIC, QUASI ELASTIC AND INELASTIC NEUTRON SCATTERING STUDIES ON HYDROGEN-BONDED SYSTEMS OF BIOPHYSICAL INTEREST*

S. Magazù and F. Migliardo†

Dipartimento di Fisica, Università di Messina, Messina, Italy

ABSTRACT

Neutron scattering has been demonstrated to be a powerful tool for investigating physical and chemical mechanisms of biophysical processes. The present paper shows Elastic (ENS), Quasi Elastic (QENS) and Inelastic Neutron Scattering (INS) findings on a class of hydrogen-bonded systems of biophysical interest, such as homologues disaccharides (i. e. trehalose, maltose and sucrose)/water mixtures as a function of temperature and concentration.

The aim of the present work is to give an overview of dynamical properties of the investigated systems in order to characterise the dynamical transition, the diffusion and the vibrational modes of disaccharides in solution and to link the observed features to their bioprotective effectiveness. The role of hydrogen bond and fragility in these bioprotectant systems is discussed in detail. The obtained findings highlight a strong destructuring effect on the hydrogen-bonded network and a marked slowing down effect on the dynamics of water by disaccharides together with a considerable cryptobiotic action and can justify the superior capability of trehalose in respect to maltose and sucrose to encapsulate biostructures in a more rigid and protective environment.

Keywords: elastic neutron scattering; quasi elastic neutron scattering; inelastic neutron scattering; hydrogen-bonded systems; bioprotection; water mixtures; mean square displacements; diffusion coefficients; vibrational properties.

---

* A version of this chapter was also published in *Chemical Physics Research Journal,* Volume 1, Issue 2-3, Nova Science Publishers. It was submitted for appropriate modifications in an effort to encourage wider dissemination of research.

† E-mail address: fmigliardo@unime.it. Tel.: +39 0906765019. Fax: +39 090395004 Federica Migliardo, Dipartimento di Fisica, Università di Messina, C.da Papardo, S.ta Sperone 31, P.O. Box 55, 98166 S. Agata, Messina, Italy. (Corresponding author)

INTRODUCTION

The role played by the solvent in the physical-chemical mechanisms that determine the conformational and dynamical properties of biostructures are today an open question in investigating biophysical systems [1-4]. Changes in the structure and internal dynamics of proteins as a function of solvent conditions at physiological temperatures have been found by using several experimental techniques [5-7], demonstrating that an increase of protein conformational flexibility, and hence of the activity, is related to hydration and that the dynamic behaviour of protein solutions follows that of the solvent [8].

Homologues disaccharides, such as trehalose, maltose and sucrose, are cryptobiotic activating substances, i. e. they allow to many organisms to undergo to cryptobiosis (hidden life) when prohibitive environmental conditions occur [9]. During cryptobiosis, undetectable levels of metabolic functions are maintained, these levels reaching normal values when external conditions become again favourable to life [9,10], but, in spite of the several formulated bioprotection hypotheses [11-13], the effectiveness mechanisms remain still cryptic.

In this frame Green and Angell [11] suggest that the higher value of the glass transition temperature of trehalose and its mixtures with water, in comparison with the other disaccharides, is the only reason for its superior bioprotectant effectiveness. In fact the higher T_g values of the trehalose/H2O mixtures, in respect to those of the other disaccharides/H2O mixtures, implies that at a given temperature the glass transition for trehalose mixtures always occurs at a higher water content. As a matter of fact such a hypothesis alone is not entirely satisfactory if one keeps in mind that other similar systems, such as for example dextran, a linear polysaccharide, present even a higher T_g value, but do not show comparable bioprotective action. Crowe and co-workers [12] formulated the hypothesis that a direct interaction between the sugars and the object of protection occurs. More specifically their "water replacement hypothesis" justifies the trehalose protective function with the existence of direct hydrogen bonding of trehalose with the polar head groups of the lipids as water does. This hypothesis is supported by a simulation of Grigera and co-workers [13], which argue that the structure of trehalose is perfectly adaptable to the tetrahedral coordination of pure water, whose structural and dynamical properties are not significantly affected by trehalose.

Many experimental findings obtained by several spectroscopic techniques [14-18] indicate that the structural and the dynamical properties of water result drastically perturbed by disaccharides, and in particular by trehalose. More specifically neutron diffraction results [14] show for all disaccharides, and for trehalose to a large extent, a strong distortion of the peaks linked to the hydrogen bonded network in the partial radial distribution functions which can be attributed to the destroying of the tetrahedral coordination of pure water. Analogously Raman scattering findings [15] show that the addition of trehalose, in respect to the other disaccharides, more deeply destroys the tetrahedral intermolecular network of water, which by lowering temperature would give rise to ice.

Furthermore ultrasonic velocity measurements [16] point out that, in respect to the other disaccharides, the trehalose-water system is characterized, in all the investigated concentration range, by both the highest value of the solute-solvent interaction strength and of hydration number. These results clearly indicate that the disaccharide-water molecule interaction strength is much higher in respect to that between the water molecules.

Furthermore viscosity measurements on trehalose, maltose, and sucrose aqueous solutions [17] highlight that trehalose shows in respect to the other disaccharides, a "stronger" kinetic character in the Angell's classification scheme. QENS and INS were also employed to investigate the low frequency dynamics across the glass transition of trehalose, maltose and sucrose water mixtures [18]. The obtained experimental findings, through the relaxational to the vibrational contribution ratio, confirm that the trehalose/H_2O mixture shows a stronger character and furnish for it a higher force pseudo-constant (resilience) value in comparison to that of the other disaccharides/H_2O mixtures.

Neutrons are an excellent probe to characterize thermal molecular motions and conformational changes in biological systems [19-21]. In particular they furnish information on mean-square fluctuations in a given time scale by elastic scattering [19], on correlation times of diffusion motions by quasi elastic scattering [20] and on vibrational modes by inelastic scattering [21].

In this work an overview of ENS, QENS and INS results on homologues disaccharides (trehalose, maltose, sucrose)/H_2O mixtures as a function of concentration and temperature is presented. As we shall see ENS findings allow to characterize the systems "flexibility" and fragility, which justify the better cryptoprotectant effectiveness of trehalose; QENS measurements have been addressed to clarify the role played by trehalose, maltose and sucrose in determining the dynamics switching off of the solvent, which is a fundamental point for clarifying the bioprotective mechanisms; and INS have been aimed to describe the effects of disaccharide concentration in water mixtures on the vibrational properties of water, linking them to the disaccharide cryoprotective action.

EXPERIMENTAL SECTION

Ultra pure powdered trehalose, maltose and sucrose, D_2O and H_2O, purchased by Aldrich-Chemie, were used for the experiments.

ENS experiments were performed by using the IN13 spectrometer at the Institute Laue Langevin (ILL) in Grenoble (France). In the used IN13 configuration the incident wavelength was 2.23 Å and the Q-range was 0.28÷4.27 Å^{-1}. Measurements were performed in a temperature range of 20÷310K on hydrogenated trehalose, maltose and sucrose in H_2O and on partially deuterated trehalose, maltose and sucrose in D_2O at a weight fraction values corresponding to 19 and 6 water (H_2O and D_2O) molecules for each disaccharide molecule. Raw data were corrected for cell scattering and detector response and normalized to unity at Q=0 Å^{-1}.

QENS measurements were carried out using the IRIS high resolution spectrometer at ISIS at the Rutherford Appleton Laboratory (RAL, UK). We used the high resolution configuration of IRIS (graphite 002 and mica 006 analyser reflections) to measure sets of QENS spectra covering a Q,ω-domain extending from $\hbar/=$-0.3 to 0.6 meV and Q=0.3 to 1.8 Å^{-1}. The detectors used give a mean energy resolution of Γ=8 μeV of Half Width at Half Maximum (HWHM) as determined by reference to a standard vanadium plate. Measurements were performed in a temperature range of 283÷320K on hydrogenated trehalose, maltose and sucrose ($C_{12}H_{22}O_{11}$) in H_2O and on partially deuterated trehalose, maltose and sucrose

($C_{12}H_{14}D_8O_{11}$) in D_2O at a weight fraction values corresponding to 19 water (H_2O and D_2O) molecules for each disaccharide molecule. The raw spectra were corrected and normalised using the standard GENIE procedures and the IRIS data analysis package [22].

INS measurements have been performed by using the TOSCA indirect geometry time-of-flight spectrometer at ISIS at the Rutherford Appleton Laboratory (RAL, UK). The high energy resolution of TOSCA ($\Delta E/E\approx1.5$-2% for energy transfers up to several hundred meV) coupled with the high intensity of the ISIS source makes TOSCA ideal for studying the dynamics of water and aqueous mixtures below 2000 cm^{-1} (250 meV). Measurements were performed at a temperature value of 27K on pure H_2O and disaccharides ($C_{12}H_{22}O_{11}$)/H_2O mixtures at different weight fraction values corresponding to 7, 10 and 14 H_2O molecules for each disaccharide molecule. For the data treatment the standard GENIE programme has been used [23].

RESULTS

A comparison among elastic incoherent neutron scattering spectra of trehalose+19H_2O, maltose+19H_2O and sucrose+19H_2O mixtures is shown in Fig. 1. It is evident that a dynamical transition occurs for the trehalose/H_2O mixture at T~238K, for maltose/H_2O and sucrose/H_2O mixtures at T~235K and T~233K, respectively. In the insert the elastic intensity versus temperature of trehalose+6H_2O and sucrose+6H_2O mixtures is reported.

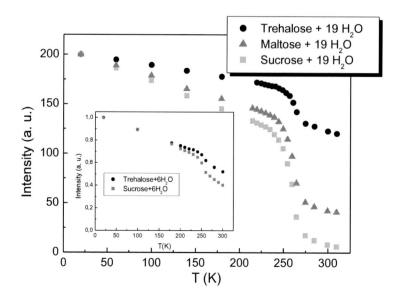

Figure 1. Elastic incoherent neutron scattering spectra $S(Q,\omega=0)$ of trehalose+19H_2O mixture (*black circles*), maltose+19H_2O mixture (*dark grey triangles*) and sucrose+19H_2O mixture (*grey squares*) as a function of temperature. In the insert the elastic intensity vs temperature of trehalose+6H_2O mixture (*black circles*), sucrose+6H_2O mixture (*dark grey squares*), trehalose+19D_2O mixture (*grey circles*) and sucrose+19D_2O mixture (*light grey squares*) are reported.

The decrease in the elastic intensity above the dynamical transition temperature can be attributed to the excitation of new degrees of freedom [20], especially at low Q, and is very less marked in the case of trehalose/water mixture than for the other disaccharide/water mixtures. This circumstance clearly indicates that trehalose shows a larger structural resistance to temperature changes and a higher "rigidity" in comparison with the maltose/H_2O mixture and the sucrose/H_2O mixture.

In Fig. 2, as an example, the derived mean square displacements for trehalose, maltose and sucrose+19H_2O mixtures as a function of temperature are shown. In the insert the derived mean square displacements of trehalose+6H_2O and sucrose+6H_2O mixtures versus temperature is reported. In the Q range in which the Gaussian model is valid and it results $S_{inc}^{el}(Q) \propto \exp\left[-\left\langle \Delta u^2/6\right\rangle Q^2\right]$, the mean square displacement behaviour can be fitted within the framework of the harmonic approximation [19-21]:

$$\left\langle \Delta u^2(T)\right\rangle = \frac{h\langle v\rangle}{2K}\coth\left(\frac{h\langle v\rangle}{2K_B T}-1\right)$$

(1)

where K_B is the Boltzman constant, K and $\langle v\rangle$ the average force field constant and the average frequency of a set of oscillators considered as an Einstein solid respectively and the term $h\langle v\rangle/2K$ the zero-point mean square displacement. The fitting procedure according to Eq. 1 furnishes for K the values of K=0.40 N/m, K=0.25 N/m and K=0.22 N/m for trehalose, maltose and sucrose+19H2O mixtures, respectively.

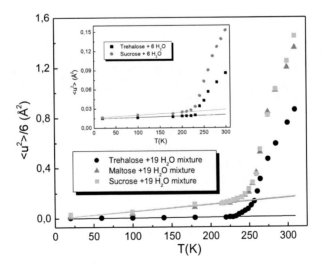

Fig. 2 Mean square displacements of trehalose/H_2O mixture (*black circles*), maltose/H_2O mixture (*dark grey triangles*) and sucrose/H_2O mixture (*grey squares*) as a function of temperature. In the insert the derived mean square displacements of trehalose+6H_2O and sucrose+6H_2O mixtures versus temperature is reported.

With the purpose of connecting the bioprotectant effectiveness of disaccharide/H2O mixtures to the "fragility degree" of these systems, we introduce a new operative definition for fragility, based on the evaluation by neutron scattering of the temperature dependence of the mean square displacement. The procedure is based on the relation between a macroscopic transport quantity, viscosity, and an atomic quantity, the nanoscopic mean square displacement.

The "fragility" is operatively defined as:

$$m = \frac{d \log \eta}{d\left(T_g/T\right)}\Bigg|_{T=T_g^+} \tag{2}$$

based on viscosity measurements.

Starting from the works on Selenium by Migliardo et al. [24] and Magazù et al. [25], a correlation between viscosity and the atomic mean square displacement has been proposed. More specifically defining $<u^2>_{loc}$ as the difference between the mean square displacements of the disordered phases (amorphous and liquid) and the ordered phase (crystalline):

$$<u^2>_{loc} = <u^2>_{disord} - <u^2>_{ord} \tag{3}$$

a linear relation is observed between the logarithm of viscosity and the inverse of $<u^2>_{loc}$.

The following interpretative model for the elementary flow process (α-relaxation) can be proposed. A given particle is jumping back and forth in the fast processes (β-relaxation) with a Gaussian probability distribution of mean square amplitude $<u^2>_{loc}$. When the amplitude of that fast motion exceeds a critical displacement u_0, a local structural reconfiguration (α-relaxation) takes place. Under the assumption of temperature independence of the time scale of the fast motion, the waiting time for the occurrence of a α process at a given particle is proportional to the probability to find the particle outside the sphere with radius u_0.

Within this picture, one can express viscosity with the following expression:

$$\eta = \eta_0 \exp\left[u_0^2 / \left\langle u^2 \right\rangle_{loc}\right] \tag{4}$$

Taking into account Eqs. 2-4, we now introduce a new operative definition to characterize the "fragility" degree by elastic neutron scattering as follows:

$$M = \frac{d\left(u_0^2 / \left\langle u^2 \right\rangle_{loc}\right)}{d\left(T_g/T\right)}\Bigg|_{T=T_g^+} \tag{5}$$

Obviously such a definition implies a fragility parameter depending on the instrumental resolution, which determines the observation time scale. From Eq. 5 we evaluate a fragility parameter M of 302 for the trehalose+19H$_2$O mixture, of 346 for maltose+19H$_2$O mixture and of 355 for the sucrose+19H$_2$O mixture, respectively. Employing a less wide viscosity data sets, for trehalose/D_2O and sucrose/D_2O mixtures at a concentration value corresponding to 19 water molecules for each disaccharide molecule we obtain a fragility parameter M of 272 and of 295, respectively, whereas for the trehalose/H_2O and sucrose/H_2O mixtures at a concentration value corresponding to 6 water molecules for each disaccharide molecule we obtain for the fragility parameter M the value of 241 and of 244, respectively. When the M parameters, obtained with a given resolution, are reported as a function of m, see Fig. 3, the data arrange themselves on a straight line whose slope depends uniquely on the instrumental resolution.

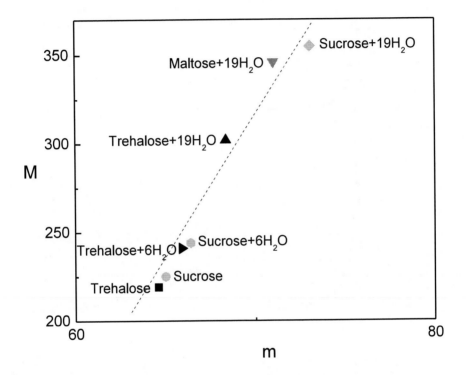

Figure 3. Linear behaviour of the M fragility parameter, as defined in the present work, versus the fragility parameter m. The solid lines indicate the best fits.

From this analysis it clearly emerges that the trehalose/H_2O mixture is characterized in respect to the maltose/H_2O and sucrose/H_2O mixtures by a lower fragility namely by a higher resistance to local structural changes when temperature decreases towards the glass transition value.

By QENS in the analysed hydrogenated systems we can distinguish among different proton populations which contribute to the QENS scattering law: i) protons of the disaccharide molecule and of its hydration shell, that follow the same diffusion law; ii) protons of the water molecules of higher hydration shells, whose diffusion is influenced by the presence of the disaccharide; iii) protons by bulk water that in our case are not present.

The spectra have been analyzed by using the fitting function:

$$S_{inc}(Q,\omega) = A(Q)\left\{ f_{Disaccharide}\left[F(Q)\frac{1}{\pi}\frac{\Gamma_1(Q)}{\Gamma_1^2(Q)+\omega^2} + (1-F(Q))\frac{1}{\pi}\frac{\Gamma_2(Q)}{\Gamma_2^2(Q)+\omega^2} \right] \right. $$

$$\left. + f_{hydr}\frac{1}{\pi}\frac{\Gamma_3(Q)}{\Gamma_3^2(Q)+\omega^2} \right\} \tag{6}$$

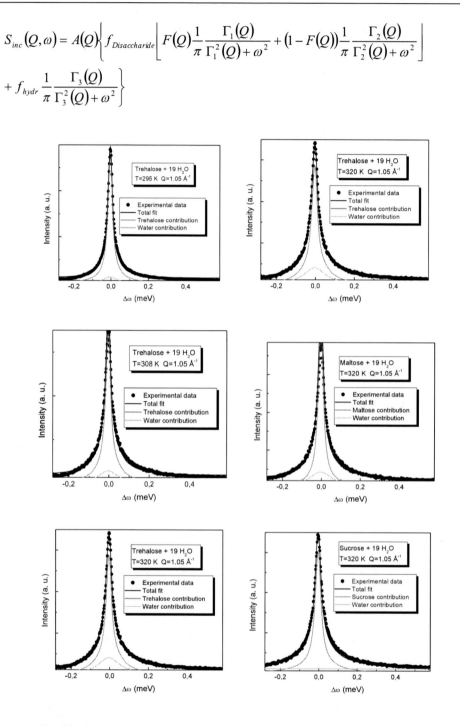

Fig. 4 Best fit of hydrogenated aqueous solutions of a) (left column) trehalose for three temperature values and b) (right column) of trehalose, maltose and sucrose at T=320K. The total fit (black line) is composed by the disaccharide contribution (grey line), resulting by the fit of disaccharide/D_2O spectra, and the water contribution (light grey line), where only the translational part is present.

where the first two terms refer to the translational and rotational contribution of hydrated disaccharide ($f_{Disaccharide}$ and f_{hydr} represent fraction factors of the total scattering from disaccharide and its strongly bonded water molecules), and the third one refers to hydration water ($f_{Disaccharide}+ f_{hydr} =1$). Therefore the dynamical information of the diffusive dynamics of disaccharide can be obtained by the analysis of disaccharide+D_2O spectra analysis for which f_{hydr} results negligible.

For trehalose aqueous solutions, the f_{hydr} parameters keep constant with Q at the values of 0.032, 0.108, 0.223 and 0.328 for T=283, 295, 308, 320 K, respectively, corresponding to 18.0, 15.7, 12.2, 9.0 water molecules bound to the trehalose molecules. These hydration number values are in excellent agreement with those obtained by ultrasonic, hypersonic, viscosity and Raman scattering techniques [14-17]. As far as the hydration number values of the three disaccharides is concerned, at T=320K the f_{hydr} parameters result 0.328, 0.348, 0.378 for trehalose, maltose and sucrose solutions, respectively, corresponding to 9.0, 8.4, 7.5 water molecules bound to the disaccharide molecules.

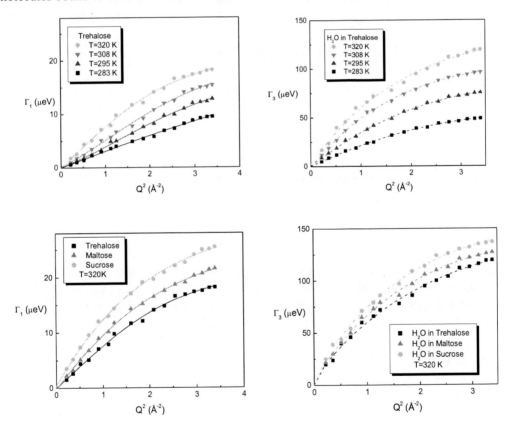

Figure 5 Linewidth of the translational contribution as a function of Q^2 for a) trehalose aqueous solutions at four temperature values and for trehalose, maltose and sucrose at T=320K and b) water in trehalose aqueous solutions at four temperature values and water in trehalose, maltose and sucrose aqueous solutions at T=320K. The solid lines are fits obtained following the RJD model.

The linewidth Γ_1 of the translational contribution of disaccharides as a function of Q^2 follows a typical Random Jump Diffusion (RJD) model [26], as shown in Fig. 5a for trehalose aqueous solutions for different temperature values and for trehalose, maltose and sucrose aqueous solutions at T=320K, respectively:

$$\Gamma_1(Q)=D_sQ^2/(1+D_sQ^2\tau) \tag{7}$$

where D_s is the self diffusion coefficient of the molecule and τ is the residence time. The RJD model furnishes the diffusion coefficient D_s value from the extrapolation to $Q \to 0$ and the residence time τ from the inverse of the asymptotic value at $Q \to \infty$.

For trehalose as a function of temperature the RJD model furnishes for the diffusion coefficient D_s and the residence time τ the values of D_s=2.83 × 10^{-7} cm²/s and τ=24.7 ps, D_s=3.82 × 10^{-7} cm²/s and τ=20.6 ps, D_s=5.35 × 10^{-7} cm²/s and τ=19.1 ps and D_s=8.50 × 10^{-7} cm²/s and τ=18.3 ps for T=283, 295, 308, 320 K, respectively. From the relation $<l^2>=6D\tau$ we obtain the values $<l^2>^{1/2}$ = 0.64 Å, 0.68 Å, 0.78 Å and 0.99 Å T=283, 295, 308, 320 K, respectively.

The RJD model furnishes for the diffusion coefficient D_s and the residence time τ the values of D_s=8.50 × 10^{-7} cm²/s and τ=18.3 ps for trehalose, D_s=1.00 × 10^{-6} cm²/s and τ=14.3 ps for maltose and D_s=1.23 × 10^{-6} cm²/s and τ=13.8 ps for sucrose at T=320K. Furthermore from the relation $<l^2>=6D\tau$ we obtain the values $<l^2>^{1/2}$ = 1.00 Å for the three disaccharides.

The linewidth Γ_3 of the translational contribution of water is reported in Fig. 5b as a function of Q^2 together with the best fit according to the RJD model for water in trehalose solutions as a function of temperature and for water in trehalose, maltose and sucrose aqueous solutions at T=320K. From the relation $<l^2>=6D\tau$ we obtain for pure water the mean jump length $<l>$ values of 1.35 Å for all the investigated temperature values. The whole water dynamics in trehalose solutions for T=283, 295, 308, 320 K resembles that of water at ~256K, ~261K, ~263K and ~268K, indicating that the water has a diffusive behaviour strongly triggered by the trehalose molecules and suffers of a noticeable frozen effect.

For the diffusion coefficient of water in the three disaccharide aqueous solutions we obtained at T=320K the value of D_w = 8.31 × 10^{-6} cm²/s for trehalose solution, D_w = 8.46 × 10^{-6} cm²/s for maltose solution and D_w = 8.60 × 10^{-6} cm²/s for sucrose solution, with the values of residence times of τ = 3.7 ps, 3.4 ps and 3.0 ps for trehalose, maltose and sucrose solutions, respectively obtaining for the mean jump length $<l>$ the value $<l^2>^{1/2}$ = 1.36 Å, 1.31 Å and 1.24 Å for trehalose, maltose and sucrose solutions, respectively. It is interesting to compare the values of the diffusion coefficient values obtained for water in presence of disaccharides with that of pure water at the same temperature, which is D_w = 3.94 × 10^{-5} cm²/s.

For the investigated systems the water dynamics resembles that of water at ~268K in the case of trehalose solution, at ~271K in the case of maltose solution and at ~277K in the case of sucrose solution. Analogously to the trehalose aqueous solutions, all the disaccharides show a slowing down effect of the water dynamics, which is stronger for trehalose than the other disaccharides.

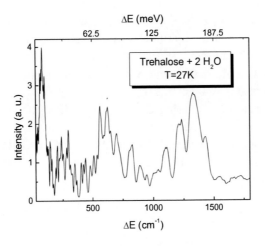

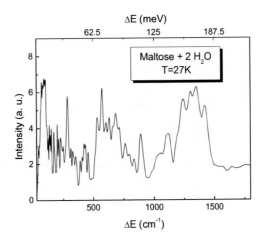

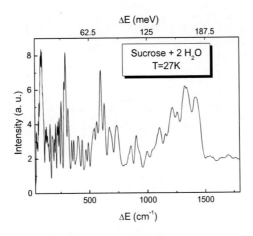

Figure 6. INS spectrum of a) trehalose/2 H_2O, b) maltose/2 H_2O and c) sucrose/2 H_2O mixture at T=27 K in the 0÷1800 cm^{-1} region. The MPNS contribution correction has been performed for all the spectra. The disaccharides mixtures profiles are shifted for clarity.

By using INS, in order to analyse the changes induced by disaccharides on the water vibrational modes, we distinguish the following vibrational modes in the H_2O spectrum: i) the region of the H_2O intermolecular vibrations up to ~1060 cm^{-1}, in which one distinguishes a "translational" part up to ~400 cm^{-1} and a "librational" part, and ii) the region of the H_2O intramolecular vibrations up to ~3600 cm^{-1}, which presents the bending modes at ~1600 cm^{-1} and the stretching modes at ~3360 cm^{-1} [27-29].

The multiphonon neutron scattering contribution (MPNS), which can be significant at high temperature and large momentum transfer, has been calculated directly from the measured spectra by using a method of sequential iterations. Since measurements were performed at low temperature, the MPNS contribution is not large at the translational modes region (i. e. at low Q region) [27-29].

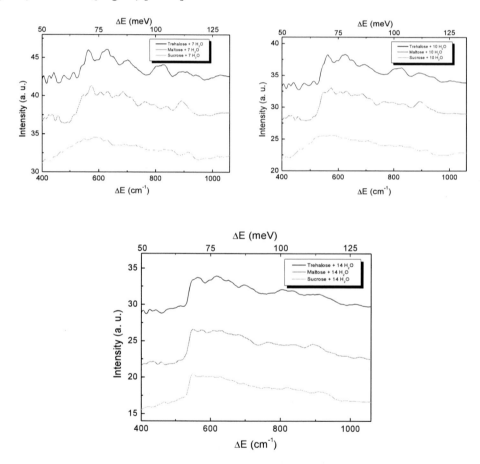

Figure 7. INS spectra of a) disaccharides/7 H_2O mixtures, b) disaccharides/10 H_2O mixture and c) disaccharides/14 H_2O mixture at T=27 K in the 400÷1060 cm^{-1} librational spectral region. The MPNS contribution correction has been performed for all the spectra. The disaccharides mixtures profiles are shifted for clarity.

Due to the spectral features observed by increasing the water content in the mixtures, in Fig. 6 the INS spectra of a) trehalose/2 H_2O (the dehydrate form is the natural state of trehalose), b) maltose/2 H_2O and c) sucrose/2 H_2O are shown as references.

It was found that the position of the librational low-energy cut-off is a characteristic value for the band in this INS spectral region for different ice forms. The observed shifts of the cut-off position are proportional to the transverse forces between the water molecules whose intensities depend on the different ice forms [27-29].

In order to point out the differences due to the trehalose, maltose and sucrose mixtures in the librational spectral region, in Fig. 7 the INS spectra for disaccharides/7 H_2O, disaccharides/10 H_2O and disaccharides/14 H_2O in the 400÷1060 cm^{-1} spectral region are shown. A depression of the intensity of the cut-off is observed and, as it can be expected, it is more evident for trehalose, maltose and sucrose/7 H_2O, the cut-off becoming sharper by increasing the water content. Concerning the shift of the cut-off position, which for H_2O at T=27 K is at ~550 cm^{-1} [27-29], for all the investigated concentration values in trehalose mixtures we observe the same shift of ~16 cm^{-1}, whereas for maltose mixtures the shift is of ~3 cm^{-1}, and for sucrose mixtures no shift occurs. This result emphasise that trehalose is more capable than maltose and sucrose to modify the H_2O spectral features by affecting the intermolecular interaction forces and the arrangements of the H_2O molecules.

In trehalose vibrational spectrum obtained by simulation [30] we note that the peak linked to the C-O stretching mode at ~1000 cm^{-1} does not appear, or at least it is not detectable, indicating that this kind of modes in the rings is constrained because of the presence of hydrogen bonds with water. In general terms, however, below 960 cm^{-1} it is not possible to assign a well defined character to all the bands [30]. The stretching of the C-O bonds occurs in a range between 970 and 1100 cm^{-1}, while the oscillations of the OH groups around the minimum of the C-C-O-H torsional potential are well localized below 950 cm^{-1}. These modes possess energy between 420 and 570 cm^{-1}, with the exception of the OH group giving rise to the intramolecular hydrogen bond, whose torsional mode has energy of 700 cm^{-1} [19]. From Fig. 7 it is evident that the peak corresponding to the torsional mode is more and more depressed by increasing the water content.

The 1060÷1800 cm^{-1} spectral region corresponds to bending vibrational modes range for ice [27-29]. The H_2O spectrum is characterised by two distinct bands centred at ~1224 cm^{-1} and ~1608 cm^{-1}, respectively. In the spectra of the investigated trehalose, maltose and sucrose/H_2O mixtures, shown in Fig. 8, these features are totally changed. The band at ~1608 cm^{-1} is absent in the disaccharide/H_2O spectra, confirming that disaccharides are able to affect the water hydrogen bond O-H···O bending modes connected to the strength and tetrahedrality of the hydrogen bonding [27-29]. In particular the role played by disaccharides is to impose to water a network which deviates from tetrahedral bonding and for which the hydrogen bonding among water molecules is diminished while that among disaccharides and water molecules is increased.

By the observation of the second band in Fig. 8 it is evident that the trehalose/H_2O spectra appear more "structured" in comparison with the other disaccharide mixtures and show more distinctly the three peaks present in the trehalose/2 H_2O spectrum (see Fig. 6). These peaks correspond to the hybridized H-C-H, C-C-H and C-O-H bending modes, as indicated by simulation [30]. By increasing the water content, the three peaks are still evident even if they are less marked. For sucrose mixtures these peaks appear larger and broader. The more ordered spectrum of trehalose/water mixture in comparison with the other disaccharide mixtures reflects the larger "cryptocrystallinity" of trehalose, i. e. the more marked presence of nano-crystallized domains in the amorphous sytem composed by trehalose and water.

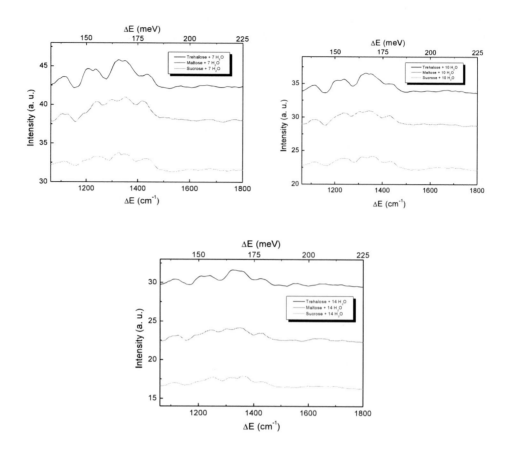

Figure 8. INS spectra of a) disaccharides/7 H$_2$O mixtures, b) disaccharides/10 H$_2$O mixture and c) disaccharides/14 H$_2$O mixture at T=27 K in the 1060÷1800 cm^{-1} bending spectral region. The MPNS contribution correction has been performed for all the spectra. The disaccharides mixtures profiles are shifted for clarity.

CONCLUSION

By the ENS results it has been shown that the glass-forming properties of disaccharides play an important role in the bioprotection mechanisms. We concluded that trehalose, besides modifying significantly the structural and dynamical properties of water, forms with H2O a less fragile entity able to encapsulate biological structures and to protect them in a more rigid environment.

The observed QENS results clearly show that, contrarily to the simulation findings obtained by Grigera [13], also the diffusive dynamics of water is significantly affected by the presence of disaccharides. More specifically al the investigated disaccharides, and trehalose with a greater extent, slow down the dynamics of water, thus showing a high "switching off" capability. This circumstance also implies a higher ability to hold volatile substances, and results to be more pronounced for trehalose than for the other disaccharides, coherently with its superior bioprotective effectiveness. Finally by the INS findings it is evident that the trehalose-water system shows a macroscopic amorphous conformation in which nano-crystallized domains are mixed with remaining liquid, so revealing a "cryptocrystalline" character. Therefore the present INS findings help to give an explanation to the previous results, because the locally most ordered structure of trehalose can justify the highest rigidity of this system.

The physical picture obtained from the performed studies shows that the highest bioprotectant effectiveness of trehalose in comparison with the other disaccharides is due to the combined effect of different co-factors. What emerges is that the biological action of disaccharides can be explained by the strong interaction with water molecules, as emphasized by the highest value of interaction strength and hydration number of the trehalose/H_2O system in respect to the other disaccharide mixtures. This capability imply that disaccharides, and in particular trehalose, promote an extensive layer of structured water around its neighborhood, which destroys the tetrahedral H-bond network of pure water, providing a structure whose spatial positions and orientations are not compatible with those of ice.

On the other hand, the marked "crptobiotic" effect of disaccharides, and in particular trehalose, is to be ascribed to the enhanced rigidity and to the "cryptocrystalline" character of the disaccharide/H_2O systems by increasing disaccharide concentration, this circumstance indicating that trehalose shows a larger structural resistance to temperature changes and a lower "fragility" in comparison with the maltose and sucrose/H_2O mixtures.

REFERENCES

[1] Frauenfelder, H.; McMahon, B. *Proc Natl Acad Sci USA* 1998, 9995, 4795-4801.

[2] Paciaroni, A.; Cinelli, S.; Onori, G. *Biophys J* 2002, 83, 1157-1162.

[3] Vitkup, D.; Dagmar, R.; Petsko, G. A.; Karplus, M. *Nat Struct Biol* 2000, 7, 34-38.

[4] Zaccai, G. *Philos Trans R Soc Lond B Biol Sci* 2004, 359, 1269-1274.

[5] Poole, P. L.; Finney, J. L. *Int J Biol Macromol* 1983, 5, 308-312.

[6] Rupley, J. A.; Gratton, E.; Careri, G. *Trends Biol Sci* 1983, 8, 18-24.

[7] Poole, P. L.; Finney, J. L. *Biopolymers* 1984, 23, 1647-1651.

[8] Réat, V.; Dunn, R.; Ferrand, M.; Finney, J. L.; Daniel, R. M.; Smith, J. C. *Proc Natl Acad Sci USA* 2000, 97, 9961-9966.

[9] Crowe, J. H.; Cooper, A. F. Jr. *Scientific American* 1971, 225, 30-36.

[10] Storey, K. B.; Storey, J. M. *Ann Rev Physiol* 1992, 54, 619-637.

[11] Green, J. L.; Angell, C. A.; *J Phys Chem B* 1989, 93, 2880-2882.

[12] Crowe, J. H., Clegg, J. S.; Crowe, L. M. In The Roles of Water in Foods; Reid, D. S.; Chapman & Hall, New York, 1998; 440-455.

[13] Donnamaria, M. C.; Howard, E. I.; Grigera, J. R. *J. Chem. Soc. Faraday Trans.* 1994, 90, 2731-2735.

[14] Branca, C.; Magazù, S.; Migliardo, F. *Rec Res Develop in Phys Chem* 2002, 6, 35-73.

[15] Branca, C.; Magazù, S.; Maisano, G.; Migliardo, P. *J Chem Phys* 1999, 111, 281-287.

[16] Branca, C.; Faraone, A.; Magazù, S.; Maisano, G.; Migliardo, F.; Migliardo, P.; Villari, V. *Rec Res Develop in Phys Chem* 1999, 3, 361-403.

[17] Branca, C.; Magazù, S.; Maisano, G.; Migliardo, F.; Romeo, G. *J Phys Chem B* 2001,

105, 10140-10145.

[18] Branca, C.; Magazù, S.; Maisano, G.; Migliardo, F. *Phys. Rev. B* 2001, 64, 2242041-2242048.

[19] Zaccai, G. *Science* 2000, 288, 1604-1607.

[20] Doster, W.; Cusack, S.; Petry, W. *Nature* 1989, 337, 754-756.

[21] Smith, J. C. *Q Rev Biophys* 1991, 24, 227-291.

[22] Telling, M. T. F.; Howells, W. S. MODES - IRIS DATA ANALYSIS (1st Edition), ISIS Facility, Rutherford Appleton Laboratory, UK, 2003.

[23] Colognesi, D.; Parker, S. F. The TOSCA Manual, Rutherford Appleton Laboratory UK, 1999.

[24] Galli, G.; Migliardo, P.; Bellissent, R.; Reichardt, W. Solid State Commun. 1986, 57, 195-198.

[25] Burattini, E.; Federico, M.; Galli, G.; Magazù, S.; Majolino, D. Il Nuovo Cimento D 1988, 10, 425-434.

[26] Bee, M. Quasielastic Neutron Scattering; Hilger, A.; Bristol and Philadelphia, 1988.

[27] Kolesnikov, A. I.; Li, J. C.; Ahmad, N. C.; Loong, C.-K.; Nipko, M.; Yocum, L.; Parker, S. F. *Physica B* 1999, 263, 650-654.

[28] Kolesnikov, A. I.; Li, J. C.; Parker, S. F.; Eccleston, R. S.; Loong, C.-K. *Phys Rev B* 1999, 59, 3569-3573.

[29] Li, J. C. *J Chem Phys* 1996, 105, 6733-6738.

[30] Ballone, P.; Marchi, N.; Branca, C.; Magazù, S. *J Phys Chem B* 2000, 104, 6313-6317.

INDEX